Technology, Work and Globalization

Series Editors

The Technology, Work and Globalization series was developed to provide policy makers, workers, managers, academics and students with a deeper understanding of the complex interlinks and influences between technological developments, including information and communication technologies, work organizations and patterns of globalization. The mission of the series is to disseminate rich knowledge based on deep research about relevant issues surrounding the globalization of work that is spawned by technology.

More information about this series at
http://www.palgrave.com/gp/series/14456

Nik Rushdi Hassan • Leslie P. Willcocks
Editors

Advancing Information Systems Theories

Rationale and Processes

Editors
Nik Rushdi Hassan
Department of Management Studies
University of Minnesota Duluth
Duluth, MN, USA

Leslie P. Willcocks
Department of Management
London School of Economics and Political Science
London, UK

ISSN 2730-6623 ISSN 2730-6631 (electronic)
Technology, Work and Globalization
ISBN 978-3-030-64886-2 ISBN 978-3-030-64884-8 (eBook)
https://doi.org/10.1007/978-3-030-64884-8

This Palgrave Macmillan imprint is published by the registered company Springer Nature Switzerland AG.
The registered company address is: Gewerbestrasse 11, 6330 Cham, Switzerland

Contents

Notes on Contributors

Pär J. Ågerfalk is a professor at Uppsala University, Uppsala, and Visby, Sweden, where he holds the Chair in Information Systems. He received his PhD in Information Systems Development from Linköping University and has held full-time positions at Örebro University, University of Limerick, Jönköping International Business School, and Lero—The Irish Software Engineering Research Centre. His work has appeared in many leading journals and conferences in the information systems field, such as *MIS Quarterly* and ICIS. Currently, he is Editor-in-Chief of the *European Journal of Information Systems*, a member of the *Journal of Information Technology* editorial board, and a member of the Distinguished Editorial Advisory Board of the *International Journal of Information Management*. He is involved in many national and international organizations and networks, such as the Academy of Management, the OR Society, and the Association for Information Systems (AIS). He was the founding chair of the AIS Special Interest Group on Pragmatist IS Research (SIGPrag).

Robert D. Galliers is the University Distinguished Professor Emeritus, formerly Provost at Bentley University, USA, Honorary Visiting Professor in the University of Loughborough's School of Business & Economics, UK, and Professor Emeritus, formerly Dean of Warwick Business School, UK. He is also an Associate Director of the European Foundation for Management Development (EFMD) Quality Services, Belgium. He was the founding editor-in-chief of the *Journal of Strategic Information Systems* up until the end of 2018. He received the Association for Information Systems (AIS) LEO Award for exceptional lifetime achievement in IS in 2012. He has published 100 articles in many leading international journals in IS and management (e.g., *MISQ, JAIS, JMIS, ISJ, EJIS, JIT, JMS, BJM, LRP*), and has authored or co-authored 13 books.

Göran Goldkuhl is Professor in the Department of Management and Engineering, Linköping University, Sweden, also associated to Campus Gotland, Uppsala University, Sweden. He holds an honorary doctorate from Örebro University, Sweden. He serves on the AIS SIGPRAG Board. His research interests cover areas such as qualitative and pragmatic research methodologies, practice research, design science, action research, innovation and change management, IS theorizing, practice theory, work and IT co-design, business process modelling, communication analysis, digital service design, user-interface design, IS evaluation, egovernment, values and law in digitalization. He has published in conference proceedings of ALOIS, ECIS, DESRIST, ICIS, ISD, LAP, POEM, among others. He has published in journals like *Australasian Journal of Information Systems, Business Process Management Journal, Communications of ACM, Communications of the Association for Information Systems, European Journal of Information Systems, Government Information Quarterly, Information and Organization, International Journal of Qualitative Methods, Journal of Information Technology Theory and Application, Semiotica, Transforming Government*, among others.

Amir Haj-Bolouri holds a Ph.D. in Informatics with a specialization on work-integrated learning at the Department of Informatics at University West in Sweden. His research interests comprise philosophy, information systems design, Design Science Research, and Action Design Research. He has published in the *European Journal of Information systems, International Journal of E-Learning* and presented at major IS conferences such as the Hawaiian International Conference on Systems Sciences (HICSS), International Conference on Design Science Research in Information Systems and Technology (DESRIST) and the European Conference on Information Systems (ECIS).

Nik Rushdi Hassan is Associate Professor of Information Systems (IS) at the Labovitz School of Business and Economics (LSBE), University of Minnesota Duluth, USA. He is currently Associate Editor for the History and Philosophy Department of the *Communications of the AIS* and Senior Editor of *Data Base Advances in Information Systems*. He has served as President of the Association of Information Systems (AIS) Special Interest Group on Philosophy in Information Systems (SIGPhil) and was one of the editors of a recent special issue of the *European Journal of Information Systems* on Philosophy and the Future of the IS Field. His research areas include the philosophical foundations of the IS field, theorizing and theory building, IS development, business analytics, social network analysis and complexity science. He has published in the *Journal of the Association for Information Systems*, *Journal of Information Technology*, *European Journal of Information Systems*, *Data Base Advances in Information Systems*, *Information Systems Management Journal*, *Communications of the AIS*, *Journal of IS Education*, *Informing Science Journal*, *Review of Accounting and Finance*, *Journal of Documentation*, *Journal of Business Analytics* and *The 2018 Routledge Companion to Management Information Systems*.

Fredrik Karlsson is a professor in Informatics at Örebro University School of Business, Örebro, Sweden. He received his PhD in Information Systems Development from Linköping University. His research on information security, tailoring of systems development methods, system development methods as reusable assets, CAME-tools, requirements engineering, method rationale, and electronic government has appeared in a variety of information systems journals and conferences including *European Journal of Information Systems*, *Journal of Strategic Information Systems*, *Scandinavian Journal of Information Systems*, *Government Information Quarterly*, and various information systems conferences. Karlsson is currently Associate Editor of the *European Journal of Information Systems* and research leader of the research environment Centre for Empirical Research on Information Systems (CERIS).

David Kreps is philosopher in information systems at National University of Ireland, Galway. Formerly he was Interim Associate Dean of Research and Innovation, Salford Business School, University of Salford, Manchester, United Kingdom. He is chair of the International Federation for

Information Processing's Technical Committee 9 on ICT and Society and was a British Academy Mid-Career Fellow in 2018. For decades David has been an 'early adopter', pioneering thinker and commentator, with a fascination for technology and its impact upon society. A web developer since 1995 (making probably the first Arts Centre Website in the UK), he did his PhD thesis on Cyborgism, and has become an expert in Web Accessibility and explorer into the philosophy of Virtuality, and Cultural Identity in the Socially Networked society. He was formerly Director of Information Systems, Organisations and Society Research Centre (ISOS) 2009–12. His books include *Understanding Digital Events: Bergson, Whitehead, and Experiencing the Digital* (Routledge); *Against Nature: The Metaphysics of Information Systems* (Routledge); *Bergson, Complexity and Creative Emergence* (Palgrave); *Gramsci and Foucault: A Reassessment* (Routledge); *This Changes Everything: ICT and Climate Change – What We Do?* (Springer); and *Technology and Intimacy: Choice or Coercion* (Springer).

Paul Lowry Ph.D. is the Suzanne Parker Thornhill Chair Professor and Eminent Scholar in Business Information Technology at the Pamplin College of Business at Virginia Tech where he serves as the BIT Ph.D. and graduate programs director. He is a former tenured Full Professor at both City University of Hong Kong and the University of Hong Kong. He received his Ph.D. in Management Information Systems from the University of Arizona and an MBA from the Marriott School of Management. He has published 235+ publications, including 130+ journal articles in *MIS Quarterly* (MISQ), *Information Systems Research* (ISR), *Journal of Management Information Systems* (JMIS), *Journal of the Association for Information Systems* (JAIS), *Information Systems Journal* (ISJ), *European Journal of Information Systems* (EJIS), *Journal of Strategic Information Systems* (JSIS), *Journal of Information Technology* (JIT), *Decision Sciences Journal* (DSJ), various *IEEE Transactions*, and others. He is on the senior editorial board of *Journal of Management Information Systems* (JMIS). He also is a Senior Editor (SE) at *Journal of the Association for Information Systems* (JAIS) and *Information Systems Journal* (ISJ) and an Associate Editor (AE) at the *European Journal of Information Systems* (EJIS). His research interests include (1) organizational and behavioral security and privacy; (2) online deviance, online harassment, and computer ethics; (3) HCI, social media, and gamification; and (4) business analytics, decision sciences, innovation, and supply chains.

M. Lynne Markus is the John W. Poduska, Sr. Professor of Information and Process Management at Bentley University and an associated researcher at MIT's Center for Information Systems Research. She is coauthor (with Daniel Robey) of "Information Technology and Organizational Change: Causal Structure in Theory and Research" (*Management Science*, 1988), the paper that inspired the current work. She has published extensively in the areas of digital business and interorganizational governance, enterprise systems and business processes, electronic communication and knowledge reuse, and organizational change management. Her current research interests include digital innovation in the financial and health sectors, the responsible use of data and algorithms, and the changing nature of work. Markus was named a Fellow of the Association for Information Systems in 2004 and received the AIS LEO Award for Exceptional Lifetime Achievement in Information Systems in 2008.

Lars Mathiassen is Georgia Research Alliance Eminent Scholar, Professor at the Computer Information Systems Department, and co-Founder of The Center for Digital Innovation at Georgia State University. His research focuses on digital innovation, IT development and management, and the use of IT for health services. He has published extensively in major information systems and software engineering journals and coauthored several books on the subject, such as *Professional Systems Development, Computers in Context: The Philosophy and Practice of Systems Design, Object Oriented Analysis and Design, and Improving Software Organizations: From Principles to Practice*. He has served as senior editor for *MISQ*, and he currently serves as senior editor for *Information and Organization* and for the *Journal of Information Technology*.

Mohammad Moeini is an Assistant Professor in Information Systems at Warwick Business School, UK. Prior to this role, he was a Senior Lecturer at the University of Sussex Business School. He obtained his PhD from HEC Montreal, Canada. His research is published in leading IS outlets including MISQ, JAIS, JSIS, and JIT. He is the recipient of two Best

Reviewer Awards from AIS SIGITProjMgmt (2013/14), an Outstanding Associate Editor award for ICIS 2019, the Prix La Relève (a recognition prize for upcoming professionals in project management) from PMI-Montréal (2015), and the Research Excellence Award for Emerging Scholars from the University of Sussex Business School (2019).

Sune Dueholm Müller is Associate Professor in the Department of Management, School of Business and Social Sciences, Aarhus University, Denmark where he teaches at both the graduate and undergraduate levels in IS strategy, IT and Organizations and Business Process Management. His research areas are in digital innovation, process innovation and business process management. He is published in the Journal of the Association for Information Systems, Business Process Management Journal, Communications of the Association for Information Systems, International Journal of Electronic Governance, Information Technology & People, Journal of Systems and Software, IEEE Transactions on Engineering Management and presented at major IS conferences such as the International Conference on Information Systems (ICIS), the Hawaiian International Conference on Systems Sciences (HICSS), Americas Conference on Information Systems (AMCIS) and the European Conference on Information Systems (ECIS).

Suzanne Rivard is a professor of information technology (IT) at HEC Montreal and is the HEC Montreal endowed chair in strategic management of information technology. She is a fellow of the *Royal Society of Canada* and a fellow of the *Association for Information Systems.* She received her Ph.D. from the Ivey School of Business, the University of Western Ontario. Her research pertains to IT project risk management, outsourcing of IT services, and user-related issues such as user resistance to IT implementation. A former senior editor of the Theory and Review department of the *MIS Quarterly*, she is currently senior editor with the *Journal of Strategic Information Systems.* Her work has been published in journals such as *Journal of the Association for Information Systems*, *Journal of Information Technology*, *Journal of Management Information Systems*, *Journal of Strategic Information Systems*, *MIS Quarterly*, and *Organization Science.*

Frantz Rowe is a professor at the Université de Nantes and SKEMA Business School. His research interests revolve around the philosophy of information systems, IS-enabled organizational transformation, interorganizational systems, and the effects of IT. He believes that our research should illuminate the complexity of the phenomena we study so that we better understand action consequences and design technology accordingly. He has published in 40 different peer-reviewed journals, has coedited five books, including *Innovation and IT in an International Context* (Palgrave Macmillan, 2014) with Dov Te'eni, and coauthored three books, two of which were awarded the FNEGE and the EFMD prize in 2016. He is Emeritus Editor of the *European Journal of Information Systems* and Board member of the *International Journal of Information Management* and of *Systèmes d'Information et Managemen*t. He is an AIS fellow and is preparing ECIS 2021 in Marrakech as cochair of the conference.

Carol Saunders is a Professor Emeritus of Management at the University of Central Florida. Dr. Saunders has received lifetime accomplishment awards from two disciplines: the LEO award in the Information Systems (IS) discipline and the Lifetime Achievement Award from the OCIS Division of the Academy of Management. She is an Association of Information Systems Fellow, a Schöller Senior Fellow and the 2020 Distinguished Educator. She served on a number of editorial boards, including a three-year term as Editor-in-Chief of MIS Quarterly. She served as General Conference Chair of the premier IS conference, ICIS and as the Program Co-chair of AMCIS 2015. She was Vice President of Publications for the Association of Information Systems from 2016–2019. Carol helped found the Organization Communication and Information Systems (OCIS) division of the Academy of Management and served in a variety of positions including its program chair and division chair. She was the Distinguished Fulbright Scholar at the Wirtschafts Universität—Wien (WU) in Austria and earlier held a Professional Fulbright with the Malaysian Agricultural Research and Development Institute. She has held research chairs in Germany, New Zealand, Singapore, and the Netherlands. Her current research interests include overload, sourcing, virtual teams, business models and control. She has published in top-ranked management, IS, computer science and communication journals.

Boyka Simeonova is an Assistant Professor in Information Management at Loughborough University, UK, and is the Director of the Knowledge and the Digital Economy Network and Deputy Director of the Centre for Information Management. Boyka has published in leading Information Systems journals including *Information Systems Journal*, *Journal of Information Technology* and is a co-editor of the book, *Strategic Information Management: Theory and Practice* (Routledge).

Jonas Sjöström is associate senior lecturer in the Department of Informatics and Media, at Uppsala University Campus Gotland, Sweden. Jonas does research in software engineering and information systems. His current work includes design issues related to online psychosocial support, IS curriculum and IS didactics, and design science methodology; he has published in the *Journal of Medical Internet Research* and various DESRIST conferences. He is currently president of the Association for Information Systems Special Interest Group on Pragmatist Information Systems Research (AIS SIGPrag).

Lars Taxén is an associate professor from Linköping University, Sweden. In addition to his academic career, he has more than 30 years' experience from the telecom industry, where he worked with system development methods, processes, information modelling, and information systems. His research interest has from the outset centered around the dialectical constitution of the individual and the social, both from a practical and theoretical point of view. Currently, his research is focused on the interaction between specialized disciplines in general, and between information systems and neuroscience in particular. He has published a book, several book chapters, and various journal and conference articles.

Leslie P. Willcocks is Professor of Work, Technology and Globalisation, Department of Management, London School of Economics, UK. His research areas include automation, digital business, the future of work, IT and business process outsourcing, organisational change, management, and global strategy. As well as being a professor in the Information Systems and Innovation Faculty Group, he is a Fellow of the British

Computer Society. For the last 30 years he has been Editor-in-Chief of the *Journal of Information Technology*. He is co-author of 67 books, including most recently *Robotic Process and Cognitive Automation: The Next Phase, Dynamic Innovation Through Outsourcing Service Automation Robots and The Future of Work* (2016, www.sbpublishing.org) and *Global Business: Strategy In Context* (2021). He has published over 240 refereed papers in journals such as *Harvard Business Review, Sloan Management Review, California Management Review, MIS Quarterly, MISQ Executive* and *Journal of Management Studies*.

Alex Wilson is Senior Lecturer in Strategy in the School of Business and Economics, Loughborough University, UK. Previously he was a Research Fellow in strategic management at Warwick Business School. He has held visiting positions as Lim Kim San Research Fellow and Visiting Professor at Singapore Management University. He was Chartered Association of Business Schools (UK) Research Fellow (2017 and 2018). He is researching the use of technologies in open strategy and continues to study the strategic development of business schools and management education. He has published in *Long Range Planning, Journal of Strategic Information Systems, Journal of Information Technology, British Journal of Management*, and *Accounting Auditing and Accountability Journal*.

List of Figures

List of Tables

1

Introduction: Why Theory? (Mis) Understanding the Context and Rationale

Nik Rushdi Hassan and Leslie P. Willcocks

Prelude

Writing an introduction to the highly discussed and most misunderstood topic of theory in information systems (IS) research circles can be challenging. The IS field has been debating the nature and role of theories for some time with intense debates regarding whether or not a theoretical core is necessary (Weber, 2003, 2006; Lyytinen & King, 2004, 2006) and disagreements concerning whether or not the field can speak of native theories (Grover, Lyytinen, & Weber, 2012; Weber, 2012), and what constitute IS theory and the role of theories in IS (Holmström,

N. R. Hassan (✉)
Department of Management Studies, University of Minnesota Duluth, Duluth, MN, USA
e-mail: nhassan@d.umn.edu

L. P. Willcocks
Department of Management, London School of Economics and Political Science, London, UK
e-mail: l.p.willcocks@lse.ac.uk

N. R. Hassan, L. P. Willcocks (eds.), *Advancing Information Systems Theories*, Technology, Work and Globalization, https://doi.org/10.1007/978-3-030-64884-8_1

2005; Gregor, 2006; Gregor & Jones, 2007; Holmström & Truex, 2011; Bichler et al., 2016; Markus & Rowe, 2018). It is therefore premature to write an integrative summary of all these discussions in this introduction. What is not up for debate is how the field undertakes its research using theories from its "reference disciplines." By borrowing from these reference disciplines "the theories and methods of these disciplines serve to set the standards by which the quality and maturity of IS research should be measured" (Baskerville & Myers, 2002, p. 1). Whether it is the theory of reasoned action (TRA) or its derivative, the theory of planned behavior from social psychology, resource based view (RBV) and absorptive capacity theory from strategic management, game theory and transaction cost theory (TCT) from Economics, innovation diffusion theory (IDT) from communications, or social cognitive theory and activity theory from psychology, the IS field has consistently borrowed (Lim, Saldanha, Malladi, & Melville, 2013), often uncritically (Markus & Saunders, 2007; Hassan, 2011), to legitimize its research. Given the background of the history of theories in the IS field, the goal of this series of volumes is to advance IS research beyond this form of borrowed legitimization and derivative research toward fresh and original research that naturally comes from its own theories—information system theories. It is inconceivable for a field so relevant to the era of the hyper-connected society, disruptive technologies, Big Data, social media, and "fake news" to not be brimming with its own theories. That is why the title of this series of volumes, "Advancing Information Systems Theories," is phrased in such way as to emphasize its intended goal. The focus of this series of volumes is on "information systems theories" and not just "theories in the information systems field."

Any advancement of theory has to begin with some form of agreement with regard to IS theories from thought leaders of the IS community. Although much progress has been made, many questions remain unanswered. The major questions that will be addressed include the following: What can we agree on with regard to theories? What constitutes theories and what doesn't? Why do we need theories? How can one go about developing theories? What does an IS theory look like? Therefore, the approach in this series of volumes is to solicit from the field's thought leaders of all levels and accomplishments how they each address these questions. For a field as diverse in its content and approaches as the

IS field, such a goal will be challenging but not insurmountable. The symptoms stemming from the ambiguity surrounding theories in the IS field manifest in various forms. It is omnipresent in panels at major IS conferences since the inception of the IS field (Keen, 1980; Mumford, Hirschheim, Fitzgerald, & Wood-Harper, 1985; Klein, Hirschheim, & Nissen, 1991; Nissen, Klein, & Hirschheim, 1991; Karahanna et al., 2002). It can be seen in editorials and special issues of major IS journals (Weber, 2003, 2006; Hirschheim, 2006; Lyytinen & King, 2006; Straub, 2012; Avison & Malaurent, 2014; Gregor, 2014; Lee, 2014; Markus, 2014). It also manifests itself in how authors often hedge their prose when introducing the theory portion of their submissions with the phrase "toward a theory" or "building theory." And it emerged as a title in the 2019 *International Conference for Information Systems* (ICIS): "How and Why 'Theory' Is Often Misunderstood in Information Systems" (Siponen & Klaavuniemi, 2019).

Much of the discussion on theory in the IS field mirrors the debates in the management research surrounding theory, and these continue unabated (Bacharach, 1989; Weick, 1989, 1995, 1999; Whetten, 1989; van de Ven, 1989; Doty & Glick, 1994; Sutton & Staw, 1995; Colquitt & Zapata-Phelan, 2007; Hambrick, 2007; Whetten, Felin, & King, 2009; Corley & Gioia, 2011; Suddaby, Hardy, & Huy, 2011; Byron & Thatcher, 2016; Shepherd & Suddaby, 2017). Often the misunderstanding about theories in IS and the management fields has to do with how the notion is framed. For example, both Hirschheim (2019) and Hambrick (2007), in criticizing their respective field's obsession with theory, traced its origins to the historical period when business schools were under pressure to draw its scholarly methods from the more established social sciences, such as psychology and economics, in order to enhance rigor and raise its level of scholarship. In the IS field, this phenomenon of drawing from the "natural science model" (March & Smith, 1995) built a tradition labeled "behavioral IS research" (Hevner, March, Park & Ram, 2004; Goes, 2013). These kinds of derivative research either prominently cite their sources of theory from these reference disciplines, or in the absence of such theories, will as Hambrick (2007) laments, pepper their articles with enough words like "theory," "theoretical framework," or "theorizing" to persuade reviewers that the article has enough theory in it. Kaplan (1964), over half a century ago, wrote about how authors would resort to similar strategies of augmenting their research with words suggesting theory, albeit in essence treating theory only as an afterthought.

One reason why researchers resort to these strategies has to do with a sense of insecurity about whether or not their research has "enough" theory to satisfy journal editors, and this insecurity is in turn exacerbated by a misunderstanding of what theory means. If the researcher is unclear whether or not the research contains theory, and is under pressure to demonstrate theory, the researcher is very likely to overcompensate and may even find instances of some theory to latch on to regardless of however strained its relevance. This phenomenon is neither a twenty-first-century tendency nor is it limited to the IS field and management. The confusion with regard to theory should not come as a surprise because even the experts on philosophy of science disagree on what exactly scientific theories look like (Suppes, 1967; Suppe, 1977, 2000; Swedberg, 2014). It is no surprise that IS and its allied management fields also experience confusion. The goal of this introduction is to begin the engagement with theories in a substantive fashion by addressing the question as to why theories are needed in the first place. We use the Aristotelian approach documented in his *Posterior Analytics* to answer this "Why Theory?" question. To Aristotle, answers to the why questions comes from answering the question about the meaning and essence of the *explananda* (Charles, 2000); that is, to answer the why question, one needs to answer the what question. The "Why theory?" question requires a clear response to the "What is theory?" question because, much like everything else, why we need something is dependent on a clear definition of that something which is needed. We need air because air contains life-giving oxygen and other elements that help us breathe and therefore live. So, this introduction sets the stage for our engagement with theory by answering two questions: What is theory and why do we need it?

What Is Theory?

Interestingly, Gregor (2006) acknowledges that the nature of theory itself has not received much attention within the IS field. This observation coincides with the struggles the field has with theory. We begin our engagement by reviewing major discussions about the definition of theory within the IS field and compare those discussions with those outside the IS field.

Theory from the IS Field

As Gregor (2006) notes, IS researchers who use "theory" in their research topics fail to give any explicit definition of theory itself and therefore work within their limited scope of their understood version of theory. For instance, when Sarker and Lee (2002) test three competing theories of business process redesign—the technocentric theory of redesign, the sociocentric, and sociotechnical theories of redesign—their version of theory follows from Markus and Robey's (1988) focus on the causal structure of theory. Their work, therefore, is restricted to a subset of theories that discusses cause and effect, which do not encompass all that could be said about theory. Gregor (2006, p. 616) defines theory in a way that is not limited to causality, but as "abstract entities that aim to describe, explain, and enhance understanding of the world and, in some cases, to provide predictions of what will happen in the future and to give a basis for intervention and action." Gregor's definition favors a more inclusive form of theory that could describe, explain, or predict. This definition also implies that although prediction is one of the goals of theories, it is not the only goal. Weber (2012), on the other hand, prefers to align with the definition of theory limited only to Gregor's Type-IV theory—theory for explanation and prediction—because only this category of theory satisfies his requirement for precision in defining theory's components, associations, states, and events.

Lee (2014) addresses the question of "What is Theory?" by suggesting that theory cannot be separated from science, and like science the nature of theory differs across the different terrains of science. According to Lee (2014), the dominance of the natural sciences has colored the conception of theory to mean the science of description and explanation, which may not be suited for the IS field. IS falls into the category of applied sciences that describe how to innovatively create better information systems to meet human needs that do not now exist or have not yet existed. For Lee, this conception of theory excludes both the definition of theory within the positivistic sciences and the interpretive kind. In summary, although Gregor (2006, p. 614) favors the more inclusive definition of theory that includes "conjectures, models, frameworks, or body of knowledge," her sentiment is not shared across the community (Hirschheim, Dennis, & Willcocks, 2019).

Theory from the Management Field

Much of the discussion about theory in the IS field is guided by discussions in the management field that continue unabated. The first wave of these discussions in 1989 defined theory, what constitutes theory, and what theoretical contribution looks like. Introducing the forum on theory, van de Ven (1989, p. 486) lauded good theory because it "advances knowledge in a scientific discipline, guides research toward crucial questions, and enlightens the profession of management." Bacharach (1989, p. 498) offers a definition of theory that is repeated by many IS researchers (Mueller & Urbach, 2017) as "a statement of relationships between units observed or approximated in the empirical world …[with the goal of answering the] how, when, and why." He made it clear that descriptions (such as categorization of data, typologies, and metaphors) do not qualify as theories because they only answer the "what" question. To this, Whetten (1989) adds that models are indistinguishable from theories. Thus, a theoretical model qualifies as a theory. To him a "complete theory" must contain four essential elements that address the what, how, and why of the theory and a fourth element consisting of who, where, and when. The what and the how elements define the domain of the theory as well as the model that could be tested, while the why element explains the logic and justification for the theory.

The second wave of discussion among management theorists also addressed what theory is, by discussing what theory is not (Sutton & Staw, 1995; Weick, 1995). They emphasize that theory has to do with answering the why question, and since literature reviews, data, and hypotheses do not address the why question, they do not qualify as representations of theory. They recommend lessening the expectations for strong theory to allow researchers to focus on data and empirical findings without necessarily demanding theory to be constructed. This argument about accepting papers that may have strong theory but not empirical data, or papers with empirics but not strong theory, is picked up in the IS field by Avison and Malaurent (2014) and by Grover and Lyytinen (2015). Weick (1995, p. 385) adds that "most products that are labelled theories actually approximate theories," so, half-done frameworks, preliminary hypotheses and discussion of concepts qualify as means to develop theory and should be at least valued as such. For Weick, the term theory should not be restricted to just good theory or grand theories and should include the interim struggles, just like how physicians theorize about what is wrong with the patient from data and symptoms without necessarily introducing any theory. In this example, equally important to the theory (or diagnosis) is the context in which the theory lives, and, thus, the theorizing process itself becomes critical because it points to the depth of the theoretical analysis provided by the researcher. Theory therefore is much richer, multidimensional, and much more nuanced than most would like it to be, and that is why "good theories" are hard to come by (DiMaggio, 1995)

The third wave of discussions on theory in management was marked by Hambrick's (2007) complaint of concerning the management field's obsession with theory at the expense of interesting empirical discoveries. In response, Corley and Gioia (2011) laid out what theoretical contributions amount to, providing some practical guidance as to what originality, utility, relevance, and prescient scholarship that influences the future of management might look like. The conclusion of all that discussion (Suddaby et al., 2011) was that by and large management researchers failed to develop their own theories and that any theory they use fails to capture the rich manifestations of organizations in society (Kilduff,

Mehra, & Dunn, 2011; Oswick, Fleming, & Hanlon, 2011; Sandberg & Tsoukas, 2011; Shepherd & Sutcliffe, 2011). One reason for this phenomenon is that the management field neglected testing to whittle down the morass of theories (Pfeffer, 2014) that had already cluttered the field since its early history (Koontz, 1961).

It would take more than a decade after Hambrick's (2007) complaint for management theorists to consider that perhaps they have restricted the meaning of theory too much, leaving no room for the varied knowledge about management. Thus, Sandberg and Alvesson (2020) propose an expansion of the meaning of theory in order to engender a more creative approach toward alternative, multidimensional, heterogenous views of organizational phenomena. Their proposal includes structural elements specifying minimal requirements for "theory" without restricting its meaning. Those structural elements specify a purpose for the theory, direct the theory to a phenomenon, embody some conceptual order, provide insights beyond common sense, hold relevance that can be evaluated, have empirical support, and are constrained by certain boundary conditions. Using primarily purpose and the target phenomenon as criteria, they build a typology of theory types very different from traditional views of theory—theories that explain, comprehend, order, enact, and provoke.

Theory from the Social Sciences

Before discussing what theory means to social scientists such as economists and sociologists, it is useful to understand what the term "theory" means. The term "theory" itself is Greek from *theoria*, which means "looking at, viewing, beholding" (Liddell & Scott, 1889), and historically, in this sense, theorists played the role of philosophers as the beholders of society. Thus, in *Nicomachean Ethics*, Aristotle saw the life of contemplation (*theoretikos*) as the only activity that could be loved for its own sake, compared to the lower lives of pleasure and politics (Liedman, 2013). Perhaps it was Francis Bacon (1620) and Isaac Newton (1687) who changed how people viewed theory from the act of beholding to understanding the precision of the universe and, ultimately, transforming the hallmark of theory toward predictability. Their legacy would be later carried by others who see theories as axiomatic systems containing sets of propositions to be tested in experiments (Hallberg, 2013).

The original meaning of theory as the act of beholding was exemplified by scholars such as Adam Smith (1776), one of the founders of present-day economics, who set out on *An Inquiry into the Nature and Causes of the Wealth of Nations.* Theory, to the classical economists, was not just about utilitarian goals as depicted by models promoted by the neoclassical economists; it was also about understanding the workings of an economy based on both market transactions and requirements of moral philosophy. Economists have always held competing views about what constitutes economic theory, but this diversity of thought in economics succumbed to positivist pressures, and in the end economic theory subscribed to either post-positivist Popperian views of theory or its more scientistic extremes (Blaug, 1997). Those following Popper (1959, p. 59) saw theory as a means of explaining the world that's out there since "theories are nets cast to catch what we call 'the world': to rationalize, to explain, and to master it. We endeavor to make the mesh ever finer and finer." Mainstream economics have largely followed this understanding of theory. Others such as Duhem (1906) and Quine (1951) argue that the world is not so deterministic and theory is necessarily underdetermined. The human sciences schools of theory (Dilthey, 1883; Gadamer, 1976) hold that social science theories differ in nature from natural science theories. For Dilthey, theories are not just descriptions of physical and natural phenomena. Our meaningful lives demand that theory should also include descriptions of lived experiences and historical statements, understanding (*verstehen*) of unique instances, and evaluative judgments and practical rules, all of which are summed up in his proposed science of the spirit (*Geisteswissenschaften*).

Going back to Greek origins that equate theory with philosophy and the love of true knowledge, Gadamer (1975, p. 454) sees theory as the highest form of practice, the "highest manner of being human." Through the act of beholding from a distance, theory gives what is being witnessed its validity by being caught up in it. To Gadamer (1998), the researcher who fulfills the intended goal of the research is not as impressive as one who is sidetracked by a puzzling outcome and as a result beholds something new and unexpected. What that researcher stumbled on outside oneself is the real theory. In the beginning, sociologists would vacillate between these forms of theories, but eventually would also take the road of scientization (Alexander, 1982). For Homans (1974), theory must be propositional, forming a deductive system following the covering-law view of explanation. Using the so-called Carnegie Project on Theory, Parsons attempted to push his idea of action theory (Parsons & Shils, 1962) as the grand framework that would unify the social sciences (Isaac, 2020). Others (Alexander, 1982) challenge this "scientization" of sociology and hold on to the view that neither the empirical observational world nor the non-empirical metaphysical world alone can determine social theory.

For Alexander (1982), scientific thought and theories can take many forms and exist on an epistemological continuum (Fig. 1.1). Larson (1973) notes that theory is "one of the most amorphous terms in science. [It may be] as broad as all thought or as narrow as a single thought [and may] vary from complete conjecture to solid confirmation, from unarticulated impression to precisely defined prediction." (p. 4). With such a wide range of possibilities, limiting theory to merely verificational dimensions of empirical practice would have an impoverishing impact on theorizing, and would limit any progress to methodological innovation and analysis. How does one reconcile these different forms of theory?

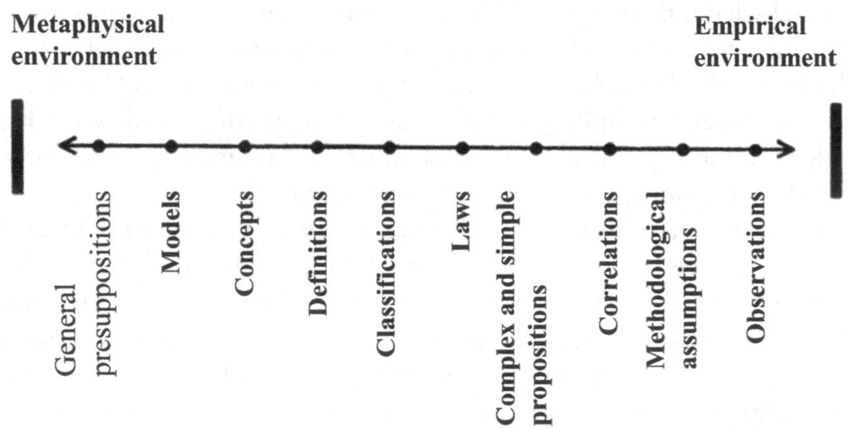

Fig. 1.1 The epistemological continuum (Reproduced from Alexander (1982) with permission from the author)

The answer might lie with the giants of the social sciences. The likes of Emile Durkheim, Max Weber, Karl Marx, and Bronisław Malinowski did not begin their research with a handy theory borrowed from some reference discipline and simple propositions with convenient constructs. They observed certain compelling facts and events, such as the different occurrences of suicide, the growth of capitalism in religious societies, and the question of class conflict and possibilities of universal culture, and began theorizing the core concerns of their phenomenon of interest in a myriad of forms. This is how giants in the past approached their research. There was no need for any ostentatious declaration of theory or elaborate justification of how their research extended other theories.

A Novel View of Theory

One way of reconciling the different meanings of theory is to redefine theory not based on its ontological or epistemological form, or even in its evaluative and teleological form, but as a communication tool like a map that points to all its discoveries (Abend, 2008). When one party insists on answering the "What is theory?" question based on its ontology, structure, composition, how good a theory it is, and what it is designed for, one is forced to take sides (Weber, 2012). At this current stage of the IS field's development, such a view of theory is at least unproductive, if not harmful (Siponen & Klaavuniemi, 2019). For example, limiting theory to its positivist epistemological form creates an unnecessary dichotomy with journal editors and reviewers between research that has or does not have theory embedded in it (Ågerfalk, 2014). This dichotomy manifests itself in how researchers become too "hung-up on theory" (Avison & Malaurent, 2014; Grover & Lyytinen, 2015; Hirschheim, 2019), causing researchers to miss out on more productive and significant areas of research to where their energies could be directed. In the IS field, this dichotomy feeds into the received view that research can be either exploratory—commonly taken pejoratively—or explanatory, the latter describing "real" research that needs to happen. As a result, certain types of qualitative research such as those that apply interviews are pigeonholed as being exploratory until such point that a "definitive" survey can be undertaken to validate the findings of the exploratory research.

Viewing theory from a communicative stance as opposed to its ontological, epistemological, evaluative, or teleological form is not unlike how formal theorists view theory. For example, Bacharach (1989, p. 496) describes a theory as "no more than a linguistic device used to organize a complex empirical world. Therefore, the purpose of theoretical statements is twofold: to organize (parsimoniously) and to communicate (clearly)." Ramsey (1965, p. 212) describes theory "simply as a language for discussing the facts the theory is said to explain" This semantic, communicative, and linguistic view of theory is proposed by Abend (2008) so that theory itself will be free of all kinds of baggage that have been piled on to it in the past century, and so that social scientists will be free to adopt and work with any form of theory in order to make epistemic progress.

By viewing theory in this way, we do not have to argue about whether or not the field has a theoretical core (Gray, 2003), what constitutes theoretical contribution (Grover, 2012; Ågerfalk, 2014), or whether we need theory-light papers (Avison & Malaurent, 2014). We only need to understand how the theory communicates its content. Doing so does not mean that we are not interested in the components of theory, or do not work toward constructing good theory. We do this so that, as a field we can all be on the same page and start theorizing about IS without arguing whether or not our discussions have theory and start making progress that will eventually lead to "good" theories. Some might say that if everything is theory, then the term "theory" itself becomes meaningless. It is not that anything can be theory; rather that theory can refer to many things. There is a big difference between saying "anything" and "many things" (Corvellec, 2013). Inspired by Abend (2008), we propose in the next section the ten semantic forms of theory that are best suited for the IS field. Following Gregor's (2006) recommendation for IS researchers to better describe the theory they are using, our ten semantic forms of theory provide the necessary detail of the communicative contents of theory that will prevent any misunderstanding that might ensue.

Ten Semantic Forms of Theory (What Theory Means)

Before describing the many semantic meanings of theory, it is important to distinguish all of them from the layman's version of theory, which is the form when someone says "that's just a theory," which normally refers to a guess or something that lacks credibility or, worse, the stuff of conspiracy theories which are not backed by any evidence. Essentially, theories are statements of claims that can be received as *knowledge* and they can take the following ten linguistic and communicative forms.

Theory 1: Theory as Proposition or Empirical Generalization

To a logician, a proposition is something that may be asserted or denied, that is, it has to be either true or false. Often they are made up of one or more declarative sentences (Copi & Cohen, 2001). These declarative sentences or general propositions represent what most scholars would call at least an "empirical generalization"(Abend, 2008), which is the minimum expected in any journal article. Merton (1945) defines empirical generalizations as taking two forms: an isolated proposition summarizing observed uniformities of relationships between two or more variables; and scientific laws, which are statements of invariance derivable from a theory. Propositions can be as simple as "all men are mortal" and in this case can be nonrelational or, as Merton (1945) defines it, can be relational and therefore relate changes of one entity to the changes of another (Fawcett, 1998). Regardless of the nature of the proposition, it generally says something about a concept or relates one concept with another. Statements of claims within this category include laws since laws explains facts and relate them to other facts. An example of such a general proposition is the law of supply and demand which is often called the theory of supply and demand. It is referred to as a law since it describes a universal increase in prices as quantity demanded increases and the decrease in prices as the quantity of supply increases. Kaplan (1964) describes at least seven different types of laws that form part of explanatory theory. The form of the law varies depending on its universality or generality. For example, Michel's (1915) "iron law of oligarchy" is a form of *genetic law* that defines the state of an organization based on its age and asserts that all complex organizations, regardless of how democratic they are when started, eventually develop into oligarchies. General propositions or empirical generalizations may or may not be laws but nevertheless yield a certain order of knowledge.

Propositions are the elements of explanation (Achinstein, 1983), which, in turn, is the goal of theories. Science is essentially about the "very process of exhibiting the systematic connections of propositions about matters of common knowledge" (Nagel, 1979, p. 6). Thus, isolated propositions trigger theory formulation by the process of restating it as a relation that extends beyond its original context. The example of Durkheim's (1895) comparison of suicide rates between Catholics and Protestant populations illustrates

(*continued*)

Theory 1: (continued)

this process. Durkheim found a lower suicide rate among Catholics which posed a theoretical problem. What could be causing such a regularity? Why are the rates different? By connecting a set of isolated propositions that included concepts of social cohesion, psychic support, unrelieved anxieties, and the proposition that Catholics had greater social cohesion than Protestants, he formulated the theory that groups with certain conceptualized attributes (e.g., social cohesion) can be expected to display different behavior (i.e., suicide) and theorized various forms of suicidal behaviors. Theory emerged from this process as a result of the cumulation of both relational propositions and research findings.

Relational propositions typically assert the patterns of covariation between two or more concepts and make up the components of both explanatory and predictive theories. These forms of explanation assert the direction, shape, or strength of specific relations. These relational propositions often begin as nonrelational propositions. Nonrelational propositions include existence propositions that assert the existence or level of existence of a phenomenon and definitional propositions that describe the characteristics of that phenomenon. The positivists' term "operational definition" is a specific type of definitional proposition that allows the researcher to use a concept in a particular manner so as to allow for the proposition to be tested (Fawcett, 1998).

Theory 2: Theory as Model

Empirical generalizations and laws are often described in mathematical language to improve its precision and to measure the magnitude of change in the relations between variables. Such interconnected collections of propositions are called models, and many theorists qualify them as theories. Ontologically, models can be created without any theoretical justification since it is essentially an imaginary material analogy, not necessarily existing in any real physical system (Hesse, 1966), but are useful because it simplifies the complex reality into an abstraction that allows the research question to be addressed at a practical level of detail (Rosenblueth & Wiener, 1945). Semantically, the model communicates useful notions of positive and neutral analogies (Hesse, 1966), albeit as an imperfect form of the phenomenon of interest, but allows the researcher to draw out new relations between the model properties and the phenomenon of interest. This process of drawing out new relations outside the original context of the model is the act of theorizing and makes models, from a linguistic perspective, theories.

(*continued*)

Theory 2: (continued)

As Harré (1970, 1976) explains, a model is no more than a putative analogue for a real mechanism, modeled on things, materials, and processes that we already understand. Harré (1970) describes several types of models distinguished according to whether the subject of the model is also the source of the model. For instance, a model airplane in a wind tunnel is constructed based on the original airplane and is called a homeomorph, which can differ in terms of scale, purity, and level of detail. Models in which the subject is not the same as the model, for example economic models, are termed "paramorphs," which are used to model processes that are unknown or yet to be investigated. One of the earliest theorizing works on magnetism by William Gilbert (1893) applied the model of the earth as a magnet to explain why compasses point north. Likewise, biologists and paleontologists apply principles of modeling each time they attempt to predict what role biological structures in animals (or dinosaurs) play in their habitat and, thus, also predict (and speculate in the case of dinosaurs) how they can be conserved (or why they went extinct).

Models are often conflated with research frameworks even though in terms of communicating what is significant about the research, they are distinct. For one, models and theories are constituents of a framework since a framework is defined as the researcher's map of the phenomenon of interest, which consists of the main concepts, constructs, intellectual traditions, models, and theories (Miles & Huberman, 1994). The conceptual framework plays the role of the mechanism that addresses why the research topic matters and why the research method is appropriate and rigorous (Ravitch & Riggan, 2012), serving as the source of stability for the research regardless of which model or theory is adopted or abandoned.

One important role of theories is to bring something novel to the "stock of current knowledge and practice" (Rynes, 2002, p. 311) or add value to other researchers and practitioners. Because models uncover what is not evident or significantly modify existing assumptions or beliefs, many consider models to be theories (Dubin, 1969; Whetten, 1989; Jaccard & Jacoby, 2010). In the IS field, two models in particular, namely the Technology Acceptance Model (TAM) (Davis, 1989; Davis, Bagozzi, & Warshaw, 1989) and the IS success model (Delone & McLean, 1992), are claimed to be the two most applied IS theories (Moody, Iacob, & Amrit, 2010; Straub, 2012). TAM was abstracted from the theory of reasoned action (TRA) (Fishbein & Ajzen, 1977), and its later versions incrementally reverted back to its origins (Benbasat & Barki, 2007). Therefore, how much TAM added to the stock of current knowledge may be put to question. The IS success model started out as a comprehensive taxonomy of factors such as system quality, information quality, use, user satisfaction, and impact from numerous studies. In other words, it too modeled what already existed and arguably add limited value to other researchers and practitioners.

Theory 3: Theory as Paradigm

Paradigms are very close to models but deserve a separate category of linguistic form of theory because of its transformative potential and unique role in the conceptual development of the IS field. As a form of heuristic, paradigms have been largely misunderstood in the IS field (Hassan & Mingers, 2018). Instead of remaining faithful to Kuhn's (1970) intended goal of positioning paradigms as concrete solutions to particular problems that can serve as exemplars for solving other problems, paradigms are understood in the IS field as philosophy or epistemology, which consequently resulted in researchers taking sides (Banville & Landry, 1989; Landry & Banville, 1992; Hassan, 2014). Paradigms draw from their communicative role of long-held theory that are employed as concrete "puzzle-solutions" (Kuhn, 1970, p. 175) or exemplars that can replace explicit rules as a basis for the solution of remaining puzzles. Paradigms as Theory 3 goes beyond models as Theory 2 because it ties together not just philosophical or epistemological elements (belief system, myths, speculation, and ways of seeing), but also sociological (recognized achievements, political bases, and accepted opinions) and conceptual elements (standard terminology, analogies, procedures, and tools).

The conceptual and artifactual sense of paradigms illustrates well how paradigms play the role of theories. Related to Theory 1, paradigms can take the form of symbolic generalizations or formal expressions that are shared unquestionably by the research community. Thus, the symbolic generalization $e = mc^2$ acts as both theory and paradigm by internally supporting researchers in their research efforts as well as explaining to others externally what those researchers (i.e., theoretical physicists) do. Examples of theory as paradigms in the IS field include the paradigm of decision support systems (DSS) as a research program that is based on Gorry and Scott-Morton's (1971) descriptive theory (Theory 8). Another example is the paradigmatic hierarchy of information largely owed to Kenneth Boulding (1955), who introduced the hierarchy of information consisting of data, information, and knowledge.

Theory 4: Theory as Worldview (*Weltanschauung*)

Closely related to paradigms and drawing from constructivist and interpretive philosophy is the notion of the worldview (*Weltanschauung*) that takes the form of theory as sets of beliefs, myths, speculation, standards, ways of seeing, and organizing principles that govern perception (Wisdom, 1972). This map that determines reality is called upon when one applies a theoretical perspective (Burton-Jones, McLean, & Monod, 2015) or experiences a

(*continued*)

Theory 4: (continued)

"worldview shift" (Dent, 1999). Dilthey (1957) notes that one constructs a worldview in order to make sense of the vagaries and mysteries of life. Typically, this form of theory takes the shape of incommensurable ways of seeing the world and practicing science in it. Masterman (1970) calls it the metaphysical paradigm. A classic example in history for this theory is the theory of phlogiston (Becher, 1703), which captured the imagination of researchers for nearly a century. This theory replaced one of the four Aristotelian elements that make up the world with a substance called phlogiston. It was an element that could not be bottled but could be transferred and was the reason for the combustibility of substances. Later, Lavoisier (1777) would help dethrone this theory and replace it with the theory of oxygen. Unlike the other theories that address the specific phenomenon, Theory 4 is about how to look at, grasp, and represent other theories. Theory 4 provides the syntax, the language, and linguistic equipment—some may say, the a priori framework (scheme, map, and grid) independent of experience, but with the power and insight to create the possibility of experience. For the IS field, a classic example of Theory 4 is General Systems Theory (Bertalanffy, 1968). Like other worldviews, it does not take the shape of a particular theory, but does help prescribe certain features of the investigation, and can be refuted with the help of other forms of theory (Wisdom, 1972).

Other examples of this form of communication commonly found in the IS field are the juxtaposition of variance theory with process theory, postmodern theory, theory of practice, actor-network theory (ANT), complexity science, and sociomateriality. Mohr's (1982) conception of variance versus process theories was introduced through Markus and Robey's (1988) work on understanding causality. Using this juxtaposition, it was possible to logically view causality as either the result of covariation among properties within a system (variance theory) or the result of necessary conditions connected in a process of events (process theory). Substantive theories can be shaped based on these two different perspectives. On the other hand, postmodern theory does not refer to causality, but to a metaphysical paradigm, a worldview that shares a tradition of mistrust of grand theories (Theory 5) and modernism (Bertens, 1993). ANT does not refer to any collection of works of different thinkers, but a deliberate elision meant to highlight the tension between the first two words—actor and network—in order to foreground the network effects of everything, social and material, in social theory (Latour, 2005). Drawing in part from ANT, sociomateriality represents the study of technology that challenges the separation of technology, work, and organizations as discrete entities or mutually dependent ensembles and takes the extreme view that people and things only exist in relation to each other (Orlikowski & Scott, 2008). All of these perspectives provided the IS field with a variety of worldviews that added to the richness of how IS field studies technology.

Theory 5: Grand Theory for Society

Related to Theory 4, which provided the framework for how theory could be constructed, Theory 5 covers instances within those perspectives that take the shape of an all-embracing unified theory in which observations about every aspect of the phenomenon, unbounded in space and time, find their ordained place. It is natural to seek grand theories to explain all phenomena as researchers are under a lot of pressure to address the vast ranges of concerns of their society. Early philosophers such as Aristotle, Kant, and Hegel experienced the same pressures and offered their grand solutions, most of which fell into deserved disuse. Merton (1968, p. 45) calls these theories "total systems of sociological theories" and argues that although scholars with any credibility felt they should follow in the footsteps of their illustrious grand theorists of the past, such efforts are premature because the field itself is not ready for such theories.

C. Wright Mills (1959, pp. 33–34), who coined the term "grand theory," went further and lamented that the insistence on a single, often normative view "are so general that its practitioners cannot logically get down to observation ... to problems in their historical and structural contexts ... [it is] drunk on syntax, blind to semantics" and called for a moratorium on grand theories. Examples of scholars attempting to offer such theories read like the who's who of social theorists. Karl Marx (1866) proposed a grand theory of class relations, social conflict and transformation, and claims that the ownership of economic modes of production manifests in exploitation that ultimately leads to revolution by those who do not own those modes of production. Talcott Parsons espoused the need for "the establishment of a general theory in the social sciences ... by unifying discrete observations under general concepts ... providing generalized hypotheses for the systematic reformulation of existing facts and insights" (Parsons & Shils, 1962, p. 1). This grand theory of social action became one of the foundations for functionalism, a Theory 4 for society that sought to achieve societal equilibrium through stable roles and functions.

Despite the alleged failures of grand theories, there has been a remarkable resurgence of new forms of these theories by their students and proponents. Returning back to Kant's Critique and Hegel's dialectical materialism, the founders of the Frankfurt School, Theodor Adorno, Max Horkheimer and others who were critical of both capitalism and Soviet socialism, established a reinterpretation of Marxism called critical theory to fill the gaps in Marxist theory to address more contemporary social issues (Held, 1980). Some of those ideas found their way into the IS field through the work of Habermas (Lyytinen & Klein, 1985). Functionalist thought espoused by Spencer and Parsons continued to be spread by Merton and later Daniel Bell and Anthony Giddens (1976, 1984). Giddens sought a middle ground between macro theories and micro theories and proposed the

(*continued*)

Theory 5: (continued)

theory of structuration, which viewed societal accomplishments as having an effect on both the people and the actions of the people shaping or structuring society. Understanding this process requires a double hermeneutic; not just interpreting what people do, but also interpreting how people interpret their world. It is this form of grand theory that IS scholars attempted to appropriate with limited success (Orlikowski & Robey, 1991; Jones, 1999). Another grand theory appropriated by IS scholars was the theory on attitude (Fishbein & Ajzen, 1975), which made its way into TAM. That version of grand theory from social psychology has had a considerable influence on IS research.

Theory 6: Theory as Methodology

Among the great thinkers that could also be considered grand theorists were those that focused on anti-positivist thought. These opened up a whole landscape of theories that cannot be strictly called a theory in the sense of providing some form of explanation (see next Theory 7) or model. This category of theory is closest to Kuhn's (1970) notion of paradigm as classical textbooks, standard illustrations and analogies, standard procedures, and techniques and tools (Masterman, 1970), which we dub theory as methodology. Thus, when sociologists discuss "Weberian theory" (Collins, 1986), they are not referring to the theory of bureaucracy or of Protestant ethics, which Max Weber (1930, 1947) is famous for, they are referring to his approach and recasting of sociology as a science that deals with meaningful social action, action that is directed to another human being. Included in that recasting of sociology are precise theoretical concepts such as different types of meaningful social actions, the focus on instrumental rationality, and the interpretive approach (from where we get interpretivism) using the German word *verstehen*, which redefined the goal of theory toward *understanding* rather than explaining (von Wright, 1971; Benton & Craib, 2001). Also included in this form of theory is Lewin's (1951) "field theory," which he defined as "a method *of analyzing causal relations and of building scientific constructs*" (p. 45, original emphasis).

A related approach that was introduced at about the same time was phenomenology, also proposed as an alternative to positivistic behaviorism and its denial of self-reflection as a source of knowledge. Although the prime source for phenomenology is credited to Edmund Husserl (1911), who would influence later philosophers and theorists such as Heidegger, Merleau-Ponty, Garfinkel, and Schütz, phenomenology was practiced,

(*continued*)

Theory 6: (continued)

albeit unnamed, by Hindu and Buddhist philosophers, Brentano's (1874) classification of mental phenomena, and William James' (1890) appraisal of stream of consciousness. It was primarily Schutz (1954, 1961) who influenced Berger and Luckmann (1966) to coin the terms "social reality" and "social construction," which have become part of an acceptable research method in the IS field (Boland, 1985; Walsham, 1995). For Husserl, a phenomenological investigation involves setting aside what we already know (bracketing) including commonsense beliefs (or any theories we might have) in order to describe using just our sense perceptions (phenomenological reduction). Heidegger (1977a, 1982) reinterprets Husserl's phenomenology as the philosophical problem of ontology and rejects Husserl's bracketing of the world, and instead views phenomenology as the problem of *being* (or being-in-the-world) rather than of the act of observing beings. Any object therefore cannot be experienced without necessarily experiencing how those objects *can be* in the world taking the form of a seamless engagement (*ready-to-hand*) with them. Riemer and Johnston (2014, 2017) apply Heidegger's phenomenological approach to reimagine the role of the IT artifact and connects Heidegger's phenomenology to the *Weltanschauung* of sociomateriality.

Heidegger's critique of Husserl provided the background for another theory of methodology to emerge—hermeneutics—which had existed for as long as text needed to be interpreted. Thus, biblical hermeneutics referred to the scriptural theory of exegesis, which was later appropriated by Schleiermacher (1978) as the science or art of understanding. Dilthey (1883) chose hermeneutics as the historically oriented theory for developing his project of *Geisteswissenschaften* (Human Science). Heidegger (1999) would expand Husserl's (1911) concept of phenomenology from its limited image of what *something is* to what it means to *being something* (or in short, its ontology). Taking hermeneutics as the manner of engaging, approaching, accessing, and explicating this ontology, Heidegger (1999) proposed the hermeneutics of facticity, which would then be refined by his student Hans-Georg Gadamer (1975) as hermeneutic phenomenology. When Gadamer (1975) talks about the theory of hermeneutic experience, he is referring to an understanding that fully realizes any presuppositions of both text and author and that distinguishes true prejudices and biases from false ones in order to experience the others' claim to truth.

At about the same time Gadamer was working out the new hermeneutics on the Continent, Barney Glaser and Anselm Strauss (1967) were working on filling the gap in the social science between grand theories and empirical data. Dissatisfied with solutions such as Merton's (1968) theory of the middle range, they needed a method that could improve the generation of

(*continued*)

Theory 6: (continued)

useful theory from qualitative data. They argue that such a process is possible by grounding the theory in and generating it directly from data—the method of grounded theory. By taking advantage of the methods of quantitative verification such as methods of sampling, coding, reliability, and validity, they placed the discovery of the concepts and hypotheses relevant to the phenomenon of interest (Hanson, 1958) at the same level as the process of verification. The results of grounded theory can take many communicative forms, including Theory 1 (Empirical generalizations), Theory 7 (Explanation), and Theory 8 (Significant description).

Theory 7: Theory as Explanation

This is probably what most IS researchers mean by theory with a capital "T." Within this category, explanatory theories can be philosophical (Lear, 1988), formal (Blalock, 1969), mathematically structured (Freese, 1980), deductive-nomological (Hempel, 1965), in the middle-range (Merton, 1968), constructive (van Fraassen, 1980), statistical (Salmon, 1998), portrayed as a mechanism (Glennan, 1996), or positivist case studies (Eisenhardt, 1989; Yin, 1989). Some of these theories take the form of "logically interconnected sets of propositions from which empirical uniformities can be derived," essentially taking the shape of a closed system of propositions that offers some kind of explanation (Merton, 1968, p. 39).

Gregor (2006) considers theories that not only explain but also predict phenomena to be the commonly held view of theories in both the natural and social sciences. What distinguishes this form of theory from other theories is the need for explanations to answer the question "Why?" (Salmon, 1998). For example, propositions or laws in Theory 1 do not necessarily contain their own explanation and therefore can be explained by Theory 7. Using Hempel's deductive-nomological form of theory, one or more Theory 1 can be part of an explanation (Theory 7)(Abend, 2008). For example, the question "Why does ice float?" can be explained with the help of a combination of laws and premises: (1) a substance with a lower density floats above the substance with a higher density, and (2) ice is less dense than water, together making up Theory 7. Another way of viewing Theory 7 as a kind of proposition is to use the term "theoretical laws" to mean that the relation concerning the concepts is not ostensibly observable and requires an elaborate "explanation" in order to make it work (Nagel, 1979). Theory 2 as models (mathematical, economic, conceptual, narrative, or otherwise) apply Theory 7 in various degrees of complexity to explain why in a limited

(continued)

Theory 7: (continued)

abstract way certain concepts or variables vary as a result of other concepts or variables. The ability to answer "why" questions are not limited to positivist approaches. All anti-positivist approaches and methods are equally if not more capable of addressing the "why" questions, especially as they pertain to meaningful social actions (Dilthey, 1883; Gadamer, 1975; Heidegger, 1977b) as we argue below.

Aristotle's classic four causes is an example of answering the "Why?" question. For Aristotle, to say that something exists by nature is to cite its cause. This understanding of causality differs from our modern Humean understanding of an antecedent cause that is sufficient to produce an effect. For Aristotle, causality is the calculus of change. A pile of bricks changes to a house because the building of the house is the "primary source of change" often mistranslated as "efficient cause." Similarly, the matter (often translated as the first cause—"material") of the house is brick, "that of which the thing comes to be," and what subsumes both material and efficient cause is form since form is the manner by which the material is organized to realize the change. The final cause, often translated as telos or "that for the sake of which," is really for the sake of maintaining form. It is not a different cause; it is referring back to our task of understanding the reasons why nature works in certain ways to maintain the final form. In other words, as Aristotle puts it, "nature is a cause, a cause that operates for a purpose" and it is the goal of science to understand that purpose (Lear, 1988, pp. 25–38).

After Hume, most of Aristotle's philosophical understanding of Theory 7 is subsumed by the discussion of causality as either deductive, probabilistic, teleological, or genetic. Deductive explanations consist of an axiomatic formal structure. Thus, natural occurrences such as moisture collecting on the outside of a glass of ice water, as well as social phenomena such as historical occurrences of suicide studied by Durkheim, communicate deductive explanations. They may contain premises of a statistical form, such as studies of heredity, but are nevertheless deductive because their premises necessarily imply their explanations. The premises of probabilistic explanations do not guarantee the explanation because they are only "probable" based on a statistical assumption. For example, the exact reasons why Cassius plotted the death of Caesar is unknown and can only be based on assumptions that are at best statistical generalizations of Theory 1 (perhaps a societal or psychological predisposition). Using Markus and Rowe's (2018) sensitizing framework with 27 explanations for causality in IS phenomena, most of the conceptions of explanation including causal mechanisms, cross-boundary and internal changes, and agency-related explanations fall within probabilistic explanations of causality. Functional or teleological explanations describe one or more functions (or dysfunctions) to realize certain traits

(*continued*)

Theory 7: (continued)

(similar to Aristotle's final cause) or the instrumental role an action plays in achieving some goal. Typically, these include language like "in order that" or "for the sake of" and are exemplified in Markus and Rowe's (2018) causal autonomy dimension that traces technological outcome as either an instrument of human social action or technology, or both.

Achinstein (1983) proposes a very intimate relationship between explanation and understanding by saying that the act of presenting knowledge is the illocutionary act (uttering sentences) performed with the intention of making something understandable, and defines someone understanding as someone in a complete knowledge-state with respect to the question. The *explanandum* (what is to be explained) need not be limited to physical entities, behaviors, or events in constant conjunction. It can include social and historical phenomena, which widens the communicative role of theory beyond Hempel and Oppenheim's (1948) deductive nomological model of explanation. It is entirely possible that such an explanation could be an understanding of the situation (*verstehen*) (Dilthey, 1883). Even Salmon (1998), who is sympathetic to the positivistic sciences, states that the goal of science is "*Understanding, comprehension, and enlightenment*" (p. 126, original emphasis). Ontologically, Gregor (2006) agrees with this assessment that anti-positivist explanations qualify as explanatory theory, and also considers some of them capable of prediction (Type IV).

Theories that are capable of prediction need not be capable of explanation and that is why Gregor (2006) and others distinguish theories that explain from theories that predict. When Gregor (2006) describe theories that predict (Gregor Type III), she was not able to find many examples in the IS field. However, with the explosion of interest in data analytics (Abbasi, Sarker, & Chiang, 2016; Hassan, 2019), theories that only predict have come into favor (Dhar, 2013; Mayer-Schönberger & Cukier, 2013). A closer look at explanation and prediction exposes the thin line between them; for, although it is possible to predict without providing any explanation, the explanation will almost certainly offer some capabilities for predicting. Therefore, prediction is really nothing more than a description of an event, albeit one in the future, whereas explanation must be more than a description of it. Distinguishing between explanation and prediction as the goal of the research sets the stage of how modeling in theorizing is done especially in analytics. For example, choices of the volume, sample size, and variety of data required depend on whether the goal of the analysis is to predict or to explain (Shmueli, 2010).

Another version of explanation that overlaps with the next category of theories (Theory as Significant Description) is explanation as mechanism, whereby explaining *why* a phenomenon occurs by explaining *how* it

(*continued*)

Theory 7: (continued)

occurred. Inspired by developments in the biological sciences, a mechanism is defined as an orchestrated functioning of component parts, their operations or activities, and organization that produces an outcome (Machamer, Darden, & Craver, 2000; Bechtel & Abrahamsen, 2005). It claims that the deductive-nomological model or statistical explanations are subject to Humean limitations of causality that do not really explain anything beyond constant conjunction. Instead of reducing the explanation to a set of premises or laws independent of the explanation itself (Achinstein, 1983), the correctness of the explanation requires the complete mechanism scheme with all its component parts and activities. Hedström and Udehn (2009) and Hedström and Swedberg (2010) tie this form of theory to Merton's theory of the middle range because, like middle-range theories, mechanisms unveil the processes in natural or social systems (Bunge, 1997) at a level that allows for empirical testing, and are employed and adapted to particular situations and explanatory tasks.

Theory 8: Theory as Significant Description

Whereas Theory 7 answers the "Why" question, Theory 8 answers the "What" and "How" questions. Theory 8 communicates the significance of the components or constituents of the phenomenon often by describing an insight or pattern that lies underneath the data. Statements that seek to communicate forms of classification such as Linneaus' (1735) classificatory system of the animal kingdom, taxonomies and typologies in management studies such as McGregor's (1960) Theory X and Y, and Blake and Mouton's (1964) *Managerial Grid*, what researchers often call "frameworks," not only built the foundations of their respective disciplines, but also impacted practice right to this very day. The organization, embedded patterns, and consistency of the patterns declare the significance of their phenomenon of interest. In the IS field, a classic example of such theories is Gorry and Scott-Morton's (1971) framework for MIS that supported the growth of and research in MIS for decades (Dickson, 1981; Keen, 1987) and spawned several technologies as a result (Benbasat & Konsynski, 1988; Rockart & DeLong, 1988). This meaning of theory is closest to Gregor's (2006) Type I Theory for Analyzing, although Gregor's ontological and epistemological categorization of Theory Type I, drawing from Fawcett and Downs (1986), excludes the findings of ethnographic, phenomenological, and hermeneutic studies that offer answers to the "what" and "how" questions. We follow Fawcett (1998), who includes the findings of these anti-positivist approaches, as

(*continued*)

Theory 8: (continued)

"descriptive theory." In other words, depending on the communicative goals of the study, anti-positivist research can take the form of any theory.

The source of Theory 8 can come from positivist case study methods (Eisenhardt, 1989; Yin, 1989) that answer the "what" and "how" questions, and survey research using both open-ended and structured interviews, from models or simulations that could represent the phenomenon (Theory 2) or paradigms (Theory 3). Models that are limited to describing the significance of the phenomenon captures the meaning of Theory 8. In the interpretive vein, phenomenological lived experience descriptions and hermeneutical interpretations (Kvale, 1983) and ethnographic thick descriptions of the expressions of cultures (Geertz, 1973) also communicate these forms of Theory 8.

Theory 9: Theory for Prescription

Theory for prescription or prescriptive theories take the form of theories that say what ought to be done. A good prescriptive theory needs to have explanatory power and must pass the test of utility, that is, it must work. What makes prescriptive theory different from "best practices" and benchmarking operations is its generalizability and analyzability to explain how and why prescriptions in one context work within their context, and the analysis reveals how subparts of the theory need to be adapted to fit new contexts (Clegg & Bailey, 2008). An example of research in IS that results in prescriptive theory is action research. As Baskerville and Myers (2004) describe it, "the action researcher is concerned to create organizational change" (p. 329), and this requires a theory by which the change can be made possible, a theory that William James proposed of "thought and action" (p. 331). This theory "must be explicit before the action is taken, otherwise there is a risk that the action is purposeless, and therefore meaningless" (p. 333). Theory in action research is not just limited to explaining or diagnosing the problem in action research (Baskerville & Wood-Harper, 1996); it is critical in guiding the intervention and is the result of the action, that is, it is theory grounded in action (Susman & Evered, 1978; Davison, Martinsons, & Kock, 2004).

Within social psychology and the organizational science, the notion of prescriptive theory came about as a result of efforts to narrow the increasing gap between theory and practice. Kurt Lewin (1951, p. 169) led such an effort as enshrined in his aphorism, "there is nothing so practical as a good theory," which he wrote in 1943–44 as part of the description of field theory (Theory 6: Theory as Methodology) from which the principles of action

(continued)

Theory 9: (continued)

research were derived. In the organizational sciences, early efforts to bridge the gap between theory and practice include Argyris and Schön's (1974) definition of prescriptive theory or a "theory of action" as taking the form "in situation S, if you want to achieve consequences C, do A" (p. 5). They relate their theory of action to theories of practice (not to be confused with Bourdieu's (1977) theory of practice which is Theory 4: Theory as Worldview), which consists of "a set of interrelated theories of action that specify for the situations of the practice the actions that will, under the relevant assumptions, yield intended consequences" (Argyris & Schön, 1974, p. 6). They also distinguish these theories of action into "espoused theories," which actors communicate to others, and "theories-in-use," which governs the actual action (p. 7). They and others (Friedman & Rogers, 2009) describe models and principles on building good prescriptive theories.

The technological version of prescriptive theory is the IS field's efforts to define a design science (Hevner et al., 2004) for artifacts. The goal of design science is to bridge the aspirations of the IS field to achieve legitimacy by adopting the approaches of the natural sciences (i.e., explanatory and predictive theory) with that of "design-related issues" (Walls, Widmeyer, & El Sawy, 1992, p. 37). Drawing from Herbert Simon's (1981) *The Sciences of the Artificial*, the IS field's goal in developing a class of design theories is to carry out a design process in a way that is both effective and feasible (Walls et al., 1992). Such a goal limits the IS field to the formalities and restrictions of logico-deductive structures of the natural sciences. As a form of prescriptive theories (Walls et al., 1992; Gregor & Jones, 2007), design theories can draw from multiple sources including those of the arts (Hassan, Mingers, & Stahl, 2018), architecture (Lee, 1991), and humanities. McPhee (1996) describes these two alternative paths of design theory: scientific approach to problem-solving following the natural sciences and the social approach following the human sciences. Design science in IS follows the former, while other disciplines such as art, architecture, and urban design follow a more intuitive, artistic, and idiosyncratic approach to design. Since design is a social activity involving communicative, cultural, political, ethical, and aesthetic issues, the logico-deductive prescriptive theories may not be well suited for the design of IS.

Theory 10: A Theory of Theories—Metatheory

Metatheory takes on different meanings depending on what the prefix "meta" means. Lewis and Grimes (1999) focus on applying the insights of multiple paradigms, especially metaphysical paradigms or worldviews (Theory 4) and dub it multitriangulation—the process of creating theory from combining multiple paradigms. Such a process of building these metaparadigm theories defies Burrell and Morgan's (1979) claim that organizational paradigms are incommensurable. Metaparadigm theory building takes place after the underlying paradigmatic assumptions are reviewed and analyzed (e.g., see Gioia & Pitre 1990) and applied to cultivate varied representations of the complex phenomena (e.g., see Hassard (1991)). Ritzer (2001) defines metatheorizing as systematic theorizing from theories and views it as the reality that has taken place in sociology since its inception. The value of these metatheories is no different from theories that emerge from theorizing directly from social phenomena because their role as an overarching theoretical perspective (Ritzer, 1990) also helps us understand, explain and make predictions. Ritzer's (1996) *McDonaldization of Society* combines Weber's theory of bureaucracy, Taylor's theory of scientific management, and Marxist and Habermasian critical theory to build a metatheory of social criticism, and predicts how McDonald's successful model for the fast-food industry has expanded into other areas of American and global culture including retail, tax services, childcare, and higher education.

In the IS field, Mingers and colleagues (1997, 2001, 2003) have always emphasized the need for combining different methods in research (Theory 6: Theory as Methodology) that would engender metaparadigm theory building. Bostrom, Gupta, and Thomas (2009) view Adaptive Structuration Theory (DeSanctis & Poole, 1994) as an example of metatheory that combines the decision-making paradigm, socio-technical theory, and institutional theory to provide the interplay between advanced IT, social structures and human interaction. The social structures of an advanced IT supplies the spirit of the technology that can be characterized by dimensions of decision processes, leadership, efficiency, conflict management, and the formal or informal atmosphere of the system and all which help predict behaviors of individuals and teams as they interact with the advanced IT system. Niederman and March (2019) view metatheory as comprising of a set of theory instances that are related by similar characteristics. Thus, they review process and variance metatheories, network metatheories, and co-evolution metatheories that fill the pages of IS journals.

The Need for Theories

The attack on theory comes from both academics and industry. In the IS field, many authors argue against a fixation on theory. Avison and Malaurent (2014) suggest that the desperate search for, and overemphasize on, IS theory has produced uninteresting research, and Grover and Lyytinen (2015) claim that scripted research strategies that domesticate theories from other disciplines have led to the lack of boldness and originality in IS. Hirschheim (2019) argues that our research cycle has been modified creating a circular dysfunction that generates a focus on "scholarly" theory only to create more problems rather than creating knowledge that resolves problems. Much of this attack comes from "*conflicting or overly narrow definitions of theory and theoretical contributions* among authors, editors, and review panelists" (Markus, 2014, pp. 342, original emphasis), and hopefully going back to Aristotle's approach to explanation that begins with definition, our in-depth description of the ten semantic meanings of theory should be sufficient to explain why these attacks require a more nuanced examination. As Gregor (2014) reminds us, it is not theory but the overemphasis on deductive logic and theory testing that slows progress toward new knowledge. And Hirschheim (2019), after his initial attack on theory, recommends focusing instead on "understanding" and "insight," which are the goals of theories in the first place.

The need of theory is in fact an intrinsically unique human need. Philosopher and behavioral scientist Kaplan (1964, p. 294) regards theorizing as "the most important and distinctive" activity for human beings: "to engage in theorizing means not just to learn by experience but to take thought about what is there to be learned … lower animals grasp scientific laws but never rise to the level of scientific theory" (p. 295). Within academia, this higher level of intellectual activity is naturally sought after. Corley and Gioia (2011, p. 12) begin their treatise on what constitutes a theoretical contribution by stating, "Theory is the currency of our scholarly realm," and Alvesson and Sandberg (2011, p. 247) note, "as researchers, we all want to produce interesting and influential theories." Academic priority is given first to those who can build original and interesting theories, second to those who can use them effectively, and third to those who understand them. Even for theory creators, only when scholars outside of their disciplines acknowledge and apply their theories can they say that their theories have been fully useful (Corvellec, 2013).

The context of research matters when we consider the role of theory in our research. Classically, the context of research can be divided into two phases: the context of discovery and the context of justification; and how these two roles interact with theory is described in more detail in Chap. 5, "The Process of Information Systems Theorizing as a Discursive Practice," by Hassan, Mathiassen, and Lowry. To add to their discussion on the context of discovery, Glaser and Strauss (1967) summarize some of the academic arguments on the role of theory in enhancing research:

1. Enabling prediction and explanation of social action and behavior
2. Providing evidence for progress in the field of study (in our case, that of IS)
3. To be useful in practical applications and provide practitioners understanding and control of situations
4. Provide a perspective on the phenomenon of interest and data to be collected
5. Guide research and provide a style for handling and conceptualizing data

Theory engenders a higher level of intellectual capability in researchers that goes beyond mimicking other people's research or approaches or following their cookie-cutter guidelines. This kind of mimicry is what Gregor and Hevner (2013) decry in referring to how IS researchers appropriate the guidelines for design science research (Hevner et al., 2004). Research is not just a matter of retrieving and translating what is available into one's own context. Even the menial process of summarizing an article, if done well and supported by theory, will communicate the author's thought more clearly than the original. Ibn Rushd's summaries of Aristotle became one of the sources that triggered the Enlightenment and the development of Western civilization out of the dark ages (Lyons, 2009). Theory is the source from which reinterpretation and corrections of previous works become possible. The continuing progress of any discipline can only happen with the guidance from and attention to theory (Levine, 2015). An example of such important work is the step of clearly defining problems in theorizing and in asking the right questions (Hassan, Mathiassen, & Lowry, 2019). The current crisis of fake news and the weaponization of information (Waltzman, 2017; Kang & Frenkel, 2018) require novel approaches capable of stemming its corruptive impact on our democracy. Such measures and interventions begin with acknowledging that we know very little about the problem and need to start with a clear problem statement. As Merton (1996, p. 53) notes: "It requires a newly informed theoretical eye to detect long-obscured pockets of ignorance as a prelude to newly focused inquiry."

Concerning the attack from industry, *Wired Magazine's* chief editor Chris Anderson (2008) argues that with big data we no longer have to settle for imperfect models, and since the scientific method relies on models from which we test hypotheses, big data has essential made the scientific method obsolete. Extending this argument, because theory is the goal of the scientific method, theory itself becomes unnecessary. Why do we need theory when big data can already help us predict? Not surprisingly this claim has attracted much attention from both industry and academia (Mayer-Schönberger & Cukier, 2013; Kitchin, 2014). Like many highly cited pieces, the *Wired* editorial has taken on a life of its own, as it is interpreted and reinterpreted by many to support their own stance on the topic of theory. Some have taken the predictive power of models to the point where explanatory theory becomes no longer necessary. The model "just needs to work: prediction trumps explanation" (Siegel, 2016, p. 90). For instance, *Wired Magazine* (Steadman, 2013) featured how big data predicted Osama bin Laden's location from publicly available data without any need for theories.

This argument from big data against theory is equally vacuous. First, theory is not irrelevant because even the process of collecting big data itself is based on some kind of theory (Mayer-Schönberger & Cukier, 2013). Without theory, big data itself is bereft of value. Our practical everyday life is supported and enhanced by the propositions or empirical generalizations that we hold, our Theory 1. For instance, our understanding of viral infections and pandemics in general guide our actions in the existing COVID-19 pandemic (World Health Organization, 2020). The search for a vaccine for this novel virus begins with various theories and empirical generalizations collected from previous infections such as the Spanish flu, zika, HIV-AIDS, Ebola, and SARS viruses (Fauci, Touchette, & Folkers, 2005). In the area of data analytics, exploratory data analysis and other models such as the "flattening-the-curve" hypothesis (Anderson, Heesterbeek, Klinkenberg, & Hollingsworth, 2020) provide effective simplifications of complex real phenomena that help public health officials determine the most effective intervention strategies to reduce infection and fatalities from the COVID-19 pandemic.

Related to the issue of such intractable problems that has IT and information embedded in it, one aspect of theory that has not received attention is its connection with values (Suddaby, 2014). This axiological nature of theory or the lack thereof in regard to theory applied in IS holds back research and progress of the field in addressing such IT and information-related crises. Maximizing profits and achieving certain level of effectiveness and efficiencies or similar utilitarian goals in systems carry with them specific sets of values that are not often highlighted in IS theory. The psychologists and neuroscientists who helped Cambridge Analytica mine the private data of millions of Facebook users vetted their research through a review board and followed the necessary ethical rules, but nevertheless were not aware that the research was used for illegal and unethical use (Cadwalladr & Graham-Harrison, 2018). Other disciplines are already cognizant and actively addressing the value-laden nature of theory. In Callon's (2007) terms, theories are performative; thus, economic theories do not just explain the economy, but they build the economy. In finance, MacKenzie (2006) showed that those Nobel-winning economic theories developed after the 1970s helped build hitherto non-existent derivative markets that became the source of several global financial crises. Theory holds the keys to addressing the crises and intractable problems society faces. Researchers and industry practitioners need to embrace the broader meaning of theory in order to realize its true potential.

Intimations of What Is to Come

The chapters in this volume are organized according to a logical sequence starting with Chap. 1, the introduction that answers the question "Why theory?" which leads to an answer to the question of "What is theory?" Chapter 2, "Theoretical, Empirical and Artefactual Contributions in Information Systems Research: Implications Implied," by Ågerfalk and Karlsson expands on the answer to what is theory by viewing theory as various types of knowledge contributions, the ultimate goal of any research effort as well as how that research is evaluated. The authors argue that theoretical contributions are not the only possible knowledge contribution, and the other two types of contributions, empirical and artifactual, also provide useful implications for both research and practice. Instead of viewing the implications for practice as the flipside of implications for research, as is common practice, research can be viewed as a combination of practices that study other practices to develop knowledge that aims to improve the studied practice itself. Such contribution the authors call empirical contributions, which go, by their account, hand in hand with theoretical contributions. The third type of contribution, artifactual contribution, presupposes the empirical contribution, but adds the results from intervention-oriented research that introduces an artifact. Regardless of the type of contribution, the value of that contribution lies in the opportunities that it provides for both the research domain and the domain of practice.

In Chap. 3, "Theoretical Diversity in IS Research: A Causal Structure Framework," Rowe and Markus complement their treatise published in the *MIS Quarterly* (Markus & Rowe, 2018), "Is IT changing the world? Conceptions of causality for information systems theorizing." They offer an explanation of that article's development and pedagogical aspects. Rowe and Markus describe how causal structure differs from the goals of theory, methods to build theory, and the techniques applied in analyzing causes. They elaborate on the need for more diversity in theoretical positions because of the multifaceted and rapidly evolving phenomena of interest to the IS field, the equally, if not more, diverse aims and goals of IS researchers, and the need to make IS research more relevant to stakeholders. The justification behind each causal structure and its ontology, trajectory, and autonomy are given a more personal treatment with regards to how they can be useful toward building substantive and middle-range theories, as opposed to borrowing from other disciplines.

The discussion on diverse approaches offered by the causal structure framework is followed by a description of the creative nature of theory building by Rivard in Chap. 4, titled "Theory Building is Neither Art nor a Science, But a Craft." In this chapter, developed from a paper published in the *Journal of Information Technology*, the process of building theory is contrasted against the romantic view of theory building as the artistic, detailed, and exhaustive explanation of a phenomenon to a more pragmatic, interim struggle toward stronger theories. Following this view, Rivard argues that theory building resembles a craft even more than an art or a science; thus, while demonstrating creativity and imagination as an artist does, and rigor as a scientist does, all these elements are folded into an effort that requires patience and perseverance, and an iterative process that combines acts of reading, reflecting, writing that develop erudition, motivation, definition, imagination, explanation, presentation, and contribution for the theory being developed. The ultimate goal of these iterations, Rivard explains, is to produce a cohesive whole of constructs, assumptions, boundaries, and relationships that represent the theory.

The detailed processes of building theory and discursive practices presented by Rivard are elaborated in Chap. 5, "The Process of Information Systems Theorizing as a Discursive Practice," by Hassan, Mingers, and Lowry. The authors reintroduce the stage of theorizing, the context of discovery, that is neglected by most researchers in IS who spend most of their efforts within the context of justification. They argue that exciting theories are the result of efforts within the context of discovery where the intuitive leaps, mistakes, and happy accidents that litter this stage result in creative and serendipitous discoveries. The authors develop a framework for this stage of theorizing consisting of foundational and generative theorizing practices. The former enacts the discourse, problematizes the phenomenon, leverages paradigms, and bridges with practice. The latter applies various processes of analogizing, metaphorizing, mythologizing, modeling, and constructing the research framework to support the constructions of creative and exciting theories. What their framework provides is freedom from the need to borrow from the reconstructions of other disciplines instead, to be inspired creatively by them, and consequently connect with other disciplines in a meaningful way.

The following four chapters offer examples of how such creative theorizing takes place in different domains, specifically in digitization (Chap. 6, "Theorizing Digital Experience: Four Aspects of the Infomaterial," by Kreps), design science (Chap. 7, "Design Science Theorizing," by Goldkuhl and Sjöström), IS Strategy (Chap. 8, "Recontextualizing IT-Rich Theories: Illustrations from IS strategy," by Moeini, Galliers, Simeonova, and Wilson), and organizational politics and business process transformation (Chap. 9, "Pluralist Theory Building: A Methodology for Generalizing from Data to Theory," by Müller, Mathiassen, and Saunders). In Chap. 6, Kreps introduces the notion of infomateriality that provides a philosophical grounding for digitizing in the twenty-first century where IT and people are mangled in an intimate relationship. Drawing from the oeuvre of philosopher Henri Bergson, Kreps argues for the condition of society where the exchange of information and the digital tools with which society undertakes it have become constitutive of the physical context in which it exists. He merges this understanding of the blurred distinction between mental and physical with Grover's four aspects of the digital—(1) embeddedness, (2) decoupling, (3) representation, and (4) generativity—to provide insights from the experiences of an ongoing digitization project aimed at capturing human digital experiences.

In the domain of design science, Goldkuhl and Sjöström lament the ambivalence that design science shows toward theory and offers a solution in the form of "practical theory," which is not limited to domain-specific knowledge but includes knowledge to support the inquiry and design process itself. Such a theory does not distinguish between utility of a theory toward building an artifact in a specific domain and the methods and the experience of building, what they call "instrumentalities," that are applied to build that same artifact. The authors eschew the practitioner-theorist dualism and consider design as a process of inquiry. They propose several types of practical theory that are not necessarily exhaustive, but nevertheless produce the elements of a theory in progress that emerge out of the alternation between situational design inquiry and theorizing, that are capable of grasping the complexity and detail of lived experience.

Chapter 8, "Recontextualizing IT-Rich Theories: Illustrations from IS strategy," by Moeini, Galliers, Simeonova, and Wilson emerged from the authors' experience of distilling theories from the IS strategy domain. To the authors, borrowing theory can be useful especially if their constructs, configurations, and logic can be modified and successfully recontextualized. Consequently, even the black-boxed IT artifact, when properly reconceptualized, will add value to IS strategy studies by highlighting its taxonomic nature, its system-specific nature, its subcomponents, its more holistic attributes, its different functionalities, and its limitations and goals, all which contribute to better specification of the strategic role that IT artifact plays. The authors demonstrate how IS strategy studies accomplish the construct and relationship recontextualization in terms of specification and distinction.

Within the domain of organizational politics and business process transformation in Chap. 9, "Pluralist Theory Building: A Methodology for Generalizing from Data to Theory," Müller, Mathiassen, and Saunders propose a pluralist theory building process that leverages the power of multiperspective inquiry. This theory building methodology involves moving between description and theory, and between single and multiple perspectives through four iterative steps with specific deliverables: create perspective accounts, synthesize multiperspective account, create theory fragments, and synthesize pluralist theory. In this chapter, they provide insights into the challenges involved in using the methodology and the activities in which researchers may engage to address these challenges. By drawing from Mingers' pragmatic approach to pluralism and Lee and Baskerville's generalization framework, they offer a useful methodology of abstracting theory from empirical studies.

The final two chapters move the discussion on theorizing into a more philosophical level. Chapter 10, "Revitalizing thoughts on Theory, Theorizing, and Philosophizing in Information Systems," by Haj-Bolouri connects theory to philosophy via the notion of kernel philosophy, drawing an analogy from the notion of kernel theories in design science. Kernel philosophy forms the integral part of theorizing by encouraging IS scholars to question the significance of data collected and the assumptions taken, underlying epistemologies and boundaries erected, and includes ethical considerations for the research. As part of the theorizing space within the space of inquiry, kernel philosophy feeds into the process where the theory is actualized out of the research project. Chapter 11, "Reviving the Individual in Information Systems Theorizing," is by Taxén. He identifies the nexus between the individual, social, and material elements of the phenomenon of interest and ties them together through the process of communalization. This process forms a communal infrastructure comprising the individual, with his or her neurobiological factors and the social, institutionalized factors, which together enable or constrain the actions performed. Using this perspective, all the under-theorized elements—the IT artifact, information, and systems—are given new avenues for advancing IS theorizing.

Conclusion

The IS field has remained in its comfort zone of borrowed legitimization, derivative research, and half-hearted defense of its identity for too long. Serious efforts are required to take the field out of its doldrums into a new phase of advancement, not for its own sake but because the world requires it. Peter Drucker once commented to Markus (1999, pp. 200–201), "The problem with your field, is that you haven't figured out that it's about information, not about technology." The role of information and systems may have been one of novelty when Daniel Bell (1973) and Peter Drucker (1969) wrote about them half a century ago. Today, information and systems are turning the world upside down, becoming both the engine of economic growth and, at the same time, helping to divide and rip our society apart. As Galliers (2003, p. 346) noted:

> Narrowly focused, reductionist … thinking that assumes that the whole is no greater than the sum of the parts, that individual components of a complex entity will interact one with another in exactly the same way when certain of those components are taken out of the equation, that systems do not exhibit emergent properties—such thinking is unlikely to lead to the kind of insights that would emerge from a more systemic approach.

The field cannot wait for another half a century to take on its responsibility as custodians of information and of systems, as well as of the associated technologies. The road ahead needs to be clear and unambiguous and the necessary tasks to be undertaken cannot be accomplished without a clear answer to the question of why it needs to exist as a field of study beyond its service to its members. The answer to this question lies in its concepts and its theories that the field declares to the world. Fortunately, its subject matter remains current and is ever changing, providing the necessary opportunities for the field to realize its true potential.

References

Abbasi, A., Sarker, S., & Chiang, R. (2016). Big data research in information systems: Toward an inclusive research agenda. *Journal of the Association for Information Systems, 17*(2), i–xxxii. https://doi.org/10.17705/1jais.00423

Abend, G. (2008). The meaning of "theory". *Sociology Theory, 26*(2), 173–199.

Achinstein, P. (1983). *The nature of explanation*. New York: Oxford University Press.

Ågerfalk, P. J. (2014). Insufficient theoretical contribution: A conclusive rationale for rejection? *European Journal of Information Systems, 23*(6), 593–599.

Alexander, J. C. (1982). *Positivism, presuppositions, and current controversies, theoretical logic in sociology*. Berkeley, CA: University of California Press.

Alvesson, M., & Sandberg, J. (2011). Generating research questions through problematization. *Academy of Management Review, 36*(2), 247–271.

Anderson, C. (2008). The end of theory. *Wired, 16*(7), 71.

Anderson, R. M., Heesterbeek, H., Klinkenberg, D., & Hollingsworth, T. D. (2020). How will country-based mitigation measures influence the course of the COVID-19 epidemic? *Lancet, 395*(10228), 931–934.

Argyris, C., & Schön, D. A. (1974). *Theory in practice: Increasing professional effectiveness*. San Francisco, CA: Jossey-Bass Publishers.

Avison, D., & Malaurent, J. (2014). Is theory king?: Questioning the theory fetish in information systems. *Journal of Information Technology, 29*(4), 327–336.

Bacharach, S. B. (1989). Organizational theories: Some criteria for evaluation. *Academy of Management Review, 14*(4), 496–515.

Bacon, F. (1620). *Novum Organum*. London.

Banville, C., & Landry, M. (1989). Can the field of MIS be disciplined? *Communications of the ACM, 32*(1), 48–60.

Baskerville, R., & Myers, M. D. (2004). Special issue on action research: Making information system research relevant to practice. *MIS Quarterly, 28*(3), 329–335.

Baskerville, R. L., & Myers, M. D. (2002). Information systems as a reference discipline. *MIS Quarterly, 26*(1), 1–14.

Baskerville, R. L., & Wood-Harper, A. T. (1996). A critical perspective on action research as a method for information systems research. *Journal of Information Technology, 11*, 235–246.

Becher, J. J. (1703). *Physica subterranea*. Spirensis, Germany: Lipsiae.

Bechtel, W., & Abrahamsen, A. (2005). Explanation: A mechanist alternative. *Studies in History and Philosophy of Biological and Biomedical Sciences, 36*(2), 421–441. https://doi.org/10.1016/j.shpsc.2005.03.010

Bell, D. (1973). *The coming of the post-industrial society: A venture in social forecasting*. New York: Basic Books.

Benbasat, I., & Barki, H. (2007). Quo vadis TAM? *Journal of the Association for Information Systems, 8*(4), 211–218.

Benbasat, I., & Konsynski, B. (1988). Introduction to special section on GDSS. *MIS Quarterly, 12*(4), 588–590.

Benton, T., & Craib, I. (2001). *Philosophy of social science: The Philosophical foundations of social thought*. New York: Palgrave.

Berger, P. L., & Luckmann, T. (1966). *The social construction of reality*. New York: Anchor Books.

Bertalanffy, L. v. (1968). *General systems theory: Foundations, development, applications*. New York: George Braziller.

Bertens, H. (1993). The postmodern weltanschauung and its relation to modernism: An introductory survey. In J. Natoli & L. Hutcheon (Eds.), *A postmodern reader* (pp. 25–70). Albany, NY: State University of New York Press.

Bichler, M., Frank, U., Avison, D., Malaurent, J., Fettke, P., Hovorka, D., et al. (2016). Theories in business and information systems engineering. *Business Information Systems Engineering, 58*(4), 291–319.

Blake, R. R., & Mouton, J. S. (1964). *The managerial grid: Key orientations for achieving production through people*. Houston, TX: Gulf Publishing Co..

Blalock, H. M. (1969). *Theory construction: From verbal to mathematical formulations*. Englewood Cliffs, NJ: Prentice-Hall.

Blaug, M. (1997). *Economic theory in retrospect*. Cambridge, UK: Cambridge University Press.

Boland, R. J. (1985). Phenomenology: A preferred approach to research on information systems. In E. Mumford et al. (Eds.), *Research methods in information systems* (pp. 193–200). North-Holland: Elsevier Science Publishers B. V.

Bostrom, R. P., Gupta, S., & Thomas, D. (2009). A meta-theory for understanding information systems within sociotechnical systems. *Journal of Management Information Systems, 26*(1), 17–47.

Boulding, K. E. (1955). Notes on the information concept. *Exploration, 6*, 103–112.

Bourdieu, P. (1977). *Outline of a theory of practice.* Cambridge, UK: Cambridge University Press.

Brentano, F. C. (1874). *Psychology from an empirical standpoint.* London: Routledge.

Bunge, M. (1997). Mechanism and explanation. *Philosophy of the Social Sciences, 27*(4), 410–465.

Burrell, G., & Morgan, G. (1979). *Sociological paradigms and organisational analysis.* London: Heinemann.

Burton-Jones, A., McLean, E. R., & Monod, E. (2015). Theoretical perspectives in IS research: From variance and process to conceptual latitude and conceptual fit. *European Journal of Information Systems, 24*(6), 664–679. https://doi.org/10.1057/ejis.2014.31

Byron, K., & Thatcher, S. M. B. (2016). Editors' comments: "What I know now that I wish I knew then"—teaching theory and theory building. *Academy of Management Review, 41*(1), 1–8.

Cadwalladr, C., & Graham-Harrison, E. (2018). How Cambridge Analytica turned Facebook "likes" into a lucrative political tool. *The Guardian.*

Callon, M. (2007). What does it mean to say that economics is performative. In D. MacKenzie, F. Muniesa, & L. Siu (Eds.), *Do economists make markets? On the performativity of economics.* Princeton, NJ: Princeton University Press.

Charles, D. (2000). *Aristotle on meaning and essence.* Oxford, UK: Oxford University Press.

Clegg, S., & Bailey, J. R. (2008). Prescriptive theory. In *International encyclopedia of organization studies.* SAGE.

Collins, R. (1986). *Weberian sociological theory.* New York: Cambridge University Press.

Colquitt, J. A., & Zapata-Phelan, C. P. (2007). Trends in theory building and theory testing: A five-decade study of the academy of management journal. *Academy of Management Journal, 50*(6), 1281–1303.

Copi, I. M., & Cohen, C. (2001). *Introduction to logic*. New York: Prentice-Hall.

Corley, K. G., & Gioia, D. A. (2011). Building theory about theory building: What constitutes a theoretical contribution. *Academy of Management Review, 36*(1), 12–32.

Corvellec, H. (2013). Why ask what theory is? In H. Corvellec (Ed.), *What is theory? Answers from the social and cultural sciences* (pp. 9–24). Copenhagen: Liber CBS Press.

Davis, F. D. (1989). Perceived usefulness, perceived ease of use, and user acceptance of information technology. *MIS Quarterly, 13*(3), 318–340.

Davis, F. D., Bagozzi, R. P., & Warshaw, P. R. (1989). User acceptance of computer technology: A comparison of two theoretical models. *Management Science, 35*(8), 982–1003.

Davison, R. M., Martinsons, M. G., & Kock, N. (2004). Principles of Canonical action research. *Information Systems Journal, 14*, 65–86.

Delone, W. H., & McLean, E. R. (1992). Information system success: The quest for the dependent variable. *Information Systems Research, 3*(1), 60–95.

Dent, E. B. (1999). Complexity science: A worldview shift. *Emergence, 1*(4), 5–19. https://doi.org/10.1207/s15327000em0104_2

DeSanctis, G., & Poole, M. S. (1994). Capturing the complexity in advanced technology use—adaptive structuration theory. *Organization Science, 5*(2), 121–147.

Dhar, V. (2013). Data science and prediction. *Communications of the ACM, 56*(12), 64–73.

Dickson, G. W. (1981). Management information systems: Evolution and status. In M. C. Yovits (Ed.), *Advances in Computers* (pp. 1–37). New York: Academic Press.

Dilthey, W. (1883). *Introduction to the human sciences*. Princeton, N.J.: Princeton University Publishers.

Dilthey, W. (1957) *Philosophy of existence: Introduction to Weltanschauungslehre* (W. Kluback and M. Weinbaum, Trans.). New York, NY: Bookman Associates.

DiMaggio, P. (1995). Comments on "What Theory Is Not". *Administrative Science Quarterly, 40*(3), 391–397.

Doty, D. H., & Glick, W. H. (1994). Typologies as a unique form of theory building: Toward improved understanding and modeling. *Academy of Management Review, 19*(2), 230–251.

Drucker, P. F. (1969). *The age of discontinuity: Guidelines to our changing society*. New York: Harper & Row.

Dubin, R. (1969). *Building theory*. New York: The Free Press.
Duhem, P. (1906). *The aim and structure of physical theory* (P. Wiener, Trans.). Princeton, NJ: Princeton University Press.
Durkheim, E. (1895). *Les règles de la méthode sociologique* (*The rules of the sociological method*). Paris: F. Alcan.
Eisenhardt, K. M. (1989). Building theories from case study research. *Academy of Management Review, 14*(4), 532–551.
Fauci, A. S., Touchette, N. A., & Folkers, G. K. (2005). Emerging infectious diseases: A 10-year perspective from the National Institute of Allergy and Infectious Diseases. *Emerging Infectious Diseases, 11*(4), 519–525. https://doi.org/10.3201/eid1104.041167
Fawcett, J. (1998). *The relationship of theory and research*. Philadelphia: F. A. Davis Company.
Fawcett, J., & Downs, F. S. (1986). *The relationship of theory and research*. Norwalk, CT: Appleton-Century-Crofts.
Fishbein, M., & Ajzen, I. (1975). *Belief, attitude, intention and behavior*. Reading, MA: Addison-Wesley.
Fishbein, M., & Ajzen, I. (1977). *Belief, attitude, intention and behavior*. Reading, MA: Addison-Wesley.
van Fraassen, B. C. (1980). *The scientific image*. Oxford, UK: Clarendon Press.
Freese, L. (1980). Formal theorizing. *Annual Review of Sociology, 6*, 187–212.
Friedman, V. J., & Rogers, T. (2009). There is nothing so theoretical as good action research. *Action Research, 7*(1), 31–47. https://doi.org/10.1177/1476750308099596
Gadamer, H.-G. (1975). *Truth and method* (2nd ed.). New York: Continuum Publishing Group.
Gadamer, H.-G. (1976). *Philosophical hermeneutics*. Berkeley: University of California Press.
Gadamer, H.-G. (1998). *Praise of theory: Speeches and essays*. New Haven, CT: Yale University Press.
Galliers, R. D. (2003). Change as crisis or growth? Toward a trans-disciplinary view of information systems as a field of study: A response to Benbasat and Zmud's call for returning to the IT artifact. *Journal of the AIS, 4*(6), 337–351.
Geertz, C. (1973). *The interpretation of cultures*. New York, NY: Basic Books.
Giddens, A. (1976). *New rules of sociological method: A positive critique of interpretive sociologies* (2nd ed.). New York: Basic Books.
Giddens, A. (1984). *The constitution of society: Outline of the theory of structuration*. Berkeley: University of California Press.

Gilbert, W. (1893). *On the loadstone and magnetic bodies and on the great magnet the earth*. New York: John Wiley & Sons.

Gioia, D. A., & Pitre, E. (1990). Multiparadigm perspectives on theory building. *The Academy of Management Review, 15*(4), 584–602.

Glaser, B. G., & Strauss, A. L. (1967). *The discovery of grounded theory: Strategies for qualitative research*. New York, NY: Aldine de Gruyter.

Glennan, S. S. (1996). Mechanisms and the nature of causation. *Erkenntnis, 44*(1), 49–71. https://doi.org/10.1007/BF00172853

Goes, P. B. (2013). Editor's comment. Commonalities across IS silos and intradisciplinary information systems research. *MIS Quarterly, 37*(2), iii–vii.

Gorry, G. A., & Scott Morton, M. S. (1971). A framework for management information systems. *Sloan Management Review, 13*(1), 55–70.

Gray, P. (2003). Introduction to the debate on the core of the information systems field. *Communications of the AIS, 12*(1), p. Art. 42.

Gregor, S. (2006). The nature of theory in information systems. *MIS Quarterly, 30*(3), 611–642.

Gregor, S. (2014). Theory—Still king but needing a revolution! *Journal of Information Technology, 29*(4), 337–340.

Gregor, S., & Hevner, A. R. (2013). Positioning and presenting design science research for maximum impact. *MIS Quarterly, 37*(2), 337–355.

Gregor, S., & Jones, D. (2007). The anatomy of a design theory. *Journal of the AIS, 8*(5), 312–335.

Grover, V. (2012). The information systems field: Making a case for maturity and contribution. *Journal of the Association for Information Systems, 13*(4), 254–272.

Grover, V., & Lyytinen, K. (2015). New state of play in information systems research: The push to the edges. *MIS Quarterly, 39*(2), 271–296.

Grover, V., Lyytinen, K., & Weber, R. (2012). Panel on native IS theories. In *Special Interest Group on Philosophy and Epistemology in IS (SIGPHIL) Workshop on IS Theory: State of the Art*. Orlando, FL, Dec 16–19.

Hallberg, M. (2013). Looking at theory in theory of science. In H. Corvellec (Ed.), *What is theory? answers from the social and cultural sciences* (pp. 65–87). Copenhagen: Liber CBS Press.

Hambrick, D. C. (2007). The field of management's devotion to theory: Too much of a good thing? *Academy of Management Journal, 50*(6), 1346–1352.

Hanson, N. R. (1958). *Patterns of discovery: An inquiry into the conceptual foundations of science*. London: Cambridge University Press.

Harré, R. (1970). *The principles of scientific thinking*. Chicago, IL: University of Chicago Press.

Harré, R. (1976). The constructive role of models. In L. Collins (Ed.), *The use of models in the social sciences* (pp. 16–43). Boulder, CO: Westview Press, Inc..

Hassan, N. R. (2011). Is information systems a discipline? Foucauldian and Toulminian insights. *European Journal of Information Systems, 20*(4), 456–476.

Hassan, N. R. (2014). Paradigm lost … paradigm gained: A hermeneutical rejoinder to Banville and Landry's "Can the Field of MIS be Disciplined?". *European Journal of Information Systems, 23*(6), 600–615.

Hassan, N. R. (2019). The origins of business analytics and implications for the information systems field. *Journal of Business Analytics*. Taylor & Francis, *2*(2), 118–133. https://doi.org/10.1080/2573234x.2019.1693912.

Hassan, N. R., Mathiassen, L., & Lowry, P. B. (2019). The process of information systems theorizing as a discursive practice. *Journal of Information Technology, 34*(3), 198–220.

Hassan, N. R., Mingers, J., & Stahl, B. (2018). Philosophy and information systems: Where are we and where should we go? *European Journal of Information Systems*. Taylor & Francis, *27*(3), 263–277. https://doi.org/10.1080/0960085X.2018.1470776.

Hassan, N. R., & Mingers, J. C. (2018). Reinterpreting the Kuhnian paradigm in information systems. *Journal of the Association for Information Systems, 19*(7), 568–599.

Hassard, J. (1991). Multiple paradigms and organizational analysis: A case study. *Organization Studies, 12*(2), 275–299.

Hedström, P., & Swedberg, R. (2010). Social mechanisms: An introductory essay. In *Social mechanisms: An analytical approach to social theory* (pp. 1–31). Cambridge, UK: Cambridge University Press.

Hedström, P., & Udehn, L. (2009). Analytical sociology and theories of the middle range. In P. Hedström & P. Bearman (Eds.), *The Oxford handbook of analytical sociology* (pp. 25–47). New York: Oxford University Press.

Heidegger, M. (1977a). *Basic writings from being and time to the task of thinking*. New York: Harper & Row.

Heidegger, M. (1977b). *The question concerning technology, and other essays*. New York: Harper & Row.

Heidegger, M. (1982). *The basic problems of phenomenology*. Bloomington, IN: Indiana University Press.

Heidegger, M. (1999). *Ontology-The hermeneutics of facticity*. Bloomington, IN: Indiana University Press.

Held, D. (1980). *Introduction to critical theory: Horkheimer to Habermas*. Berkeley, CA: University of California Press.
Hempel, C. G. (1965). *Aspects of scientific explanation and other essays in the philosophy of science*. New York: Free Press.
Hempel, C. G., & Oppenheim, P. (1948). Studies in the logic of explanation. *Philosophy of Science, 15*(2), 135–175.
Hesse, M. B. (1966). *Models and analogies in science*. Notre Dame, IN: University of Notre Dame Press.
Hevner, A. R., March, S. T., Park, J., & Ram, S. (2004). Design science in information systems research. *MIS Quarterly, 28*(1), 75–105.
Hirschheim, R. (2006). Special research perspectives issue on the IS core/identity debate. *Journal of the Association for Information Systems, 7*(10), 700–702. https://doi.org/10.17705/1jais.00105
Hirschheim, R. (2019). Against theory: With apologies to feyerabend. *Journal of the Association for Information Systems, 20*(9), 1340–1357. https://doi.org/10.17705/1jais.00569
Hirschheim, R., Dennis, A. R., & Willcocks, L. (2019). Panel presentation. In *SIGPHIL@ICIS Workshop on the Death of Theory in IS and Analytics, Munich, Germany, Dec 15–16*. Special Interest Group on Philosophy in Information Systems (SIGPHIL).
Holmström, J. (2005). Theorizing in IS research: What came before and what comes next? *Scandinavian Journal of Information Systems, 17*(1), 167–174.
Holmström, J., & Truex, D. (2011). Dropping your tools: Exploring when and how theories can serve as blinders in IS research. *Communications of the Association for Information Systems, 28*(1), 283–294, Article 19.
Homans, G. C. (1974). *Social behavior: Its elementary forms*. New York: Harcourt Brace Jovanovich, Inc..
Husserl, E. (1911). Philosophy as rigorous science. In Q. Lauer (Ed.), *Phenomenology and the crisis of philosophy*. New York: Harper and Row.
Isaac, J. (2020). Theorist at work: Talcott Parsons and the Carnegie Project on theory, 1949–1951. *Journal of the History of Ideas, 71*(2), 287–311.
Jaccard, J., & Jacoby, J. (2010). *Theory construction and model-building skills: A practical guide for social scientists*. New York, NY: The Guilford Press.
James, W. (1890). *The principles of psychology*. New York: Henry Holt and Company.
Jones, M. (1999). Structuration theory. In W. Currie & B. Galliers (Eds.), *Rethinking management information systems* (pp. 103–135). Oxford: Oxford University Press.

Kang, C., & Frenkel, S. (2018). Facebook says Cambridge Analytica harvested data of up to 87 million users. *New York Times*. Retrieved March 3, 2019, from https://www.nytimes.com/2018/04/04/technology/mark-zuckerberg-testify-congress.html.

Kaplan, A. (1964). *The conduct of inquiry: Methodology for behavioral science*. San Francisco: Chandler Pub. Co..

Karahanna, E., Davis, G. B., Mukhopadhyay, T., O'Keefe, B., Watson, R. T., and Weber, R. (2002). Information systems's voyage to self-discovery: Is the First stage the development of a theory? In *International Conference on Information Systems (ICIS)*. Barcelona, Spain.

Keen, P. G. W. (1980). MIS research: Reference disciplines and a cumulative tradition. In E. McLean (Ed.), *International Conference on Information Systems (ICIS)* (pp. 9–18). Philadelphia, PA: ACM Press.

Keen, P. G. W. (1987). MIS research: Current status, trends and needs. In R. Buckingham et al. (Eds.), *Information systems education: Recommendations and implementation* (pp. 1–13). Cambridge: Cambridge University Press.

Kilduff, M., Mehra, A., & Dunn, M. B. (2011). From blue sky research to problem solving: A philosophy of science theory of new knowledge production. *Academy of Management Review, 36*(2), 297–317.

Kitchin, R. (2014). *The data revolution: Big data, open data, data infrastructures & their consequences*. Thousand Oaks, CA: SAGE Publications.

Klein, H. K., Hirschheim, R. A., & Nissen, H.-E. (1991). A pluralist perspective of the information systems research arena. In H. K. Klein, R. A. Hirschheim, & H.-E. Nissen (Eds.), *Information systems research: Contemporary approaches and emergent traditions*. North Holland: Elsevier Science Publishers.

Koontz, H. (1961). The management theory jungle. *The Journal of the Academy of Management, 4*(3), 174–188.

Kuhn, T. (1970). *The structure of scientific revolutions* (2nd ed.). Chicago: University of Chicago Press.

Kvale, S. (1983). The qualitative research interview: A phenomenological and a hermeneutical mode of understanding. *Journal of Phenomenological Psychology, 14*(2), 171–196.

Landry, M., & Banville, C. (1992). 'A disciplined methodological pluralism for MIS research', *Accounting. Management and Information Technology, 2*(2), 77–97.

Larson, C. J. (1973). *Major themes in sociological theory*. Philadelphia, PA: D. McKay Co..

Latour, B. (2005). *Reassembling the social: An introduction to actor-network-theory*. Oxford, UK: Oxford University Press.

Lavoisier, A. L. (1777). *Mémoires de l'Académie Royale des Sciences*. Paris: Royal Academy of Sciences.

Lear, J. (1988). *Aristotle: The desire to understand*. New York: Cambridge University Press.

Lee, A. S. (1991). Architecture as a reference discipline for MIS. In H.-E. Nissen, H. K. Klein, & R. Hirschheim (Eds.), *Information systems research: Contemporary approaches and emergent traditions* (pp. 573–592). Amsterdam: Elsevier North-Holland.

Lee, A. S. (2014). Theory is king? But first, what is theory? *Journal of Information Technology, 29*(4), 350–352.

Levine, D. N. (2015). *Social theory as a vocation: Genres of theory work in sociology*. New York: Transaction Publishers.

Lewin, K. (1951). *Field theory in social science: Selected theoretical papers*. New York, NY: Harper & Row.

Lewis, M. W., & Grimes, A. J. (1999). Metatriangulation: Building theory from multiple paradigms. *Academy of Management Review, 24*(4), 672–690.

Liddell, H. G., & Scott, R. (1889). *An intermediate Greek-English Lexicon*. Oxford, UK: Clarendon Press.

Liedman, S.-E. (2013). Beholding, explaining, and predicting—The history of the concept of theory. In H. Corvellec (Ed.), *What is theory? Answers from the social and cultural sciences* (pp. 25–47). Copenhagen: Liber CBS Press.

Lim, S., Saldanha, T., Malladi, S., & Melville, N. P. (2013). Theories used in information systems research: Insights from complex network analysis. *Journal of Information Technology Theory and Application, 14*(2), 5–46.

Linnaeus, C. (1735). *Systema naturea*. Paris: M.A. David.

Lyons, J. (2009). *The house of wisdom: How the Arabs transformed western civilization*. New York: Bloomsbury Press.

Lyytinen, K., & King, J. L. (2004). Nothing at the center? Academic legitimacy in the information systems field. *Journal of the AIS, 5*(6), 220–246.

Lyytinen, K., & King, J. L. (2006). The theoretical core and academic legitimacy: A response to professor Weber. *Journal of the AIS, 7*(10), 714–721.

Lyytinen, K., & Klein, H. K. (1985). The critical theory of Jurgen Habermas as a basis for a theory of information system. In E. Mumford, et al. (eds.), *Research Methods in Information Systems, Proceedings: IFIP WG 8.2 Colloquium, Manchester, 1–3 September, 1984, Amsterdam: North Holland*. North Holland: Elsevier Science Publishers B. V.

Machamer, P., Darden, L., & Craver, C. F. (2000). Thinking about mechanisms. *Philosophy of Science, 67*(1), 1–25.
MacKenzie, D. (2006). *An engine, not a camera: How financial models shape markets*. Boston, MA: MIT Press.
March, S. T., & Smith, G. F. (1995). Design and natural science research in information technology. *Decision Support Systems, 15*, 251–266.
Markus, M. L. (1999). Thinking the unthinkable: What happens if the IS field as we know it goes away? In W. Currie & B. Galliers (Eds.), *Rethinking management information systems*. New York: Oxford University Press.
Markus, M. L. (2014). Maybe not the king, but an invaluable subordinate: A commentary on Avison and Malaurent's advocacy of "theory light" IS research. *Journal of Information Technology, 29*(4), 341–345.
Markus, M. L., & Robey, D. (1988). Information technology and organizational change: Causal structure in theory and research. *Management Science, 34*(5), 583–598.
Markus, M. L., & Rowe, F. (2018). Is IT changing the world? Conceptions of causality for information systems theorizing. *MIS Quarterly, 42*(4), 1255–1280.
Markus, M. L., & Saunders, C. S. (2007). Editorial comments: Looking for a few good concepts…and theories…for the information systems field. *MIS Quarterly, 31*(1), iii–vi.
Marx, K. (1866). *Capital* (Vol. 1). Hamburg: Otto Meissner.
Masterman, M. (1970). The nature of a paradigm. In I. Lakatos & A. Musgrave (Eds.), *Criticism and the growth of knowledge: International colloquium in the philosophy of science (Bedford College, 1965)* (pp. 59–89). London: Cambridge University Press.
Mayer-Schönberger, V., & Cukier, K. (2013). *Big data: A revolution that will transform how we live, work, and think*. New York: Houghton Mifflin Harcourt.
McGregor, D. (1960). *The human side of enterprise*. New York: McGraw-Hill.
McPhee, K. (1996). *Design theory and software design*. Edmonton, Alberta, Canada: Department of Computing Science, The University of Alberta.
Merton, R. K. (1945). Sociological theory. *The American Journal of Sociology, 50*(6), 462–473.
Merton, R. K. (1968). Sociological theories of the middle range. In *Social theory and social structure* (pp. 39–72). New York: Free Press.
Merton, R. K. (1996). *On social structure and science* (P. Sztompka, ed.). Chicago, IL: The University of Chicago Press.

Michels, R. (1915). *Political parties: A sociological study of the oligarchical tendencies of modern democracy* (E. Paul and C. Paul, Trans.). New York: The Free Press.

Miles, M. B., & Huberman, A. M. (1994). *Qualitative data analysis: An expanded sourcebook*. Thousand Oaks, CA: Sage Publications.

Mills, C. W. (1959). *The sociological imagination*. New York: Oxford University Press.

Mingers, J. (2001). Combining IS research methods: Towards a pluralist methodology. *Information Systems Research, 12*(3), 240–259.

Mingers, J. (2003). The paucity of multimethod research: A review of the information systems literature. *Information Systems Journal, 13*(3), 233–249.

Mingers, J., & Brocklesby, J. (1997). 'Multimethodology: Towards a framework for mixing methodologies', *Omega. International Journal of Management Science, 25*(5), 489–509.

Mohr, L. B. (1982). *Explaining organizational behavior*. San Francisco CA: Jossey-Bass Publishers.

Moody, D., Iacob, M.-E., & Amrit, C. (2010). In search of paradigms: Identifying the theoretical foundations of the IS field. In *European Conference on Information Systems*. June 6–9, Pretoria, South Africa.

Mueller, B., & Urbach, N. (2017). Understanding the why, what, and how of theories in IS research. *Communications of the Association for Information Systems, 41*(Art 17), 349–388.

Mumford, E., Hirschheim, R., Fitzgerald, G., & Wood-Harper, A. T. (1985). Research methods in information systems. In *Proceedings: IFIP WG 8.2 Colloquium, Manchester, 1–3 September, 1984*. North Holland: Elsevier Science Publishers B. V.

Nagel, E. (1979). *The structure of science: Problems in the logic of scientific explanation*. Indianapolis, IN: Hackett Publishing Company.

Newton, I. (1687). *Philosophiæ Naturalis Principia Mathematica* (*The mathematical principles of natural philosophy*). London: Joseph Streater for the Royal Society.

Niederman, F., & March, S. T. (2019). Broadening the conceptualization of theory in the information systems discipline: A meta-theory approach. *Data Base for Advances in Information Systems, 50*(2), 18–44. https://doi.org/10.1145/3330472.3330476

Nissen, H.-E., Klein, H. K., & Hirschheim, R. A. (1991). *Information systems research: Contemporary approaches and emergent traditions*. North-Holland: Elsevier Science Publishers B. V.

Orlikowski, W. J., & Robey, D. (1991). Information technology and the structuring of organizations. *Information Systems Research, 2*(2), 143–169.
Orlikowski, W. J., & Scott, S. V. (2008). Sociomateriality: Challenging the separation of technology, work and organization. *The Academy of Management Annals, 2*(1), 433–474.
Oswick, C., Fleming, P., & Hanlon, G. (2011). From borrowing to blending: Rethinking the processes of organizational theory-building. *Academy of Management Review, 36*(2), 318–337.
Parsons, T., & Shils, E. A. (1962). *Toward a general theory of action*. Cambridge, MA: Harvard University Press.
Pfeffer, J. (2014). The management theory morass: Modest proposals. In J. A. Miles (Ed.), *New directions in management and organization theory* (pp. 457–468). Newcastle, UK: Cambridge Scholars Publishing.
Popper, K. R. (1959). *The logic of scientific discovery*. New York: Basic Books.
Quine, W. V. (1951). Main trends in recent philosophy: Two Dogmas of empiricism. *The Philosophical Review, 60*(1), 20–43.
Ramsey, F. P. (1965). Theories. In R. B. Braithwaite (Ed.), *The foundations of mathematics and other logical essays* (pp. 212–236). London: Routledge & Kegan Paul.
Ravitch, S. M., & Riggan, M. (2012). *Reason and Rigor: How conceptual framework guides research*. Thousand Oaks: Sage.
Riemer, K., & Johnston, R. B. (2014). Rethinking the place of the artefact in IS using Heidegger's analysis of equipment. *European Journal of Information Systems, 23*(3), 273–288.
Riemer, K., & Johnston, R. B. (2017). Clarifying ontological inseparability with Heidegger's analysis of equipment. *MIS Quarterly, 41*(4), 1059–1081.
Ritzer, G. (1990). Metatheorizing in sociology. *Sociological Forum, 5*(1), 3–15.
Ritzer, G. (1996). *The McDonaldization of society*. Thousand Oaks, CA: Pine Forge Press.
Ritzer, G. (2001). *Explorations in social theory: From metatheorizing to rationalization*. London, UK: Sage.
Rockart, J. F., & DeLong, D. W. (1988). *Executive support systems*. Homewood, IL: Dow Jones-Irwin.
Rosenblueth, A., & Wiener, N. (1945). The role of models in science. *Philosophy of Science, 12*(4), 316–321.
Rynes, S. (2002). From the editors. *Academy of Management Journal, 45*(2), 311–313.

Salmon, W. C. (1998). *Causality and explanation.* New York: Oxford University Press.

Sandberg, J., & Alvesson, M. (2020). Meanings of theory: Clarifying theory through typification. *Journal of Management Studies.* https://doi.org/10.1111/joms.12587

Sandberg, J., & Tsoukas, H. (2011). Grasping the logic of practice: Theorizing through practical rationality. *Academy of Management Review, 36*(2), 338–360.

Sarker, S., & Lee, A. S. (2002). Using a positivist case research methodology to test three competing theories-in-use of business process redesign. *Journal of the AIS, 2*(1), p. Art. 7.

Schleiermacher, F. (1978). *Hermeneutics: The handwritten manuscripts* (H. Kimmerle, Ed., J. Duke, Trans.). New York: Oxford University Press.

Schutz, A. (1954). Concept and theory formation in the social sciences. *Journal of Philosophy, 51*(9), 257–273.

Schutz, A. (1961). *Collected papers vol. 1: The problem of social reality* (M. Natanson, ed.). The Hague: Martinus Nijhoff.

Shepherd, D. A., & Suddaby, R. (2017). Theory building: A review and integration. *Journal of Management, 43*(1), 59–86.

Shepherd, D. A., & Sutcliffe, K. M. (2011). Inductive top-down theorizing: A source of new theories of organization. *Academy of Management Review, 36*(2), 361–380.

Shmueli, G. (2010). To explain or to predict? *Statistical Science, 25*(3), 289–310. https://doi.org/10.2139/ssrn.1351252

Siegel, E. (2016). *Predictive analytics: The power to predict who will click, buy, lie, or die.* New York: Wiley.

Simon, H. (1981). *The sciences of the artificial.* Boston, MA: The MIT Press.

Siponen, M., & Klaavuniemi, T. (2019). How and why "theory" is often misunderstood in information systems literature. In *Proceedings of the International Conference on Information Systems (ICIS 2019), Munich, Germany Dec 15–18.* Association for Information Systems (AIS).

Smith, A. (1776). *An inquiry into the nature and causes of the wealth of nations.* Glasgow: Edwin Cannan.

Steadman, I. (2013). Big data, language and the death of the theorist (Wired UK). *Wired UK.* Retrieved September 27, 2019, from http://www.wired.co.uk/news/archive/2013-01/25/big-data-end-of-theory.

Straub, D. (2012). Editorial: Does MIS have native theories. *MIS Quarterly, 36*(2), iii–xii.

Suddaby, R. (2014). Editor's comments: Why theory? *Academy of Management Review, 39*(4), 407–411.

Suddaby, R., Hardy, C., & Huy, Q. N. (2011). Where are the new theories of organization? *Academy of Management Review, 36*(2), 236–246.

Suppe, F. (1977). *The structure of scientific theories*. Urbana, IL: University of Illinois Press.

Suppe, F. (2000). Understanding scientific theories: An assessment of developments, 1969–1998. *Philosophy of Science, 67*(Supplement), S102–S115.

Suppes, P. (1967). What is scientific theory? In S. Morgenbesser (Ed.), *Philosophy of science today* (pp. 55–67). New York: Basic Books.

Susman, G. I., & Evered, R. D. (1978). An assessment of the scientific merits of action research. *Administrative Science Quarterly, 23*, 582–603.

Sutton, R. I., & Staw, B. M. (1995). What theory is not. *Administrative Science Quarterly, 40*(3), 371–384.

Swedberg, R. (2014). *The art of social theory*. Princeton, NJ: Princeton University Press.

van de Ven, A. H. (1989). Nothing is quite so practical as a good theory. *Academy of Management Review, 14*(4), 486–489.

Walls, J. G., Widmeyer, G. R., & El Sawy, O. (1992). Building an information systems design theory for vigilant EIS. *Information Systems Research, 3*(1), 36–59.

Walsham, G. (1995). Interpretive case studies in IS research: Nature and method. *European Journal of Information Systems, 4*(2), 74–81.

Waltzman, R. (2017). *The weaponization of information: The need for cognitive security*. Santa Monica, CA: The RAND Corporation.

Weber, M. (1930). *The protestant ethic and the spirit of capitalism* (R. H. Tawney, ed.). London: G. Allen & Unwin, Ltd.

Weber, M. (1947). *The theory of social and economic organizations*. New York, NY: Free Press.

Weber, R. (2003). Editor's comments: Theoretically speaking. *MIS Quarterly, 27*(3), iii–xii.

Weber, R. (2006). Reach and grasp in the debate over the IS core: An empty hand? *Journal of the AIS, 7*(10), 703–713.

Weber, R. (2012). Evaluating and developing theories in the information systems discipline. *Journal of the AIS, 13*(1), 1–30.

Weick, K. E. (1989). Theory construction as disciplined imagination. *Academy of Management Review, 14*(4), 516–531.

Weick, K. E. (1995). What theory is not, theorizing Is. *Administrative Science Quarterly, 40*(3), 385–390.

Weick, K. E. (1999). Theory construction as disciplined reflexivity: Tradeoffs in the 90s. *The Academy of Management Review, 24*(4), 797–806. https://doi.org/10.1503/cmaj.1032006

Whetten, D. A. (1989). What constitutes a theoretical contribution? *Academy of Management Review, 14*(4), 490–495.

Whetten, D. A., Felin, T., & King, B. G. (2009). The practice of theory borrowing in organizational studies: Current issues and future directions. *Journal of Management, 35*(3), 537–563.

Wisdom, J. O. (1972). Scientific theory: Empirical content, embedded ontology, and weltanschauung. *Philosophy and Phenomenological Research, 33*(1), 62–77.

World Health Organization. (2020). *Coronavirus disease 2019 (COVID-19) situation report—73.*

von Wright, G. H. (1971). *Explanation and understanding*. Ithaca, NY: Cornell University Press.

Yin, R. K. (1989). *Case study research: Design and methods*. Newbury Park, CA: Sage Publications.

2

Theoretical, Empirical, and Artefactual Contributions in Information Systems Research: Implications Implied*

Pär J. Ågerfalk and Fredrik Karlsson

Introduction

When evaluating research, a strong theoretical contribution is typically assumed. In fact, insufficient theoretical contribution is a very common, if not the most common, reason for prestigious journals in our field to reject article submissions (Venkatesh 2006; Straub 2009; Ågerfalk 2014). To qualify for publication in a top-tier information systems (IS) journal, an article must make a substantial contribution to knowledge. However, a theoretical contribution is not the only possible knowledge contribution. In this chapter, we expand on Ågerfalk's (2014) idea that empirical contributions also can be important to knowledge advancement. We

*An earlier version of this chapter appeared as Ågerfalk, P. J., & Karlsson, F. (2020). Artefactual and empirical contributions in information systems research. *European Journal of Information Systems, 29*(2), 109–113.

P. J. Ågerfalk (✉)
Uppsala University, Uppsala, Sweden
e-mail: par.agerfalk@im.uu.se

N. R. Hassan, L. P. Willcocks (eds.), *Advancing Information Systems Theories*, Technology, Work and Globalization, https://doi.org/10.1007/978-3-030-64884-8_2

expand Ågerfalk's idea by discussing three types of contributions—theoretical, empirical, and artefactual—and relate these to their possible implications for research and practice. Acknowledging the value of artefactual contributions is particularly relevant to IS since design and evaluation of digital artefacts is an accepted feature of research in the field, although non-constructive (i.e. behavioural) research is also common.

Understanding the implications of a contribution is important for understanding the value of the contribution itself. As authors, reviewers, and editors, we have found that the relationship between a contribution and its implications is often treated incidentally, using standards that compromise a fair review of any particular research. For example, an implication for theory is often treated as the flip side of the theoretical contribution coin and appears to be less useful concept to understand the value of a piece of research. Similarly, when authors claim a practical contribution, they often are not referring to a contribution to practice but to an implication for practice.

In this chapter, we follow Ågerfalk's (2014) lead by arguing that a paper might have a strong contribution to knowledge without a strong theoretical contribution. For some papers, the emphasis should not be on their theoretical contributions but rather on their empirical or artefactual contributions, followed by possible implications for research, including implications for theorizing. Our starting point is that scientific knowledge development can be seen as a practice—that is, we think of research as practices that develop academic knowledge. We focus on the reporting of research results (a research practice) to present (our contribution) a more informed and precise discussion of contributions and implications, and, as such, help researchers craft and evaluate manuscripts for publication (an implication for research).

The chapter proceeds as follows. In the next section, we expand on the notion of research practice and its relationship to contributions and implications. Using this discussion, we elaborate on contributions and implications. This analysis ends up with a proposed framework for how authors, reviewers, and editors can think about contributions and implications when writing and evaluating research for publication. Finally, we provide a brief summary and discussion of the usefulness of our contribution.

F. Karlsson
Örebro University, Örebro, Sweden
e-mail: fredrik.karlsson@oru.se

Research Actors and Practices

When speaking of research, one often resorts to poorly defined dichotomies, such as research versus practice, theory versus practice, and researcher versus professional. To set the stage for a more precise discussion of research concepts such as these, we view research as a combination of practices (Schatzki 2001; Goldkuhl and Röstlinger 2003). Specifically, we view research as a practice where somebody uses resources to do something for somebody in some context with some purpose, giving rise to some implications.

With this definition, *somebody* can be either a researcher or a practitioner. Somebody can also refer to a combination of researchers and practitioners collaboratively working. The *resources* used in research practices typically consist of existing theories, empirical data, research methods, and various tools (e.g. analysis and simulation software). The *doing* is the act of actually making a contribution, such as developing and reporting new knowledge. This contribution is aimed at somebody, typically other researchers and other practitioners. The practice is situated in some *context*, which is typically confined by the current study or research project. The construction of the contribution has some purpose: typically, to promote improved understandings or practices. The doing of the practice has *implications* for somebody as it changes the resources and doings of other practices.

Research practices study other practices. Sometimes researchers intervene in a studied practice for the purpose of learning and improving. These other practices can also be research practices where the aim is to develop knowledge about research or to improve research. However, researchers usually study other practices and it is common to distinguish between research, on the one hand, and practice, on the other. To avoid confusion, we use the terms research practices and domain practices to distinguish between the studying of practices and the studied practices. These two practices overlap to some extent, especially in collaborative projects that involve both researchers and practitioners, who are not researchers. Here, we bow to customary usage and use the term 'practitioner' to refer to actors conducting domain practices and the term 'researcher' to refer to actors conducting research practices. When the studied domain is research, the research practice and domain practice are of the same kind, but typically not the same (the exception being when researchers study themselves such as when producing autobiographies). For convenience, we use the term 'practice' also as a generic construct to refer to any set of practices of a certain kind. Thus, when there is no need to identify or enumerate particular research practices, we use the term 'research practice' to refer to the combination of research practices as a whole. In IS research, the domain practices studied are typically different kinds of digital practices (i.e. practices that rely on digital technologies to some extent such as digital somebodies, digital resources, digital doings, and digital results).

As this chapter focuses on contributions and implications, we will not cover resources explicitly. Much has been said about research resources, and we refer the interested reader to the existing research methods literature and other relevant chapters in this book.

Contributions and Implications

In the previous section, we conceive of research as a practice in which researchers study phenomena in other practices or intervene in those practices to make desirable changes (Ågerfalk and Wiberg 2018). A research contribution can then be seen as something that advances our understanding of a studied practice that might improve the studied practice itself. Although improvement is not always required, advancement is critical—a contribution is a contribution by virtue of its originality and novelty. Although different aspects of research, contributions and implications are often conflated. An implication exists in relation to a contribution: it is that to which the research contribution is leading us with regard to a particular practice. Therefore, when researchers make claims about their contributions, they presume or anticipate certain other claims. In the following, we explore this distinction between contributions and implications and explain how such a distinction can be useful when discussing research outcomes.

Contributions

To claim that some research makes a contribution is to claim that it adds to something. One might consider Walsham's question: 'What does the piece of written work *claim to offer* that is new to the audience and the literature?' (2006). This question emphasizes a comparison to what is already known and how research makes a difference rather than focusing on a specific contribution. However, in an IS research paper, the expected contribution is typically a theoretical contribution.

Theories and theorizing are often, for good reasons, viewed as the core of research practice (Weick 1995; Hassan et al. 2019): 'Developing theory is what we are meant to do as academic researchers and it sets us apart from practitioners and consultants' (Gregor 2006). Indeed, the theoretical contribution is the received notion of what constitutes a research contribution. However, the accumulated and structured knowledge resulting from research can also enlighten domain practices (Lewin 1943; Van de Ven 1989; Gregor 2006; Van de Ven and Johnson 2006). Lee and Baskerville (2003) argue that researchers formulate empirical statements and theoretical statements both of which serve important, but different, purposes. For example, empirical statements can help ground theoretical statements and theoretical statements can help sort data into empirical statements (Niederman and March 2019). From this, it follows that researchers could make contributions other than theoretical ones: 'A contribution to knowledge is not necessarily a theoretical contribution' (Ågerfalk 2014, p. 593) and 'Journal editors and reviewers need to reconsider their single-minded focus on contribution to theory' Tsang (2013, p. 199).

With this backdrop, it seems worthwhile to engage in a more comprehensive treatment of what constitutes a contribution in IS research—that is, a treatment of what advances our understanding of a studied domain practice. Theoretical and empirical contributions serve different but complementary purposes in research practices. In addition to theoretical and empirical contributions, there is a third type of contribution that seems to be less well recognized—the artefactual contribution. Distinguishing among these three types of contributions can be useful especially because the three types are closely related and easily conflated.

What constitutes a theoretical contribution is closely related to how one defines theory. We chose to take a broad view of what constitutes theory by drawing on Lee and Baskerville's view that theoretical statements posit 'the existence of entities and relationships that cannot be directly observed' (2003, p. 230). Similarly, Corley and Gioia, drawing on Sutton and Staw (1995), define theory as 'a statement of concepts and their interrelationships' but added that such a conceptual structure needs to show also 'how and/or why a phenomenon occurs' to qualify as theory (2011, p. 12). However, 'how and/or why' is not necessarily foregrounded in IS (Gregor 2006). Instead, 'concepts and their interrelationships' can also deal with what is, what will be, and what ought to be. A theoretical contribution can be viewed as something that advances our understanding of a specific set of entities and relationships (Ågerfalk 2014).

To pin down a particular theoretical contribution, one needs to consider the nature of the problem the research addresses. This problematization can be illustrated by the well-recognized taxonomy of theory proposed by Gregor (2006). In Gregor's taxonomy, different research problems call for theories with different primary goals and different contributions (Table 2.1). For example, advancing a theory for analyzing would mean something different than advancing a theory for design and

Table 2.1 Contributions and theories (based on Gregor 2006)

Theory type	Contribution to the theory's distinguishing attributes
Analysis	Advancing the description of a phenomenon—that is, saying something new about what is.
Explanation	Advancing the explanation of a phenomenon— that is, saying something new about what is, how, why, when, and where.
Prediction	Advancing the prediction and the testable propositions regarding a phenomenon— that is, saying something new about what is and what will be.
Explanation and prediction	Advancing the explanation, the prediction, and the testable propositions regarding a phenomenon— that is, saying something new about what is, how, why, when, where and what will be.
Design and action	Advancing the prescriptions regarding a phenomenon— that is, saying something new about how to do something

action. The former would say something new, something that previous research has not acknowledged. For example, Schultze and Leahy suggest using a theoretical framework to conceptualize the avatar–self relationship as 'the interaction between a communicator and his/her virtual (re) presentation' (2009). This approach allows them to identify and organize eight types of relationships that communicators enact with their avatars. Advancing a theory for design and action would include new normative statements about how to do something, prescriptions that have not previously been suggested and evaluated by research. For example, Germonprez et al. (2007) propose a theory of tailorable technology design (i.e. technology a user can modify for a specific context). The theory includes nine design principles for how to design this class of information system.

An empirical statement refers to 'data, measurements, observations, or descriptions about empirical or real-world phenomena' (Lee and Baskerville 2003, p. 232). Thus, empirical statements are accounts of something in the studied domain practice. Empirical statements constitute an empirical contribution if they provide 'a novel account of an empirical phenomenon that challenges existing assumptions about the world or reveals something previously undocumented' (Ågerfalk 2014, p. 594). Although containing descriptions, an empirical contribution, which says something about what is, should not be confused with an analytical theory (Gregor 2006). An analytical theory is still a generalization made beyond what is directly observable (Lee and Baskerville 2003). Empirical contributions identify phenomena that theory cannot explain or they uncover new details of existing phenomena. In addition, empirical contributions can corroborate theories, in whole or in part, by providing additional support.

As noted above, there is a well-recognized relationship between theoretical statements and empirical statements, and the two often go hand in hand. This relationship is typically described in terms of a generalization of empirical findings to theoretical statements (Lee and Baskerville 2003), a validation of theoretical statements through empirical findings (Tsang 2013), or a combination of both. On the one hand, theories are tools that help us analyse, explain, predict, and inform design and actions (Gregor 2006). On the other hand, empirical data are required to develop, validate, and evaluate these theories. However, advancing insight into a phenomenon does not necessarily require a priori conceptualization. Nor is the originality or novelty of an empirical contribution intrinsically tied to the need or possibility to theorize about the findings: 'There is no intrinsic reason for each piece of published research to go full circle and claim a substantial theoretical contribution' (Ågerfalk 2014, p. 594). Hence, researchers can settle on an empirical contribution providing a rich account of an empirical phenomenon by means of data, measurements, observations, or descriptions. However, they need to show the implications for research practice in terms of implications for theorizing, a concept developed in the next section.

The third type of contribution, the artefactual contribution, is neither a theoretical nor an empirical contribution, although closely related to both. An artefactual contribution results from using intervention-oriented research, such as design science research (Hevner et al. 2004), action design research (Sein et al. 2011), and action research (Baskerville and Wood-Harper 1998). In relation to artefactual contributions, originality and novelty refer to the change as such, which could be the introduction of a particular artefact. Because research that involves action or design is empirical by nature, it often presupposes an empirical contribution as part of motivating and validating the artefactual contribution (Gregor and Baskerville 2012). In design science research (DSR), for example, empirical data might be generated to prove the usefulness and value of an artefact and, as such, is part of the evidence of 'satisficeability' of the artefactual contribution (Simon 1961). Whether or not it is also a bona fide empirical contribution depends on the situation: 'The contribution of design-science research is the artefact itself' (Hevner et al. 2004, p. 87). This claim, however, should not be confused with the empirical account of the achievement to instantiate the artefact. For example, a new DSR artefact must be understood in relation to its intended organizational and social context. The artefact itself as described in a research article is an artefactual contribution. An empirical contribution in relation to the artefact could be a rich description of the design process leading to the artefact or the actual use of the artefact. Acknowledging artefactual contribution as a contribution in its own right suggests that the empirical contribution captures data, measurements, observations, or descriptions regarding the artefact; however, this is not an artefactual contribution. Nonetheless, intervention research, such as DSR artefacts or changes made to a studied domain practice through action research, should draw on existing theory or have an evolving design theory as a rationale (Walls et al. 1992). Thus, earlier theoretical contributions provide the rationale for the artefactual contribution. Similarly, developing artefacts and studying their design and use might lead to new or extended theories of design and action.

According to Sein et al., IS researchers believe that researchers should 'make theoretical contributions and assist in solving the current and anticipated problems of practitioners' (2011, p. 38). Our discussion so far has deliberately avoided the question of when a contribution comes into existence: Is it when the research is performed or is it when the research is published? Since we are focusing on presenting and evaluating contributions in articles, we have implicitly adopted the latter interpretation. Contributions, in this context, are linguistic. That is, contributions require language. Contributions are expressed by theoretical and empirical statements (Lee and Baskerville 2003) or artefactual statements describing an artefact or a change (e.g. a programme or an intervention). Indeed, this expression surfaces as a kind of Berkeley conundrum. If researchers solve a domain practice problem but never expose it to peer review, does their contribution actually exist? It follows that the solving of problems in the domain practice, sometimes referred to as a practical contribution, is not a contribution at all but either a part of the empirical data generation required to ground an empirical, theoretical, or artefactual contribution or an implication for domain practices of such a contribution.

Regardless of the type of contribution (theoretical, empirical, and artefactual), there is a sensible corollary question: When is a contribution sufficient to warrant publication? If I add an additional construct to the Technology Acceptance Model and establish statistical significance, have I established a sufficient theoretical contribution? What if I present a rich account of an unheard of use of a particular digital technology that helped emancipate a marginalized group that everybody thought was resisting modern society? What if I present an artefact that solves a novel problem using novel technology? Indeed, these questions cannot be answered unless we know more about the potential implications of the claimed contribution. The truth is in the consequences, pragmatically speaking.

Research Implications

Research can result in a theoretical, empirical, or artefactual contribution or a combination of the three. However, a contribution of any sort has limited value unless its potential implications are clearly communicated. As stated above, when researchers claim a contribution, they imply their contributions have a certain degree of utility or usefulness: 'The idea of contribution rests largely on the ability to provide original insight into a phenomenon by advancing knowledge in a way that is deemed to have utility or usefulness for some purpose' (Corley and Gioia 2011, p. 15). Although utility and usefulness are rather vague concepts, they indicate that the claimed new knowledge needs to lead us somewhere. Corley and Gioia (2011) claimed that a (theoretical) contribution must have the potential to lead to either the improvement of the current research practice or the improvement of how practitioners work (i.e. an improvement of the domain practice). Similarly, Baskerville suggests that we should ask ourselves 'what will my reader do differently, day-to-day, after reading my article?' (2004, p. 1). What these authors are speaking to is the notion of an implication—that is, where a research contribution is leading us. An implication is directed to the future; it is oriented towards what a specific practice, such as the domain practice, might do differently in the future as a consequence of learning about the contribution.

However, not all research comes with implications, or at least not with significant ones. Avison and Malaurent (2014), for example, express a fear that papers ticking all the necessary boxes regarding theoretical contributions are published, although the implications of those findings are likely to be minor. The fact that something is statistically significant does not imply that it is important and interesting. Indeed, there are studies that do not detail the implications of their findings and studies that have ineffective strategies for presenting implications—that is, papers that provide vague, sweeping, or generic implications. Geletkanycz and Tepper discuss the importance of not treating implications as something secondary or as an afterthought; instead, implications should be viewed as a reflection after having completed the research work, where the paper's 'larger, underlying value' (2012) is presented. This requires a clear understanding of what an implication is and how implications are distinct from contributions.

Over the years, there seems to have been more emphasis on discussing how to recognize and frame the contribution of research and less attention devoted to discussing the implication these contributions could have. Indeed, sometimes even the discussion of these concepts has been rather confusing. For example, Walsham (1995, 2006) views implications as a result of generalizations, together with concepts, theories, and rich insights. This use of the implication concept is unfortunate because it limits the possibilities about discussing where the results from generalizations lead us.

Implications are to some extent linked to the relevance concept and the long-lasting discussion about this concept (Benbasat and Zmud 1999; Davison et al. 2004; Sein et al. 2011; Siponen and Vance 2014). When discussing relevance, Benbasat and Zmud argue that '[u]nless an article's implications are implementable, i.e., prescribed in a manner that could be put to use (to some extent) in practice to exploit an opportunity or to resolve a problem, practitioners are unlikely to characterize it as being of interest' (1999, p. 5). Furthermore, they show that implications and relevance are not equivalent and that relevance is determined by implementability. However, Lee (1999) has criticized this 'instrumental model of practice' view of relevance where researchers produce research that practitioners consume. He argues that researchers do more than just hand practitioners contributions for them to apply. For example, research contributions can also criticize the studied domain practice; for example, studies might lead to an increased awareness or changed opinions in society despite not being implementable in a domain practice. Additionally, Davison et al. discuss the importance of recognizing the research domain as a practice: 'The nature of the selected topic and the implications of the results can be used to assess the relevance of a study' (2004, p. 67). Importantly, what counts as an implication depends on the scope of the current study. If, for example, data are collected to validate an artefactual or theoretical contribution based on assumed effects deduced from the contribution, these effects are not implications since they are within the confines of the current study. In such a case, the effects are part of the artefactual or theoretical contribution. Remember, implications are oriented to the future and outside the researcher's direct control. An implication (no pun intended) of this lack of control is that implications 'cannot be specified in full detail' (Ågerfalk 2014, p. 594). Nevertheless, researchers should carefully draw out potential implications for both research practice and domain practice.

By joining the building blocks discussed above (i.e. different types of practices and different types of contributions), we can construct the 3 × 2 matrix shown Table 2.2, where different implications are discerned. It

Table 2.2 Types of contributions and their implications

	Implications	
Type of contribution	Implications for research practice	Implications for domain practice
Theoretical—new or modified theory, or parts thereof	Opportunities for • analysing, explaining, or predicting a phenomenon differently, • using alternative theories or using theory differently, and • for executing research differently.	Opportunities for (improved) theory-informed action: • analysing, explaining, or predicting a phenomenon differently in a specific context and • designing differently in a specific context.
Empirical—an account of a phenomenon or details of a phenomenon not previously covered by research or covered to a limited extent	Opportunities for • generalizing to theory and to other empirical representations and • empirically informed design of artefacts.	Opportunities for • getting informed about a phenomenon, • benchmarking, and • empirically informed change.
Artefactual—description of proposed new or modified artefact(s)	Opportunities for • creating empirical accounts and • theorizing about class of artefact.	Opportunities for taking advantage of new or modified artefacts.

should be noted that the presentation is not meant to be exhaustive; rather, it serves to distinguish the different types of implications.

The artefactual contributions at the bottom of Table 2.2 are interventions made in the world, such as the introduction of new or modified artefacts. Implications for research practice are future opportunities for empirical accounts regarding, for example, a particular class of artefact. If one discusses the instantiation of a particular class of artefact that has not existed before, such an account would not have been possible to give before. Even if one considers less ground-breaking instantiations, such as modifications of existing artefacts, there are opportunities to explore how the modified aspects work in the intended context. Implications for the domain practice are opportunities for taking advantage of the introduced artefact and ways of working. For example, Smolander et al. (1991) describe the computer-aided method engineering tool MetaEdit, a tool for modelling of systems development methods using an interactive graphical notation. By introducing this tool, it was possible to give empirical accounts of computer-aided method engineering and to evaluate the usefulness of specific support of systems engineering that were generated using MetaEdit, including the Oinas-Kukkonen's (1998) evaluation of the possibility and usefulness of capturing design rationale in systems engineering projects.

Empirical contributions are accounts of a phenomenon or details of a phenomenon not previously covered by research. This type of contribution could open future possibilities for researchers to theorize: 'The outcome may or may not lead to theory creation' (Tsang 2013, p. 199). Theorizing can, at least in principle, materialize outside the immediate research context through the use of empirical accounts (Ågerfalk 2014). This means that researchers could discuss potential theoretical implications by pointing to future avenues for research. Practitioners could also benefit from empirical contributions because they can lead to opportunities for rich understanding of a novel phenomenon and to benchmarking against their own practices.

Theoretical contributions include wholly or partly new or modified theories. Different theories make different implications possible. From a tool perspective, new or modified theories offer possibilities to analyse, explain, and predict a phenomenon differently in future research. For example, Yang and Maxwell (2011), in a review of research, propose a summative framework for explaining successful inter-organizational information sharing in the public sector. The suggested framework provides a means for theory-informed description and explanation of future information-sharing activities between public sector organizations. Other implications include how to use a theory in a new area. Karlsson and Hedström (2013) discuss such implications in the context of evaluation of end-user development as a requirement engineering (RE) technique. They operationalize the design boundary object into an evaluation framework, making it possible to use this theory to evaluate other techniques used in RE. Implications could also be related to methodology, an area that could influence the direction of future studies. For example, Riemer and Vehring's study uses socio-materiality to study the adoption, use, and perception of a telephone solution to argue that 'software can only be understood in context and in a non-individualistic way' (2010, p. 2). Based on this contribution, they argue that the implication is the need to rethink the methods used for assessing usability, where the use of laboratory experiments 'might actually be counterproductive', emphasizing instead the direction of contextual usability methods (2010, p. 13).

Theoretical contributions can have significant implications for practitioners (Lewin 1945). New and modified theories can aid practitioners in solving context-specific tasks. A theoretical contribution can be an addition to the analytical toolbox that can be used to explain and predict a phenomenon differently in a specific context. Cronen (2001) expanded on the idea of a practical theory as a theory that is useful for identifying a situation and constructing judgements that implicate actions leading to improving a situation. Similarly, Yang and Maxwell (2011) studied factors that explain successful interorganizational information sharing in the public sector, a tool also useful for practitioners. The identified factors could raise awareness and support in the analysis of specific information-sharing initiatives. Therefore, theoretical contributions can enlighten practitioners' practice (Gregor 2006) by providing advice on future designs of artefacts and how to act in a specific context. For example, Muntermann (2009) addresses how individual investors react to unforeseen market events by drawing on behavioural finance research. His findings open opportunities for including new forecasting techniques in financial decision support systems. In addition, Morschheuser et al. (2018) identify 13 principles of how to engineer gamified software. Using these principles, they suggest a systems development method that includes steps that had not been previously acknowledged. For practitioners, both the principles as such and the new method makes it possible to develop theory-informed actions in gamification.

Importantly, research can have multiple contributions of different kinds; however, these contributions must be clearly and systematically discussed along with associated implications. Research publications often have main contributions as well as secondary contributions. Main contributions concern the domain practice, which is the practice being researched. As a consequence, the main contribution should also direct the main implications of the paper and researchers need to detail implications for the domain practice. Of course, this does not exclude the main contribution having implications also for research and other practices. Secondary contributions, being by-products of the performed study, can target the research practice or the domain practice or both. Thus, implications of secondary contributions do not necessarily need to address the domain practice—that is, a secondary contribution could lead to something that is only relevant to research practice.

Kolkowska et al.'s DSR study (2017) serves as an illustrative example as their study provides an analytical tool that information security managers can use to understand what 'rationality conflicts exist and the impact they have on employees' compliance' with information security policies (2017, p. 40). Thus, their domain practice is information security management. The study's primary contribution is a new method for analysing the reasons behind employees' compliance and non-compliance with information security policies, together with a set of design principles for this class of methods and an empirical demonstration of a practical method. In this case, the primary contribution covers all three types of contributions in Table 2.2, and they primarily have implications for information security management practices. First, the method itself is an artefactual contribution that provides an opportunity for information security managers to analyse employees' work with information security in a novel way. Second, the empirical contribution provides information security managers with a way to gain knowledge about how to use the method. Also, the identified rationality conflicts serve as a possible benchmark for other organizations. Finally, the set of design principles could act as a general starting point for devising a method similar to the suggested method if information security managers do not want to adopt the suggested one. The proposed method also provides opportunities for research practice to create empirical accounts of why employees comply or do not comply in different contexts. As a secondary contribution, Kolkowska et al. (2017) show how observation of employees' day-to-day work can provide valuable complementary data. These observations make it possible to identify activities that would have been overlooked had only self-reporting been used because employees tacitly know what are compliance and non-compliance activities. Therefore, Kolkowska et al. (2017) suggest that observation should be used as a complementing data collection method in future information security policy compliance research (i.e. this is an opportunity for executing research differently).

Implications are complex and lead to speculation. Some contributions might, for example, appear to have application beyond the demarcations of the current study, which might well be the case. Implications can range from concrete short-term consequences to more long-term and/or potential consequences. In addition, implications depend heavily on the strength of evidence presented and the study's limitations. Geletkanycz and Tepper (2012) focus on theoretical implications and present many suggestions for how to express these. In doing so, they identify pitfalls that seem valid across all types of implications: rehashing results, meandering, and overreaching. Thus, meticulous judgmeent is called for in order to avoid these pitfalls and ensure that more speculative parts are contained within the boundaries of the arguments developed in the research in question.

Conclusion

In this chapter, we have explored the notion of research contribution in IS. We have paid particular attention to the distinction between research contributions and research implications. Furthermore, we have explained why such a distinction can be useful when discussing research outcomes.

We have shown that contributions and implications are two different aspects of research, although often these aspects are conflated and implications often receive less attention than contributions. A research contribution can be seen as something that advances our understanding of a studied practice or improves the studied practice itself. An implication exists in relation to a contribution—that is, research contribution leads to implication with regard to a particular practice. When researchers make claims about their contribution, they necessarily presume or anticipate other claims. Therefore, understanding a study's implications is fundamental to understanding the value of its contribution.

We have distinguished between two types of practices that are important when discussing contributions and implications: research practice and domain practice. A research practice is a practice that investigates other practices. A domain practice is a practice investigated by a research practice. These two types of practices overlap to some extent, especially in collaborative projects that involve both researchers and practitioners or when researchers study research.

We began our exploration by conceptualizing contribution. By taking an incremental view on knowledge development, where each piece of research does not necessarily need to go all the way from empirical (data collection) to theoretical (generalization) to empirical (validation), we identified three types of research contributions:

- Theoretical—new or modified theory, or parts thereof;
- Empirical—an account of a phenomenon or details of a phenomenon not previously covered by research or covered to a limited extent; and
- Artefactual—description of proposed new or modified artefact(s).

Because an implication always exists in relation to a contribution, we explored what the implications these contributions could lead to in the two practices. As result, we arrived at the 3 × 2 matrix in Table 2.2. We do not claim that these implications are exhaustive; rather, the discussed implications serve as an illustration of the implications.

If we were to situate our own work in the contribution/implication framework in Table 2.2, it is both a theoretical contribution and an artefactual contribution. We acknowledge that our conceptualization of contribution and implication is not entirely new. Nonetheless, acknowledging an artefactual contribution as a contribution in its own right goes beyond how contributions have been conceptualized and integrated in previous research. In addition, our conceptualization of the relation between contribution and implication advances existing conceptualizations since implication per se has received very little attention. The framework in Table 2.2 is also an artefactual contribution as it creates a hands-on analytical framework—that is, an artefact for writing and assessing research papers. Although previous research has offered frameworks that focus on analysing research contributions, this research has not explicitly integrated research contributions and implications.

The implications of our work are oriented towards research practice (in our case, the studied domain practice is research practice, more specifically the practice of writing and publishing an article). An intended implication is to provide guidance to authors in their efforts to detail research contributions and implications of such contributions. We hope that the analytical framework will increase the quality of researchers' discussions about research contributions and implications, enabling them to pinpoint these aspects with greater accuracy. Another intended implication is also to provide guidance to reviewers and editors in their efforts to scrutinize and develop manuscripts that are sent to journals and conferences. As we have shown, not all papers need to include all three types of contributions. The analytical framework could help reviewers and editors find ways to appreciate research despite a perceived lack of theoretical contribution. Moreover, in the case of rejecting a paper, the framework could help reviewers and editors detail their specific concerns in a constructive manner and developmental spirit.

References

Ågerfalk, P. J. (2014). Insufficient theoretical contribution: A conclusive rationale for rejection? *European Journal of Information Systems, 23*(6), 593–599.

Ågerfalk, P. J., & Wiberg, M. (2018). Pragmatizing the normative artefact: Design science research in Scandinavia and beyond. *Communications of the Association for Information Systems, 43*(4), 68–77.

Avison, D., & Malaurent, J. J. (2014). Is theory king? Questioning the theory fetish in information systems. *Journal of Information Technology, 29*(4), 327–336.

Baskerville, R. (2004). An editor's values. *European Journal of Information Systems, 13*(1), 1–2.

Baskerville, R., & Wood-Harper, A. T. (1998). Diversity in information systems action research methods. *European Journal of Information Systems, 7*(2), 90–107.

Benbasat, I., & Zmud, R. W. (1999). Empirical research in information systems: The practice of relevance. *MIS Quarterly, 23*(1), 3–16.

Corley, K. G., & Gioia, D. E. (2011). Building theory about theory building: What constitutes a theoretical contribution? *Academy of Management Review, 36*(1), 12–32.

Cronen, V. (2001). Practical theory, practical art, and the pragmatic-systemic account of inquiry. *Communication Theory, 11*(1), 14–35.

Davison, R. M., Martinsons, M. G., & Kock, N. (2004). Principles of canonical action research. *Information Systems Journal, 14*(1), 65–86.

Geletkanycz, M., & Tepper, B. J. (2012). Publishing in AMJ-Part 6: Discussing the implications. *Academy of Management Journal, 50*(2), 256–260.

Germonprez, M., Hovorka, D. I., & Collopy, F. (2007). A theory of tailorable technology design. *Journal of the Association for Information Systems, 8*(6), 21.

Goldkuhl, G., & Röstlinger, A. (2003). Towards an integral understanding of organizations and information systems: Convergence of three theories. In H. W. M. Gazendam, R. J. Jorna, & R. S. Cijsouw (Eds.), *Dynamics and change in organizations* (pp. 133–161). Dordrecht: Springer.

Gregor, S. (2006). The nature of theory in information systems. *MIS Quarterly, 30*(3), 611c642.

Gregor, S., & Baskerville, R. (2012). *The fusion of design science and social science research*. Paper presented at the Information Systems Foundation Workshop, Canberra, Australia.

Hassan, N. R., Mathiassen, L., & Lowry, P. B. (2019). The process of information systems theorizing as a discursive practice. *Journal of Information Technology, 34*(3), 198–220.

Hevner, A. R., March, S. T., Park, J., & Ram, S. (2004). Design science in information systems research. *MIS Quarterly, 28*(1), 75–105.

Karlsson, F., & Hedström, K. (2013). Evaluating end user development as a requirements engineering technique for communicating across social worlds during systems development. *Scandinavian Journal of Information Systems, 25*(2), 1–26.

Kolkowska, E., Karlsson, F., & Hedström, K. (2017). Towards analysing the rationale of information security noncompliance: Devising a Value-Based Compliance analysis method. *Journal of Strategic Information Systems, 26*(1), 39–57.

Lee, A. S. (1999). Rigor and relevance in MIS research: Beyond the approach of positivism alone. *MIS Quarterly, 23*(1), 29–33.

Lee, A. S., & Baskerville, R. (2003). Generalizing generalizability in information system research. *Information Systems Research, 14*(3), 221–243.

Lewin, K. (1943). Psychology and the process of group living. *Journal of Social Psychology, 17*(1), 113–131.

Lewin, K. (1945). The research centre for group dynamics at Massachusetts Institute of Technology. *Sociometry, 8*, 126–135.

Morschheuser, B., Hassan, L., Werder, K., & Hamari, J. (2018). How to design gamification? A method for engineering gamified software. *Information and Software Technology, 95*(March), 219–237.

Muntermann, J. (2009). Towards ubiquitous information supply for individual investors: A decision support system design. *Decision Support Systems, 47*(2), 82–92.

Niederman, F., & March, S. (2019). The "theoretical lens" concept: We all know what it means, but do we all know the same thing? *Communications of the Association for Information Systems, 44*(1), 1.

Oinas-Kukkonen, H. (1998). Evaluating the usefulness of design rationale in CASE. *European Journal of Information Systems, 7*(3), 185–191.

Riemer, K., & Vehring, N. (2010). *It's not a property! Exploring the sociomateriality of software usability*. Paper presented at the 31st International Conference of Information Systems (ICIS 2010), St. Louis, USA.

Schatzki, T. R. (2001). Introduction: Practice theory. In T. R. Schatzki, K. Knorr Cetina, & E. von Savigny (Eds.), *The practice turn in contemporary theory* (pp. 1–14). London: Routledge.

Schultze, U., & Leahy, M. M. (2009). *The avatar-self relationship: Enacting presence in second life*. Paper presented at the 30th International Conference of Information Systems (ICIS 2009), Phoenix, AZ, USA.

Sein, M. K., Henfridsson, O., Purao, S., Rossi, M., & Lindgren, R. (2011). Action design research. *MIS Quarterly, 35*(1), 37–56.

Simon, H. A. (1961). *The sciences of the artificial*. Cambridge, MA: MIT Press.

Siponen, M., & Vance, A. (2014). Guidelines for improving the contextual relevance of field surveys: The case of information security policy violations. *European Journal of Information Systems, 23*(3), 289–305.

Smolander, K., Lyytinen, K., Tahvanainen, V. P., & Marttiin, P. (1991). MetaEdit – A flexible graphical environment for methodology modelling. In R. Andersen, J. A. Bubenko, & A. Sølvberg (Eds.), *Advanced information systems engineering. CAiSE 1991. Lecture notes in Computer Science, vol 498* (pp. 168–193). Berlin: Springer.

Straub, D. W. (2009). Editor's comments: Why top journals accept your paper. *MIS Quarterly, 33*(3), iii–ix.

Sutton, R. I., & Staw, B. M. (1995). ASQ forum: What theory is not. *Administrative Science Quarterly, 40*(3), 371–384.

Tsang, E. W. K. (2013). Case study methodology: Causal explanation, contextualization, and theorizing. *Journal of International Management, 19*(2), 195–202.

van de Ven, A. H. (1989). Nothing is quite so practical as a good theory. *Academy of Management Review, 14*(4), 486–489.

van de Ven, A. H., & Johnson, P. E. (2006). Knowledge for theory and practice. *Academy of Management Review, 31*(4), 802–821.

Venkatesh, V. (2006). Where to go from here? Thoughts on future directions for research on individual-level technology adoption with a focus on decision making. *Decision Sciences, 37*(4), 497–518.

Walls, J. G., Widmeyer, G. R., & El Sawy, O. A. (1992). Building information system design theory for vigilant EIS. *Information Systems Research, 3*(1), 36–59.

Walsham, G. (1995). Interpretive case studies in IS research: Nature and method. *European Journal of Information Systems, 4*(2), 74–81.

Walsham, G. (2006). Doing interpretive research. *European Journal of Information Systems, 15*(3), 320–330.

Weick, K. E. (1995). What theory is not, theorizing is. *Administrative Science Quarterly, 40*(3), 371–384.

Yang, T.-M., & Maxwell, T. A. (2011). Information-sharing in public organizations: A literature review of interpersonal, intra-organizational and inter-organizational success factors. *Government Information Quarterly, 28*(2), 164–175.

3

Theoretical Diversity in IS Research: A Causal Structure Framework*

Frantz Rowe and M. Lynne Markus

Introduction

> A number of philosophers have argued that the consideration of a diverse range of alternative theories is one of the most important methodological norms for inquiry. … Theoretical pluralism—whether in the quest for a unified or a disunified science, a single truth or a range of instrumental truths, or simply as a good ground rule for general inquiry—has had a wide range of advocates. (CJ Preston, "Pluralism and Naturalism," *Philosophical Psychology*, 2005)

*An earlier version of this chapter appeared as Markus, M. Lynne & Rowe, Frantz (2018) Is IT changing the world? Conceptions of causality for information systems theorizing. *Management Information Systems Quarterly, 42*(4), pp. 1255–1280.

F. Rowe (✉)
Université de Nantes, Nantes, France
e-mail: frantz.rowe@univ-nantes.fr

SKEMA Business School, Sophia-Antipolis, France

M. L. Markus
Bentley University, Waltham, MA, USA
e-mail: mlmarkus@bentley.edu

N. R. Hassan, L. P. Willcocks (eds.), *Advancing Information Systems Theories*, Technology, Work and Globalization, https://doi.org/10.1007/978-3-030-64884-8_3

Theoretical diversity is desirable for academic fields because all empirical studies necessarily examine only a small slice of the researched phenomenon. This is true not only because of practical limitations of time, money, and personal interest, but also because of the inherent limits on human knowledge. Only through an extended program of research on a phenomenon, drawing on the work of numerous scholars employing multiple theories and methods, is it possible to come anywhere close to comprehensive knowledge of a phenomenon.

Without theoretical diversity, a field limits its focus, scope, and relevance. Embracing theoretical diversity stimulates intellectual communities, enriching their bodies of knowledge, and minimizing the tendency to think that what they happen to study is all they need to know. Theoretical diversity increases the relevance of scholarly fields to practitioner communities (Lyytinen, Rowe, & McGuire, 2018) and to other epistemic communities. But theoretical diversity also threatens incoherence in a field's research agenda and cumulative research output. Avoiding chaos requires a map of the major patterns of causal reasoning in a field.

Many years ago, Markus and Robey (1988) presented a framework of *causal structure* in research on information technology and organizational change. Causal structure refers to theorists' explicit arguments or implicit assumptions about the nature and direction of causal influences related to their phenomena of interest. Causal structure is distinct from theory *purpose* (e.g., analysis, explanation, and prediction (Gregor, 2006)), in that causal arguments can support diverse research goals. In addition, causal structure is distinct from *method* (e.g., narrative construction and process tracing (Poole et al. 2000) and from *causal analysis* (e.g., regularity analysis, counterfactual analysis, and the instrumental variables approach (Gregor & Hovorka, 2011), even though causal arguments invariably have methodological implications.

Markus and Rowe (2018) substantially rearchitected Markus and Robey's (1988) framework, setting it on a foundation of philosophy and social theory and expanding its focus beyond organizational change so that it is more comprehensive of contemporary IS and sociotechnical research. The framework presents ideal-typical positions on three dimensions: ontology, trajectory, and autonomy. The focus of the *ontology* dimension is: Does the theorist conceive of causality as existing in the human mind or in reality? The concern of the *trajectory* dimension is: How does the theorist conceive the "social space" related to technology and envision causal movements across it? The remit of the *autonomy* dimension is: How does the theorist apportion causal effects for processes or outcomes across human or social actors and technology? The first issue is a critical question for *all* scientists; the second issue is common to all *social* scientists; and the third issue is unique to scientists who study *socio-technical* phenomena.

Diversity of theoretical positions on these three dimensions is needed and valuable for information systems (IS) and sociotechnical scholars for at least three specific reasons: the diversity and evolving nature of the phenomenon, the need for specificity in theorizing, and value of clarity for intervention. First, the phenomenon of interest to IS and sociotechnical scholars is multifaceted and rapidly evolving. At the time Markus and Robey (1988) published their framework, the use of information technology was confined to a narrow range of data-processing activities within large organizations. Today, information technology is used by vast numbers of individual consumers, not just organizations, in many forms, including social media, smart devices, and autonomous technologies. And, whatever else we know about information technology, we know that it will continue to proliferate and evolve. Theoretical diversity is critical in light of this evolution.

Second, theoretical diversity in IS and sociotechnical research is needed because research aims are so diverse. Whereas IS research in the 1980s centered on the organizational use, management, and outcomes of information technology, today's aims also include understanding and explaining the lived experience of technology users, the embedding of technology into social practices, the origins of technology features, and the consequences of technology for stakeholders far removed from hands-on users. If one wants to theorize how a technological innovation came about, one probably should not adopt the same causal positions that one would if theorizing the innovation's consequences. If one wants to know what makes for the safe operation of a class of information technology (e.g., automated securities trading), one needs a different causal logic than if one wants to explain the 2010 Flash Crash. Different research aims demand different theories. A scholar's position on the dimensions of causal structure need not be a lifetime intellectual commitment but should instead be specific to the needs of a particular investigation.

Third, theoretical diversity is also pragmatically useful. Lincoln and Guba (1985) noted that being specific about causality can help scholars obtain research funding. Although they did not elaborate, we know that external stakeholders expect sociotechnical studies to provide insights that can make a difference to organizations and society. In light of growing public recognition of IT's significant societal consequences, we are at risk as a field if we cannot talk intelligibly, which often means causally, with stakeholders about how and why IT's consequences come about and where and how to intervene in them (Leonardi, Nardi, & Kallinikos, 2012).

In this chapter, we represent Markus and Rowe's framework of causal structure[1] and discuss its utility for promoting theoretical diversity while maintaining coherence of research agenda and mutual intelligibility in the field.

Three Core Aspects of Causality

This section of the chapter represents the basic dimensions and positions of Markus and Rowe's (2018) causal structure framework.[2] The starting point for the framework is the observation that causality has disparate and sometimes irreconcilable definitions in the literatures of philosophy and social theory. It is not possible to reconcile or unify the divergent definitions of causality, but, at the same time, one does not have to choose among them once and for all. Instead, embracing the diversity of views for what each has to offer in light of particular research questions or phenomena is a viable option for IS scholars.

A theorist's views about causation as a real phenomenon is a fundamental, if not *the* fundamental, theoretical choice in academic inquiry. Hence, we called the first dimension of our framework **Causal Ontology**

[1] Reproduced with permission of the copyright holders.

[2] For a discussion of the need to clarify the nature of IS theory, an exposure to the diversity of causality/causation conceptions, a justification of the dimensions of our framework, and an exposition of how to use the framework, please see Markus and Rowe (2018). For a systematic comparison with Markus and Robey (1988), please see Rowe and Markus (2018).

to denote a theorist's beliefs about the reality of causality. Adopting a position on the Causal Ontology dimension is a critical theoretical choice for scientists, whether natural or social, as well as for many humanists, including lawyers and historians.

We called our second dimension **Causal Trajectory**, because the word trajectory denotes the movement of an entity in space and time. This dimension foregrounds the entity that the theorist believes to be moving (i.e., changing or being affected) and shifts the effect of the movement into the background. The affected entity could be an individual, a group, a sociotechnical system or infrastructure, an organization, an organizational field, a community, or an actor-network. We defined the Causal Trajectory dimension as the theorist's views about the causal movements of an affected entity. Adopting a position on the Causal Trajectory dimension represents a fundamental theoretical choice for scholars who study humans and social entities.

The third dimension of our framework foregrounds the movement of effects in space and time. Here "effects" refer to causal influences, which are sometimes referred to under the label of agency (human and/or material). We called this third dimension **Causal Autonomy**, because the word "autonomy" means self-governing or freedom from external influences, both of which meanings have been applied, not only to humans, but also to technology. We defined Causal Autonomy as the theorist's views about the movement of causal effects between human (or social) actors and technology. Adopting a position on the Causal Autonomy dimension is a theoretic choice of special importance for information systems scholars.

Issues related to the Causal Ontology, Causal Trajectory, and Causal Autonomy dimensions have been extensively discussed in the philosophical and social theoretic literatures under various names, such as agency, structure, holism, individualism, materialism, and idealism. We chose our own names for our dimensions to avoid confusion or conflation (Barley, 1998), because many similar concepts are discussed in the literature under different labels (e.g., materialism and determinism), and because some terms are used with different meanings or referents (e.g., human agency in relationship to social structure vs. in relationship to material objects).

Table 3.1 Three critical dimensions of theoretical causal structure

Dimension	Definition	Basic positions
Causal Ontology	A theorist's views about the reality of causality	• Causality is a convenient **metaphor** for a logical or metaphysical association • Causality implies a real **mechanism**, that is, a process that connects inputs to outputs • Causality is a **misnomer**, because it incorrectly implies unidirectional, deterministic, external forces
Causal Trajectory	A theorist's views about causal movements of the affected entity	• Causality occurs across the boundaries of a **stratified** entity • Causality occurs within (**internal** to) an undifferentiated entity • Causality occurs through the **accretion** (growth and complexification) of a heterogeneous entity
Causal Autonomy	A theorist's views about movement of causal effects between human (or social) actors and technology	• Causal effects move from people (or social actors) to technology—technology as **instrument** • Causal effects move from technology to people (or social actors)—technology as **influencer** • Causal effects move back-and-forth between people (or social actors) and technology—technology as **interactant**

Table 3.1 presents the definitions of the three dimensions and, for each, a set of basic positions. Below, we flesh out the dimensions and positions of our framework. We discuss the philosophical and social theoretic warrants for each dimension, and we elaborate on the basic positions, providing their warrants and one or more IS-relevant examples. We conclude each section with additional observations about the dimension and positions.

Dimension I: Causal Ontology

Causal Ontology concerns a theorist's views about the reality of causality. The Causal Ontology dimension has roots in several long-standing debates in the philosophy and social science literatures (De Pierris & Friedman, 2013; Falcon, 2012; Juarrero, 2011; Salmon, 1998; Woodward, 2003). Within these debates, three basic positions can be discerned. The first is that causality is a metaphor. Causation is a concept that refers to something that is unobservable, hence metaphysical and nonscientific. This position originates from David Hume (1711–1776). Different rationales for this position have been offered by Kant (1724–1804) and Meillassoux (2010).

The second basic position is that causation refers to something happening in the real world, that is, a process that connects inputs to outputs, such as a transfer of matter, energy, or information or a human or social dynamic, such as the self-fulfilling prophecy. This position is most commonly discussed under the headings of scientific (Bunge, 1996), speculative (Harman, 2010), or critical (Archer, Bhaskar, Collier, Lawson, & Norrie, 1998) realism.

The third basic position is that causation is a misnomer, because it incorrectly implies unidirectional, deterministic, and/or external influences and thereby radically oversimplifies and distorts what matters most in understanding human affairs. Because humans reflect on their own experiences, theorizing about people and social phenomena must necessarily differ from theorizing about physical and biological phenomena. This basic position has its origins in the philosophy of Dilthey (1833–1911) and exemplifies theorizing in numerous interpretive traditions. Some leading proponents of this position believe that a concept *like* causality is needed, both theoretically and in order to make practical recommendations, but that the concept needs complete redefinition (Lincoln & Guba, 1985).

Below, we elaborate on the basic positions presented above to bring them closer to the kinds of causal arguments one finds in the IS and related literatures. We call the elaborated positions Directional Association, Causal Mechanism, and Constitutive Causality.

Position I.A: Directional Association

The position statement for Directional Association can be phrased as follows: Causality refers to regular associations among certain hypothetical or observed conditions, possibly including temporal precedence (hence the qualifier "directional"). Directional Association theorizing is concerned with general causation only (i.e., causation in populations of similar entities or events) and does not apply to specific instances (i.e., individual cases).

Philosophical or Social Theoretic Foundations

The warrants for the Directional Association position derive from the philosophical writings of Hume (1711–1776) and Kant (1724–1804). Both scholars believed that causality cannot be directly observed. Hume argued that it is only habit or custom that allows humans to infer causality from observations about regular associations among events. Among other contributions, Hume is remembered for the "three principles" formula by which an attribution of a causal relationship is to be considered logically sound: (1) contiguity in place and time of events, (2) temporal precedence of the cause relative to the effect, and (3) constant (regular) conjunction (association) between hypothesized cause(s) and effect(s).[3] Kant agreed with Hume that causality could not be observed through sensory perception, but he believed that causality was a pure concept (like time and space) that is common to all humans and allows us to experience causal relationships in the sequences of events we observe in empirical reality (De Pierris & Friedman, 2013).

[3] In his famous example of the motion of the billiard ball hitting another and putting it in motion after the shock, Hume (1955: 186–187, 1737) explained: "Let us try any other balls of the same kind in a like situation, and we shall always find that the impulse of the one produces motion in the other. Hence there is a *third* circumstance, viz. that of *constant conjunction* betwixt the cause and the effect. Beyond these three circumstances of contiguity, priority and conjunction, I can discover nothing in this cause." Because the connection between cause and effect remains hidden, "causality" can be considered a metaphysical concept.

The Directional Association position is very much alive today in the statistical relevance (SR) model of scientific explanation (Durand & Vaara, 2009; Hovorka, Germonprez, & Larsen, 2008).

An Example from IS Research Literature

Agarwal, Animesh, and Prasad (2009) questionned the marked geographic variation in household internet usage in the United States. They reasoned on the basis of social influence theories that people whose peers used the internet would use it also. Put differently, they hypothesized that peers exert a causal influence on individuals' technology use decisions. This paper differentiated between concepts that *describe* some aspect of the phenomenon or its context (such as number of children in a household, which is statistically correlated with internet use) and concepts that can be plausibly claimed to be *causal* (e.g., peer influences). However, the paper did not address *how* the causal effect might have come about.

Position I.B: Causal Mechanism

The position statement for Causal Mechanism can be phrased as follows: Causality involves real physical, psychological, and/or social processes that connect inputs and outputs under certain conditions. Causal Mechanism theorizing applies to both general (population) and singular (case) causation.

Philosophical or Social Theoretic Foundations

The Causal Mechanism concept evolved somewhat independently in two intellectual communities, scientific realism and sociology, that appear to be converging under the banner of critical realism (Hedström & Ylikoski, 2010). Whereas adherents of the Directional Association position effectively give up on trying to specify what Hume called the "secret [hidden] connection" between causes and effects, scientific realists embrace the task of explaining why things happen (Bunge, 1996; Machamer, Darden,

& Craver, 2000; Salmon, 1998). The name they give to the hidden connection is *mechanism*, referring to a *physical* causal process that is hypothesized to "[transmit] energy, as well as information and causal influence" (Salmon, 1998, p. 146). Mechanisms are conceptualized as (1) ontologically real, even if they are unobservable, (2) reasonably inferable as causal (through abduction), and (3) able to produce effects that would not happen otherwise.

Sociologists frustrated with the Directional Association position evolved the concept of a *social* mechanism (Avgerou, 2013b; Hedstrom & Swedberg, 1998a) to theorize, rather than merely to associate or describe, the connections between social context, human belief-formation, social interaction, and social outcomes. Not to be confused with a purely descriptive chain of unique events that lead from one situation to another (Hedstrom & Swedberg, 1998b), social mechanisms are more or less *general* sets of social events or processes that, under certain circumstances, bring about changes in human social relations without necessarily being reducible to the actions of individuals (Gross, 2009). Familiar examples of social mechanisms include the self-fulfilling prophecy ("an initially false definition of a situation [that] evokes behavior that eventually makes the false conception come true" (Hedstrom & Swedberg, 1998b, p. 18) and Van de Ven and Poole's (1995) four "motors of change."

Early writings represented social mechanisms as abstract models, bits of theory, and analytic constructions or interpretations (Hedstrom & Swedberg, 1998a). This understanding of causality is metaphorical, as is the Directional Association position. More recently, however, proponents of social mechanisms appear to have embraced the philosophy of critical realism (Archer et al., 1998; Bhaskar, 1998; Bunge, 1996; Sayer, 2000; Sorensen, 1998; Stinchcombe, 2002) as a foundation for their theorizing (Hedström & Ylikoski, 2010).

In critical realist philosophy, social mechanisms, like physical mechanisms, are seen as seen as enduringly at play in the real realm independently of our perception or awareness of them (Bhaskar, 1975). Mechanisms generate actual events, only some of which can be perceived in the empirical realm. Mechanisms do not always produce actual events, because they may counteract each other or otherwise depend on the presence of various conditions (Mingers, 2004).

The Causal Mechanism concept is often employed in general causal theorizing (e.g., about the population of evolving infrastructures (Henfridsson & Bygstad, 2013) or of routines (Goh, Gao, & Agarwal, 2011)). However, the Causal Mechanism concept is also applicable to the study of individual cases and unique events (Avgerou, 2013a; Avgerou, 2013b).

An Example from IS Research Literature

Volkoff, Strong, and Elmes (2007) inquired into why enterprise systems, which are expected to make organizations more flexible, in fact make them more rigid. They inferred that embedding (of enterprise software into the organization) is the mechanism through which routines, roles, and data become rigid (Volkoff et al., 2007) with effects such as misfits (Strong & Volkoff, 2010). However, they argued that rigidification effects depend on conditions that do not always occur. In a follow-up study based on the same data, Volkoff and Strong (2013) concluded that the affordances of enterprise systems are also mechanisms that are actualized under certain conditions.

Position I.C: Constitutive Causality

The position statement for the Constitutive Causality position can be articulated as follows: "Causality" is a human belief about how meanings are enacted in highly situated social interactions (e.g., practices) and interactively construct such interactions. Proponents of this position hold that beliefs about causality have real consequences. Constitutive Causality theorizing applies to singular (case-specific) causation only.

Philosophical or Social Theoretic Foundations

We borrowed the label for this position from Schwartz-Shea and Yanow (2013), who used the term "constitutive causality" to describe how actors use language rules and interaction to make sense of a situation or phenomenon as they enact and transform it. This position has its origin in Dilthey's (1833–1911) philosophy. In opposition to Comte's (1798–1857) unified view of the sciences from physics to sociology, Dilthey argued against an overarching "universal explanatory typology for all historical facts" (Dilthey, 1989, p. 141). Instead, Dilthey's philosophy celebrated the uniqueness of experience in human life arising from complex interactions and interpretations. Dilthey's philosophy led the way for interpretivism, including the phenomenology of Husserl, Heidegger, and Merleau-Ponty, as discussed in Introna and Ilharco (2004), and the hermeneutics of Gadamer and Ricoeur, as discussed in Myers (2004).

A key warrant for the Constitutive Causality position is the late Wittgenstein (1899–1951), who argued that beliefs and reasons are deeply implicated in subsequent human actions (Sandis & Tejedor, 2017). Humans fundamentally create, share, and contest situations based on their interpretations of the meanings of those situations and on how firmly they hold to those interpretations (Wittgenstein, 1958). Causality becomes real through "languages games" in various situations. The constitution of cause-effect language games results from expressing our immediate reactions to what has affected us (Wittgenstein, 1976). The game starts with a reaction to some phenomenon and complicates itself through language. Cause-effect is the belief in our minds when we imagine the important possibility that some event undoubtedly has a particular cause (Wittgenstein, 1976). Thus, theorizing according to Wittgenstein is not about how or why things happen but rather about how situations and human meanings co-constitute each other. Furthermore, his stance that practices enacted through language games matter means that theorizing mainly involves *past* situations and relies on the identification and description of practices.

Examples from IS Research Literature

Ngwenyama and Lee (1997) investigated how and why email can support communicative action by enabling actors to move beyond their possible understanding of a situation as derived from the face validity of messages. The authors theorized that, with technologies such as email, human beings create or enact the meanings they come to hold and probe the validity claims advanced by others in order to emancipate themselves from distorted communications. Social actors can interpret messages as distorted by testing the claims made by others, thanks to the content feedback they receive in interaction. What happens makes sense to social actors when they can refer to their lifeworlds to pre-interpret patterns of actions. In another illustration of this position, Fayard and DeSanctis (2010) showed how, through their engagement in a language game, members of online communities produced five discursive practices that defined and enacted their sense of we-ness.

Interrogating how to interpret passing or failing a Turnitin (plagiarism detection software) test, Introna and Hayes (2011) theorized that human and technological actors iteratively co-constitute their attributes (e.g., fairness) and roles (e.g., copy detection and plagiarism detection) in unexpected and uncertain ways due to interpretive frames that they hold of each other. Put differently, interactions between human and technological actors had the performative outcome of subverting intentionalities. Greek students typically learn to prepare for exams by memorizing large blocks of texts. When they study in UK universities where plagiarism detection software is used, duplicated text in Greek students' exam is often interpreted as plagiarism rather than preparation.

Additional Observations about Causal Ontology Positions

The discussion above can be summarized as follows. Strikingly different, even incommensurable, positions on the Causal Ontology dimension can be articulated. Each position has authoritative warrants in the philosophical or social theory literatures. Each position can be identified in some published research in IS and other fields. What may not be immediately clear is that each position produces a different *form* of theoretical statement. To illustrate that point, we took two IS research exemplars of the Constitutive Causality position (Hayes & Introna, 2005; Introna & Hayes, 2011) and crafted theoretical statements about the same phenomenon, use of plagiarism detection software, consistent with the other two positions. As can be seen in the structured comparison in Table 3.2, the forms of these statements diverge considerably. But all of them are *causal* statements, and all are statements of *theory*. Metaphorical illustrations by our colleague Nicolas Antheaume convey the gist of each position and illustrate how they differ (see Fig. 3.1).

Table 3.2 Positions on Dimension I, Causal Ontology

Basic position	Elaborated position statement	Illustrations from the IS field
Metaphor Causality is a convenient metaphor for a logical or metaphysical association	**I.A. Directional association** Causality refers to regular associations among certain hypothetical or observed conditions, possibly including temporal precedence (hence the qualifier "directional").	• The greater the use of plagiarism detection software by professors, the greater are the number of students referred to disciplinary hearings for suspected plagiarism. • Use of several plagiarism detection packages results in fewer false positives and negatives.
Mechanism Causality implies a real mechanism, that is, a process that connects inputs to outputs	**I.B. Causal mechanism** Causality involves real physical, psychological, and/or social processes that connect inputs and outputs under certain conditions.	• Professors' use of plagiarism detection software can change the way professors understand plagiarism through the process of commensuration: the software redefines plagiarism (a complex phenomenon with many manifestations) and equates it with text duplication. • Professors' use of plagiarism detection software can improve the educational process by providing opportunities (not always realized) for discussions with students about the writing process and the meaning of original work.

(*continued*)

Table 3.2 (continued)

Basic position	Elaborated position statement	Illustrations from the IS field
Misnomer Causality is a misnomer, because it incorrectly implies unidirectional, deterministic, and external forces	**I.C. Constitutive causality** "Causality" is a human belief about how meanings are enacted in highly situated social interactions (e.g., practices) and interactively construct such interactions.	• Misinterpretations by both teachers and students emerged through the use of plagiarism detection software with Greek students in a UK educational system. (Introna & Hayes, 2011). • Use of plagiarism detection software with Chinese students who have learned by "patch writing" resulted in identity conflict and inappropriate student behavior (Hayes & Introna, 2005).

Causal Ontology refers to a theorist's views about the reality of causality. The illustrations in Table 3.2 were crafted by the authors to parallel the position taken by (Introna & Hayes, 2011) and (Hayes & Introna, 2005)

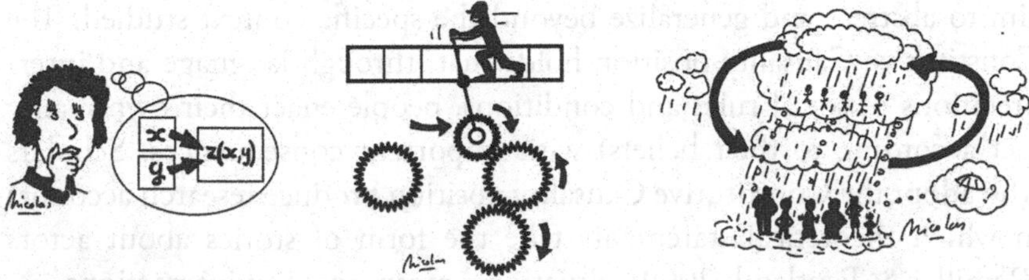

Fig. 3.1 Causal Ontology positions: directional association, causal mechanism, constitutive causality. (Illustrations by Nicolas Antheaume)

Theoretical statements produced by scholars adopting the Directional Association position emphasize causal *effects*. Their chief concern is the regularity and strength (Ziliak & McCloskey, 2014) of the associations (described as "robust dependence" by Goldthorpe (1999)) among abstract theoretical properties of individuals, social entities or concepts, and technologies. This position often results in "stories about variables" rather than in "stories about actors" (individuals and collective actors like organizations) (Ramiller & Pentland, 2009). Directional Association theorists tend to "generalize" (unduly, according to Lee and Baskerville (2003)), across contexts.

Whereas Directional Association theories hypothesize about causal *effects*, the Causal Mechanism position focuses on the contingently general causal *processes* by which causal effects come about (Avgerou, 2013b; George & Bennett, 2005). For example, in the previously cited study by Agarwal et al. (2009), hypothetical mechanisms (not explored by the authors) by which peer influence could affect household internet use include the following: (1) individual observation and imitation of others' internet use, (2) direct social pressure by peers on individuals ("if you want to contact me, use Skype"), and (3) shared cultural influences via mass and social media. Causal Mechanism theorizing produces a different kind of generalizability argument than that found in Directional Association theorizing because of the emphasis placed by Causal Mechanism theorizing on the conditions and processes by which hypothesized outcomes occur, when they do.

Theoretical statements produced by scholars holding the Constituted Causality position generally remain highly "situated," that is, they do not aim to abstract and generalize beyond the specific context studied. The Constitutive Causality position holds that, through language and interpretations of social rules and conditions, people enact their own reality (act according to their beliefs) with important consequences. Scholars who adopt the Constitutive Causality position produce research accounts in which theoretical statements take the form of stories about actors (Ramiller & Pentland, 2009), situations, meanings, and interactions.

How does the Causality Ontology dimension of our framework differ from earlier ways of characterizing causal theorizing? The Causal Ontology dimension steers clear of Markus and Robey's (1988) simple distinction between process and variance theory, because process theories can take two distinct forms (George & Bennett, 2005; Van de Ven & Poole, 2005): (1) unique sequences of empirical events (as in a historical explanation) and (2) contingently general abstract mechanisms, singly or in combination.

Dimension II: Causal Trajectory

The Causal Trajectory dimension refers to a theorist's assumptions about the causal movements of an affected entity, whether the entity is an individual, an infrastructure, an organization, or a nation. Our Causal Trajectory dimension encompasses and translates several fundamental philosophical and social science debates about the existence of society as an entity independent of the people who make it up, about the possibility of social influences on people, and about the conceptualization of society (e.g., as a social construction or a natural system).

Three basic positions on Causal Trajectory can be differentiated. The first position views affected entities as stratified, that is, consisting of layered or nested systems with relative clear (if permeable) boundaries between systems and their environments. Change is theorized as movements across boundaries. This view can be traced back to Aristotle, who believed that causes had to be external to their effects. Thus, he could only explain the self-movement of animals by positing the existence of an external soul that acted on the body. Modern philosophers like Descartes rejected the idea of bodies or substances animated by their souls (Gnassounou & Kistler, 2007), but the idea that hypothesized causes were necessarily external (i.e., exogenous) forces persisted from Newtonian era until the mid-twentieth century (Juarrero, 2002).

The widespread belief in causes as external to effects promoted the development of theories—in sociology, economics, and anthropology—about society as a whole that is greater than the sum of its parts and about "social forces"—institutions such as marriage, religion, and law—that act on individuals. This philosophy, often called *holism* (Hollis, 1994), was attacked by critics for denying the existence of human free will and was branded as cultural or environmental "determinism." By contrast, other philosophers argued that what is called society is merely the resultant of (and reducible to) individual actions and interactions—a philosophy often called *individualism* (Hollis, 1994). An example is the view of financial markets as the aggregate of individual investors' behavior.

The holism versus individualism debate, also known as the (social) structure versus agency debate, continues to vex social theorists to this day (Bunge, 1999). Holists charge that individualists' theories are undersocialized; individualists counter that the people in holists' theories are "social dopes." Numerous attempts have been made to resolve the different views. Some attempts reflect a view of society as a social construction (e.g., Alexander's (1985) neofunctionalism and Giddens' (1979) structuration theory), while at the same time conceptualizing "micro" and "macro" levels with inter-level influences. Other integrations adopt the critical realist view of a stratified natural and social world (e.g., Bunge's (1999) systemism and Archer's (1995) morphogenetic approach).

Systems theory and complex adaptive systems theory also rely on the idea of realist stratification. The phenomenon of self-organization is conceptualized in complex adaptive systems theory (Choi, Dooley, & Rungtusanatham, 2001) as involving multiple causal mechanisms: (1) interactions among the components of a social system (such as a group, an organization, a community, and a nation) and (2) between the system and its environment that (3) create, or generate changes in, more inclusive social units, which (4) subsequently constrain or enable the behavior of the lower-level components (Ellis, Murphy, & O'Connor, 2009; Juarrero, 2002). In self-organization theory, the "system" is seen to change its boundaries over time, as in neo-institutionalism (Thornton, Ocasio, & Lounsbury, 2012) and co-evolution theory (Baum & Singh, 1994).

A second basic position on Causal Trajectory, more recent in origin, rejects the idea of stratification, regardless of whether boundaries are theorized as social constructions or real. This position is characteristic of situated action theory (Suchman, 1987) and certain practice theories (Orlikowski, 2000; Orlikowski, 2002). This position eschews both micro-macro distinctions and the existence of boundaries between an affected entity and its environment. This theoretical bracketing denies or excludes from focus any external forces that might impinge on human actors and practices. Put differently, this view of causal dynamics has no "outside"—everything that matters is seen as occurring inside or within an entity (e.g., an organization) that is viewed as an analytic construct only, not as a natural system.

A third position on Causal Trajectory makes no significant use of the concept of stratification, but nevertheless implicitly assumes an inside and an outside. The affected entity is viewed as heterogeneous, consisting of mixed components such as people, organizations, ideas, and objects.

These components are not theorized as belonging to different levels (e.g., micro and macro). However, the composition of the affected entity is seen to change, sometimes radically, over time, from the accretion (i.e., importation, modification, and rearrangement) of new components that did not originally belong to it or from loss of preexisting components. This position is most clearly seen in actor-network theory (Callon, 1986; Latour, 2005). Although this view bears superficial similarity to ideas in complexity theory (Choi et al., 2001), neo-institutional theory (Thornton et al., 2012), and co-evolutionary theory (Arthur, 2009; Baum & Singh, 1994), the emphasis of the third position is on the changing *composition* of the affected entity, rather than on *influences on the entity from within or from without*.

Below we elaborate on these three basic Causal Trajectory positions. We name the positions Cross-Boundary Change, Indwelling Change, and Evolving Interlinkage.

Position II.A: Cross-Boundary Change

The position statement for Cross-Boundary Change can be phrased as follows: Change involves influences across the boundaries of natural systems with environments or embedded units. The Cross-Boundary Change position can be observed in a great majority of IS and related writings (see Winter, Berente, Howison, and Butler (2014) for a particularly rich example), and it has a number of distinct variants, including these three:[4]

1. Upward Initiative: Change involves movement from internal (lower level) to external (higher level).
2. Downward Influence: Change involves movements from external (higher level) to internal (lower level).
3. Self-organization: Change involves movements back-and-forth across system boundaries.

[4] Arguably, the most frequent Cross-Boundary Change variant in IS research might be called "same-level change," in which the idea of natural system entity (e.g., individuals or organizations) is implicit but the locus of causality is entirely internal to the entity. We omitted discussing this variant in the belief that it adds limited value to the chapter.

Philosophical or Social Theoretic Foundations

The major warrants for the Upward Initiative variant of Cross-Boundary Change are provided by philosophers considered to be individualists (Hollis, 1994) and by social scientists who insist that macro social phenomena must be explained in terms of micro-level behavior (Coleman, 1986). The position is well captured in this quotation from Weber:

> Collectivities such as states, associations, business corporations, foundations [...] must be treated as solely the resultants and modes of organization of the particular acts of individual persons, since these alone can be treated as agents in a course of subjectively understandable action. (Weber, 1968, p. 13)

In classic statements of individualism, the lower-level units are always individual humans. However, many social scientists view Weber's "collectivities" as stratified in themselves (e.g., corporations within nations), and much theorizing nowadays is devoted to the behavior of collective actors. This means that the position we call Upward Initiative can be adopted at several levels of analysis to theorize, for instance, the influences of departments on organizations or the influences of corporations on nations (Kaidesoja, 2013).

The primary warrants for the Downward Influence variant of Cross-Boundary Change come from philosophers and social theorists labeled "holists" or "institutionalists" (Hollis, 1994). For holists, the behavior of collective social entities cannot be explained solely by, or reduced to, the behavior of individuals. For instance, Durkheim's sociology aimed to establish "social facts" (e.g., the suicide rate and its causes in social integration and regulation) that are specific to each society. Social facts affect and explain individual behavior, but do not result from individual behavior. "A social fact is any way of acting, whether fixed or not, capable of exerting over the individual an external constraint [or] which is general over the whole of a given society whilst having an existence of its own, independent of its individual manifestations" (Durkheim, 1982, p. 59).

Holists allow that social structures grow initially out of human interactions, but they focus on the ways that social structures, once established, subsequently enable or constrain individual behavior through language, resources, and the material arts. A familiar example of the holist position is neo-institutional theory with its emphasis on the mechanisms (normative, mimetic, and coercive) by which social structures limit or support the choices of individuals and organizations (DiMaggio & Powell, 1983). Thus, institutional theorists often focus on downward influences (e.g., from nation states to organizations) (Wilber & Harrison, 1978). As with the Upward Initiative position, the Downward Initiative position can be adopted at any level to theorize, for example, the effects of government regulation on the actions of organizations.

The major warrants for the Self-organization variant of the Cross-Boundary Change Position come from systems theory (Mingers, 2014) and complex adaptive systems theory (Choi et al., 2001; Merali, 2004). Self-organization is often misunderstood as an exclusively bottom-up causal process, but detailed analysis (Juarrero, 2002) makes it clear that Self-organization comprises a sequence of mechanisms of different types, that is, same-level interactions among system components, bottom-up influences by a system on its environment, resulting in a new or modified higher-level configuration, and top-down influences by the emergent system on its components.

Examples from IS Research Literature

In an example of the Upward Initiative variant, Jarvenpaa and Leidner (1998) theorized that a group culture of "swift trust" can result unexpectedly in virtual teams composed of individuals with a limited history of working together and limited prospects of working together in the future. The authors argued that team members imitated the behavior that other members communicated in their first few keystrokes and that responses in kind enabled members to trust their colleagues in the face of unreliable technologies, differing views of the tasks performed, and extra-team obligations.

The Downward Influence variant can be seen in Ramiller and Swanson's (2003) discussion of how IS executives respond to media "buzz" about various IS innovations like reengineering and client-server architectures.

Ramiller and Swanson assumed that IS executives and their organizations act within a broader social milieu in which they become exposed to the "organizing visions" (created by IT vendors, academics, consultants, and the media) associated with IT innovations. These organizing visions have "careers" (i.e., they change over time), and Ramiller and Swanson showed that IS executives' responses change over time as well, offering evidence of enduring or repeated downward influences. Had Ramiller and Swanson (2003) chosen to focus on the *origins* of IS organizing visions, instead of their *effects*, they may well have adopted the Upward Initiative variant, as did Robertson, Swan, and Newell (1996), in a study of institutional entrepreneurs seeking widespread social acceptance for a particular IS organizing vision.

The Self-organization variant is exemplified by Nan and Tanriverdi's (2017) analysis of IS Strategy. Component IT innovation and architectural IT innovation lead to bottom-up effects of hyperturbulence in the competitive environment, which in turn redefine the opportunity for IT to contribute to organizational advantage in a top-down causal path (Nan & Tanriverdi, 2017). In such instances, self-organization refers to the formation of order out of organizational interactions without external or higher-level control or coordination (Tanriverdi & Lim, 2017). While the Self-organization position is multilevel by definition (Nan & Lu, 2014), most instances of multilevel IS theorizing (Burton-Jones & Gallivan, 2007; Lapointe & Rivard, 2005) also exhibit the Cross-Level Change position.

Position II.B: Indwelling Change

The position statement for the Indwelling Change position on the Causal Trajectory dimension can be phrased as follows: Change occurs through interactions within an analytically defined entity (i.e., an entity not viewed as a natural system). The entity is not differentiated into levels. There are no relevant external influences.

Philosophical or Social Theoretic Foundations

The Indwelling Change position rests heavily on the philosophical and social theoretic ideas that all social activity consists in the enacted communicative practices of individuals in idiosyncratic situations. Social structures (defined as abstract ideations like strategies, norms, and rules) are created and reproduced in action (Garfinkel, 1967) and believed to exist only as "memory traces" in the mind (Giddens, 1979).

From this point of view, it is meaningless to theorize about external environments and forces acting on humans; the very idea of something "outside" the social activity of interest does not arise. This "inside-only" view of the social world characterizes several important articulations of practice theory (Orlikowski, 2000; Orlikowski, 2002).

An Example from IS Research Literature

Mazmanian, Orlikowski, and Yates (2013) studied the use of mobile email devices in a workplace, theorizing that knowledge professionals enacted "a norm of continual connectivity and accessibility that produced a number of contradictory outcomes."

> [A]s participants [in the workplace] individually managed their mobile email, they began producing and sharing assumptions regarding how professionals should be using mobile email to get their work done. Over time, these shared assumptions were reinforced and reproduced in practice, further raising expectations about when and where participants should be engaging with their email communications. These heightened expectations led participants to feel increasing stress. (Mazmanian et al., 2013, p. 1345)

The authors portrayed this process visually via an ascending spiral mediating between the "collective" on the one side and the "individual" on the other.

Position II.C: Evolving Interlinkage

The Evolving Interlinkage position statement might be articulated as follows: Change occurs through the importation of heterogeneous new elements (e.g., ideas, actors, and resources) into an entity (or loss of elements preexisting in the entity) and by the creation of new linkages among the entity's elements. The affected entity changes qualitatively in composition.

Philosophical or Social Theoretic Foundations

The major warrant for the Evolving Interlinkage position is actor-network theory (Callon, 1986; Latour, 1996; Law & Callon, 1992). The term "heterogeneous" connotes actors and activities of different types (e.g., individuals, collectives, and nonhumans), domains (e.g., business, government, and academia), and/or geography (e.g., local, national, and international). Whereas the Indwelling Change position theorizes about the emergence of change occurring within an analytically defined setting, the Evolving Interlinkage position focuses on the changed composition of the entity through the incorporation of new (or loss of old) elements.

An Example from IS Research Literature

Heeks and Stanforth's (2007) study of e-government projects presented a framework derived from actor-network theory, in which the major causal claim is that e-government projects cannot advance unless there is both a high degree of mobilization of local network actors and a high degree of attachment of actors in a global network. Their theory contributes insights on e-government success and failure by viewing e-government projects as an unfolding trajectory rather than as a snapshot in time.

Additional Observations about Causal Trajectory Positions

Elaborated positions on the Causal Trajectory dimension are summarized in Table 3.3. Each position has authoritative warrants in the philosophical or social theory literatures. Each position can be identified in some published research in IS and other fields. The table contains constructed examples, based on Heeks and Stanforth (2007), to illustrate the differences in the positions more clearly. (See Fig. 3.2 for metaphorical illustrations of the positions by Nicolas Antheaume.)

To our knowledge, the distinctions between the positions on the Causal Trajectory dimension have not been clearly discussed in the IS literature in a way that speaks to our entire community. The fundamental distinction among the positions lies in the conceptualization of affected entities and how change happens in or to them. In the Cross-Boundary Change position, observable in the majority of IS research, change is understood as moving across the boundaries of real natural systems (e.g., from nation states to organizations and from workers to their employing organizations). By contrast, the Indwelling Change position occurs entirely within an entity whose boundaries are analytic only. Put differently, the entity is *not* viewed as a natural system that is embedded in an "environment." In the Evolving Interlinkage position, the entity (whether viewed as analytic or real) is seen as mobile: It changes qualitatively in composition as it imports new elements or diminishes through the loss of adherents.

Consider how exemplars of the Indwelling Change and Evolving Interlinkage positions contrast with the familiar Cross-Boundary Change position. Above, we described the Mazmanian et al. (2013) study as an exemplar of the Indwelling Change position. When discussing whether their findings might be transferrable (cf. Lincoln & Guba, 1985) to other settings, Mazmanian et al. wrote:

Table 3.3 Positions on Dimension II, Causal Trajectory

Basic position	Elaborated position statement	Illustrations from the IS field
Stratification Causality involves influences across the boundaries of a stratified entity	**II.A. Cross-boundary change** Change involves influences across the boundaries of natural systems with environments or embedded units. **1. Upward initiative** Change involves movement from internal (lower level) to external (higher level). **2. Downward influence** Change involves moves from external (higher level) to internal (lower level). **3. Self-organization** Change involves movements back-and-forth across system boundaries.	• E-government initiatives succeed when champions in government organizations persuade external institutions to support their plans (Upward Initiative). • E-government initiatives fail because they encounter divergent public values and opinions spread by media, consultants, and vendors (Downward Influence). • E-government initiatives succeed when champions persuade external institutions to support their plans and when their interactions create a favorable ecosystem, which in turn increases the motivation of internal stakeholders (Self-organization).
Internalization Causality occurs through interactions within an undifferentiated entity	**II.B. Indwelling change** Change occurs through interactions within an analytically defined entity. The affected entity is not differentiated into levels. There are no relevant external influences.	• E-government initiatives succeed when creative interactions unfold among stakeholders.

(*continued*)

Table 3.3 (continued)

Basic position	Elaborated position statement	Illustrations from the IS field
Accretion Causality occurs through the growth and complexification of a heterogeneous entity	**II.C. Evolving interlinkage** Change occurs through the importation of heterogeneous new elements (e.g., ideas, actors, and resources) into an entity (or loss of elements preexisting in the entity) and by the creation of new linkages among elements. The affected entity changes qualitatively in composition.	• E-government success emerges through the progressive achievement of high levels of local network mobilization and global network attachment (Heeks & Stanforth, 2007).

Causal Trajectory refers to a theorist's views about causal movements of an affected entity. The illustrations in Table 3.3 were crafted by the authors to parallel the position taken by (Heeks & Stanforth, 2007)

Fig. 3.2 Causal Trajectory positions: cross-boundary change, indwelling change, evolving interlinkage. (Illustrations by Nicolas Antheaume)

> Given that mobile technologies are increasingly pervading the lives of many contemporary workers, the shifts in practices and norms identified here ... may well extend beyond the knowledge professionals we studied. (Mazmanian et al., 2013, p. 1353)

This passage implies that conditions similar to the ones they observed might occur independently (Indwelling Change) in other settings. The possibility of influences between the organization studied and other organizations or with something like (macro-level) "society" is theoretically excluded in this account.

By contrast, others scholars observing similarities across organizations might adopt a Cross-Boundary Change position, arguing that organizations acquire common norms of behavior because of "social contagion" (i.e., contact with peers (Burt, 1987)), because of Downward Influences, for example from technology vendors and the media, or because of "structural equivalence" (Burt, 1987), that is, "common fate," or shared similar environmental circumstances.

As another example, consider the Heeks and Stanforth (2007) paper we took as our exemplar of the Evolving Interlinkage position. Many other writings on the topic of e-government adopt the Cross-Boundary Change position. For example, a paper cited by Heeks and Stanforth proffered the argument that e-government projects flounder when governments fail to specify their own objectives, relying too heavily on external vendors with conflicting interests:

> The first and the gravest mistake made by public institutions is their failure to specify the objectives of the project. The project is usually not sufficiently related to the overall goals of the organization and it is often, somewhat irrationally, expected that an ICT company will be objective in assessing public needs. A company given the job of conducting the system study, performs it in a way that suits its own interests, rather than those of the public purchaser, i.e., it subordinates informational needs to technology, sells the most expensive product or sells a product that makes the public institution dependent on the private provider for many years. (Pawłowska, 2004, p. 177)

This is an example of Cross-Boundary Change theorizing, and it differs critically in its assumptions from that of the Heeks and Stanforth (2007) study.

Dimension III: Causal Autonomy

The dimension of Causal Autonomy concerns the theorist's views about the direction of causal influences between human (or social) actors and technology. This dimension has roots in philosophical debates over "idealism versus materialism" as well as "determinism versus voluntarism" (Bunge, 1996). Three basic positions on this dimension can be differentiated.

The first basic position on the Causal Autonomy dimension is that all outcomes of technology use can ultimately be traced back to people's goals, intentions, and actions, because technology is the instrument of human (social) action and a human (social) accomplishment. This position is illustrated by the observation that "Self-driving cars can never be autonomous as long as someone tells them where to go." Consistent with this position are functional and teleological theories (Falcon, 2012) and the philosophy of (human) action (Juarrero, 2002). In this view, human motives—human ideas, meanings, and reasons, that is, goals or intentions—can properly be theorized as causes of human behavior (Juarrero, 2002) and social outcomes. In contrast, the attribution of responsibility to nonhuman entities is not appropriate.

A second basic position on Causal Autonomy is that technology, like other external physical conditions (e.g., geography and climate), influences human ideas and behaviors. The materialist point of view is usually attributed to Karl Marx (1818–1883), but this position also has contemporary appeal (Braverman, 1974; Smith & Marx, 1994; Winner, 1978).

To the extent that technology creates necessities and constraints that influence or even compel human behavior—a situation of interest in law, as well as social science—human behavior can be said to be externally influenced or determined. The assignment of legal liability (i.e., responsibility) requires the attribution of causation to someone or something (Dahiyat, 2010; Hoekstra & Breuker, 2007). A particularly interesting question today is how theorists should deal with modern technologies and systems that can operate and even self-modify, post development, without continuous human intervention. Examples include automated securities trading, driverless cars, artificial intelligence algorithms, and so on. Increasingly, lawyers and moral philosophers (Grodzinsky, Miller, & Wolf, 2008; Stahl, 2004) are mooting the possibility of attributing responsibility (or "quasi-responsibility") for certain sociotechnical outcomes to algorithms instead of their designers or users (Stahl, 2006). Ultimately, a theorist's assignment of responsibility for outcomes is a causal position. As Stahl (2006) noted, "A first condition of responsibility is causality" (p. 208).

A third basic position on Causal Autonomy attempts to synthesize the materialism versus idealism dualism (Bunge, 1996) by emphasizing the close and possibly inextricable interactions between humans and technology. Actor-network theorists (Callon, 1986; Latour, 2005) have posited the symmetry of humans and technologies in the course of social change. Other scholars have defended this position by arguing the ontological inseparability of technology (as the product of social action) and the social realm (Orlikowski & Scott, 2008). Still others adopt this position for the purpose of theorizing about "collaborative" technologies (e.g., communication tools, groupware, and social media).

Below we discuss elaborations of these basic positions, which we call Human Sovereignty, Technology Autonomy, and Relational Synergy.

Position III.A: Human Sovereignty

The Human Sovereignty position can be stated as follows: Technology is an inanimate product of intentional human action and, therefore, only people (or social actors) can be viewed as causal (i.e., agents of change). People are ultimately responsible for the consequences attributable to technology use.

Philosophical or Social Theoretic Foundations

The idea that technology doesn't *do* anything and that only what humans believe about or do with technology matters is common in everyday reasoning ("Guns don't kill people. People kill people"; "We don't have a technology problem here, we have a people problem"). Ancient Greek philosophers provided some of the earliest recorded warrants for the Human Sovereignty position. Final causes (goals, intentions, or functions) were among the types of causes known to the Greeks. For Aristotle, it was important to identify a phenomenon's primary cause, which he believed in many instances was a final cause (Falcon, 2012).

Champions of positive science attempted to eliminate human reasons from scientific explanation on the grounds that they cannot be observed and therefore should be ascribed to the realm of the nonscientific or metaphysical.[5] By contrast, idealist philosophers like Dilthey (1833–1911) insisted that human ideas exist independently of the material world, and that human intentions drive action and the consequences of action. Among contemporary philosophers of human action, intentionalists recognize that intentions do not always result in the hoped-for outcomes (Von Wright, 1971). Nevertheless, intentions are important, not only because they are distinctly different from Aristotelian material or efficient causes, but also because humans use their beliefs about intentions to ascribe responsibility for outcomes. Contemporary law, for instance, differentiates between manslaughter and murder on the basis of attributions about a perpetrator's intentions. While material and efficient causes can only be conjectured, intentions can be known by the agent (Wittgenstein, 1958), and thus revealed by the latter.

A central concern of the contemporary philosophy of action is to distinguish between voluntary, involuntary, and compulsory human behavior (Juarrero, 2002). "Voluntary actions are both purposive and appropriate to the situation. Appropriate behavior *whose principle of movement or cause is within the agent, who is aware of what he or she is doing*, is paradigmatically 'action'" (Juarrero, 2002, p. 16, emphasis added). To assign responsibility for action, Juarrero (2002) asserted, one must examine both the material conditions and the influential ideas that

[5] Auguste Comte asserted that "Le caractère fondamental de la philosophie positive est de regarder tous les phénomènes comme assujettis à des lois naturelles invariables, dont la découverte précise et la réduction au moindre nombre possible sont le but de tous nos efforts, en considérant comme absolument inaccessible et vide de sens la recherche de ce qu'on appelle les causes soit premières, soit finales" (Comte, 1830–1842). Translation (ours): "The fundamental character of positive philosophy is to regard all phenomena as subject to invariant natural laws, of which the valid discovery and reduction to the least possible number are the goal of all our efforts, while considering the study of what one calls primary or final causes as absolutely unattainable and devoid of meaning."

impinge on a person. In contemporary social sciences, Giddens (1979), among many others, has argued that human action is characterized by agency, that is, by the possibility of doing otherwise.

Examples from IS Research Literature

In a study of Enterprise Resource Planning (ERP) system use, Boudreau and Robey (2005) argued:

> humans are relatively free to enact technologies in different ways. They can use it minimally, invoke it individually or collaboratively, improvise in ways that produce novel and unanticipated consequences. This perspective advises against treating technology as a determinant of social change. Rather, technology is implicated in social change at the discretion of human agents, even with automated manufacturing technologies and especially with computer-based information systems…. As users enact technologies in response to their local experiences and needs, significant organizational changes may result over time. From [a human] agency perspective, such changes are not realized from the embodiment of social structures within the technology. (Boudreau & Robey, 2005, pp. 4–5)

Boudreau and Robey's case study of evolving patterns of use of an ERP system called Compass in a US governmental agency illustrates the Human Sovereignty position particularly well, because ERP systems are generally believed to be quite inflexible. Thus, one would have expected to see relatively little freedom for Compass users to act differently than the system's designers and implementers had intended. However, Compass users deliberately and knowingly exercised their autonomy by *not* attending training sessions and by continuing to employ "shadow systems" long after Compass was implemented.

Position III.B: Technology Autonomy

The Technology Autonomy position can be stated as follows: Technologies can affect humans (or social actors) and (post development) can sometimes operate with limited human intervention.

Philosophical or Social Theoretic Foundations

The major warrants for the Technology Autonomy position come from philosophical writings on materialism. For materialists, explanations of phenomena, including human mental activities, originate fundamentally in physical conditions. The position that technology exerts causal influences on human behavior and societal outcomes is sometimes called "technological determinism" and is attributed (incorrectly, many experts say) to the writings of Karl Marx (1818–1883). More recently, the French philosopher and sociologist Jacques Ellul (1912–1944) deeply engaged the question of technology and social change and challenged the notion that technology is subservient to humans (Ellul, 1980). Industrial and labor relations expert Harry Braverman (1974) attributed to technology the deskilling of craftwork. Philosopher of science and technology Langdon Winner (1978) argued that technology embodies power relations and becomes a way in which societies settle political questions, thereby transcending the simple categories of intended or unintended consequences.

From the perspective of this position, technologies can take on a life of their own, once they have been developed and deployed, by virtue of ideology (Ellul, 1980), economic interests (Marx, 1887; Winner, 1978), or operational independence from continuous human intervention (Sutherland, 2008). Indeed, some classes of technology today are designed to actively *prevent* human intervention (Sutherland, 2008).

Examples from IS Research Literature

"Most research on [decision support systems], knowledge management systems, and expert systems, focus [*sic*] on how these systems assist humans in decision-making" (Blue & Andoh-Baidoo, 2010, p. 46). However, certain information systems, called "directive decision devices (Sutherland, 2008) have the ability to not only gather data for input and determine the decision, but also enact the choice—all absent of human intervention" (Blue & Andoh-Baidoo, 2010, p. 46). An example of directive decision devices is financial portfolio management systems that are intended to act when human investors fail to act within a predetermined time frame (Fan, Jan, & Andrew, 2004). Similarly, Xiao and Benbasat (2007) observed that users of artificial intelligence-based product recommender systems cannot know whether or not these systems are actually doing the users' bidding.

Position III.C: Relational Synergy

We named a third common position on the Causal Autonomy dimension Relational Synergy. The position statement for Relational Synergy can be phrased as follows: The outcomes of technology use are the product of interaction between people (or social actors) and technologies.

Philosophical or Social Theoretic Foundations

The best-known warrant for the Relational Synergy position comes from actor-network theory (Latour, 1996). Latour argued that technology studies have traditionally emphasized technical attributes (i.e., an Aristotelian formal cause) in what he refers to as an "idealized materialism" (Latour, 2007). Latour countered that technologies are not inert objects that social actors can manipulate at will, nor are they autonomous agents that exert their (functional) goals onto human actors (Howcroft, Mitev, & Wilson, 2004, p. 345). Instead, Latour argued for a "material materialism," in which assemblages of humans and technology "actants" are theorized as concretely real actor-networks, not just social constructions (Latour, 2007).

Today, the leading theoretical statement on the material role of information technology in conjunction with human agency is sociomateriality (Leonardi et al., 2012). Orlikowski and Scott (2008) have advanced a sociomaterial perspective in which technology and society, while analytically separable, are ontologically "inextricably intertwined." Other IS and sociotechnical theorizing about the close synergy between humans and information technology include Schultze's (2014) discussion of "cyborgs," Ekbia and Nardi's (2014) "heteromated systems," and technology affordances and constraints theory (Majchrzak & Markus, 2013; Markus & Silver, 2008; Zammuto, Griffith, Majchrzak, Dougherty, & Faraj, 2007).

An Example from IS Research Literature

Leonardi employed the term "imbrication" to convey his theory that technology and the social world (i.e., social systems, social dynamics, and social processes) are ontologically distinct—but deeply interrelated—entities (Leonardi, 2012). In his study of automotive crash testing simulation, Leonardi (2011) theorized that perceptions of constraint invite people to change their technologies, while perceptions of affordance lead people to change their routines. For instance, because the suite of finite element tools used by engineers to build simulation models did not produce consistent results (thereby constraining the engineers), developers "recognized implicitly that their goals to make crashworthiness simulations more credible (human agency) could be fulfilled if they created code that would automatically aggregate simulation results into standard reports (material agency)" (Leonardi, 2011, p. 158).

Additional Observations about Causal Autonomy Positions

Table 3.4 summarizes the positions on the Causal Autonomy dimension and offers constructed examples based on Xiao and Benbasat (2007) for contrast. Each position has authoritative warrants in the philosophical or social theory literatures. Each position can be identified in some published research in IS and related fields, and each position can be advantageous for some theoretical purposes. (For metaphorical illustrations of the positions by Nicolas Antheaume, see Fig. 3.3.)

The Human Sovereignty position figures prominently in the literature on the social construction of technology (Howcroft et al., 2004). This position is highly relevant for theorizing about the origins of technological innovations (Van de Ven, Polley, Garud, & Venkataraman, 1999) and

Table 3.4 Positions on Dimension III, Causal Autonomy

Basic position	Elaborated position statement	Illustrations from the IS field
Technology as instrument Causal effects move from people (or social actors) to technology	**III.A. Human sovereignty** Technology is an inanimate product of intentional human action and, therefore, only people (or social actors) can be viewed as causal.	• Recommendation systems improve the efficiency of human decision-making, but do not constrain or manipulate human decision-makers.
Technology as influencer Causal effects move from technology to people (or social actors)	**III.B. Technology autonomy** Technologies can affect humans (or social actors) and (post development) can sometimes operate with limited human intervention.	• Recommendation systems screen and evaluate products for people; as a result, people cannot know whether recommendation systems are serving their interests or those of product vendors and platform owners (Xiao & Benbasat, 2007).
Technology as interactant Causal effects move back and forth between people (or social actors) and technology	**III.C. Relational synergy** The outcomes of technology use are the product of interaction between people (or social actors) and technologies.	• It is impossible to differentiate human and technological contributions to changed organizational practices such as recommendation systems.

Causal Autonomy refers to a theorist's views about movement of causal effects between human (or social) actors and technology. The illustrations in Table 3.4 were crafted by the authors to parallel the position taken by (Xiao & Benbasat, 2007)

Fig. 3.3 Causal Autonomy positions: human sovereignty, technology autonomy, relational synergy. (Illustrations by Nicolas Antheaume)

about the course of technology evolution over time (Pollock & Williams, 2009). It is also advantageous for highlighting the importance of managerial decisions and actions when implementing technology (e.g., the application of complementary technologies and process changes) in shaping subsequent technology uses and outcomes.

For an increasing number and type of technologies, including automated decision-making (e.g., automated trading and insurance underwriting), robotic automation (e.g., self-driving cars and self-piloting drones), machine learning algorithms, and embedded systems (e.g., software in refrigerators), the assumption that technology use involves continuous human intervention is no longer appropriate. It is estimated that well over 60% of financial securities and currency trading is now fully automated; the progress of machine learning is such that some experts predicted the arrival of self-programming automated trading algorithms within a decade (Beddington, 2010). Fundamentally unlike historical factory automation, in which capital substitutes for physical human labor (Marx, 1887), smart machines today are not only "informating" human cognition (Zuboff, 1988), they are also removing human thought processes from the coordination and control of physical and knowledge work (Sutherland, 2008).

In the IS literature, Kallinikos (2011) featured and developed the theme of technological autonomy, and Markus (2005) proposed a "technology shaping" perspective on the consequences of decision-technology use. The Technology Autonomy position does not deny either the human and social nature of technology or the role of human agency in technology conception, development, and deployment. Rather, the Technology Autonomy position bounds out, for theoretical reasons, the social origins of technology and brackets the subsequent and consequential influences of autonomous technologies on human or collective actors. Although discredited in the minds of many, the Technology Autonomy position still attracts the attention of historians and scholars of large-scale technological systems (Smith & Marx, 1994) and it offers needed theoretical possibilities to IS design scientists (Markus, 2005).

The Relational Synergy position has deep appeal for theorists interested in information technologies that rely on continuous human intervention and input. Examples include electronic communication systems, social media websites, gaming technology, augmented reality systems, and the like. This position highlights the dynamics by which humans and their machines emergently co-create evolving joint activity patterns.

Discussion

In this section of the chapter, we address two questions. First, what value does the Markus and Rowe (2018) causal structure framework provide over and above that provided by the not-dissimilar framework of Markus and Robey (1988)? Our answer is: It promotes theoretical diversity. Second, how can the causal structure framework of Markus and Rowe (2018) be used and useful? Our answer is: It can be used and useful in theory and research synthesis, theory building, and research design.

Promoting Theoretical Diversity

As explained in Rowe and Markus (2018), our confessional tale about the making of Markus and Rowe (2018), our motivation for tackling causal structure anew was our belief that the framework in Markus and Robey (1988) had overlaps and gaps. The single most important difference of the Markus and Rowe (2018) framework was the attempt to make the dimensions logically distinct and independent, such that a scholar's position on one dimension did not automatically predefine her position on the others. This attempt was based on careful definition and detailed reference to philosophical and social theory literatures.

Rowe and Markus (2018) provides a detailed, side-by-side comparison of the two frameworks. Both frameworks have three dimensions, but the dimensions are not commensurate, and the positions are different. The Causal Ontology dimension in Markus and Rowe (2018) (henceforth M&R18) has two positions not found in Markus and Robey (1988) (henceforth M&R88)—Constitutive Causality and Causal Mechanism—either of which are options for people who self-identify as interpretivists. The Causal Trajectory dimension of M&R18 breaks away from the levels of analysis dimension of M&R88, including two positions—Indwelling

Change and Evolving Interlinkage—that do not depend on natural or living systems theory and that characterize the positions of many process theorists, practice theorists, and scholars of science and technology studies. Assumptions about emergence pervade all three positions on Causal Trajectory, whereas M&R88 considered emergence to be an alternative position to technological and organizational imperatives. The Causal Autonomy position of M&R18 reimagines the technological imperative of M&R88 as an alternative to Human Sovereignty and Relational Synergy. It does not deny human agency in technology design and implementation, but focuses on situations in which systems operate, post-deployment, with limited human intervention. And it does not, as M&R18 did, necessarily view technology as an external force.

In retrospect, however, none of these differences between M&R88 and M&R18 represents the single most important advantage, in our view, of the latter. M&R88 took sides. They explained why they preferred a particular position on each dimension to the others. When we began our project, we expected that we would again arrive at a set of preferred positions. Instead, we gradually arrived at what is now our committed opinion: All of the positions we articulated in M&R18 have sound warrants in philosophy or social theory, and all of them have advantages for investigating particular research questions or phenomena and disadvantages for others. Theoretical diversity is functional for a field and a phenomenon as diverse as ours. No single combination of positions is best for all research purposes. Theoretical monism would restrict our future choices.

In that spirit, we also acknowledged that the M&R18 framework is not the last word. We allowed that our positions are fuzzy and need to be made specific depending on the researcher's purposes. We granted that new positions, even new dimensions, might be needed and are likely to be crafted over time. As our phenomena and collective research interests continue to proliferate and differentiate, our theories should also. As that happens, an evolving framework of causal structure can help us advance as a field without degenerating into incoherence or unintelligibility.

Using the Framework for Synthesis, Theory Building, and Research Design

For the sake of provocation, we assert that progressing as a field involves, at minimum, synthesizing what we know for various stakeholder groups, setting new research directions based on what remains unknown, and crafting new strategies and methods for acquiring new knowledge. We believe that the revisable and extensible framework of causal structure in Markus and Rowe (2018) can help advance the IS and sociotechnical communities in all three ways.

Synthesis

One major way in which the M&R18 framework of causal structure can be used is for theoretical and research syntheses (Rowe, 2014). Here, we are not talking about the type of "review paper" that tabulates the number of articles by year exhibiting a particular philosophical perspective or adopting a particular method, but rather about research that attempts to integrate or meta-triangulate (Jasperson et al., 2002) empirical findings or hypothesized causal mechanisms across the diverse literature on a phenomenon or research question. By "carving up" (Pearl, 2009) theoretical arguments in ways different from the conventional (e.g., philosophical perspective, grand theory, research method) and by not privileging any particular view, the causal structure framework of M&R18 may help a synthesizer see substantive patterns—for example, "what we know and what we don't know"—in the body of knowledge on particular sociotechnical phenomena and research questions.

Theory building

A second important way that the M&R18 framework can be useful scholars is in the process of building substantive or middle-range IS theories, in lieu of importing them from other fields (Avgerou, 2013b). Avgerou expressed dissatisfaction with the limited cumulative results of qualitative research based on interpretive social theories. She traced the source of the problem to use of an "epistemic script" in which scholars interpret case studies through the lens of concepts derived from foundational social theories.

> Analysis both demonstrates the validity of the … concepts developed in prior research and adjusts them to construct conceptual refinements suitable to the specific issues or social settings under study … However, the explanations developed are not new concepts: they convey the insights of a chosen general … theory to the particular case under investigation. [This approach] does not generate suggestions 'of relationships and connections that had previously not been suspected' [citing Karl Weick (Weick 1898)]. (Avgerou, 2013b, p. 405)

The situation Avgerou described strikes us as analogous to the elaboration of variable-oriented theories by adding new variables. This activity may increase explanatory power in statistical analysis, but it rarely breaks away from the mental frame imposed by the original theory. Avgerou's (2013b) proposal for building novel and uniquely IS middle-range theories from interpretive cases is to use process tracing to identify and label social mechanisms and use these mechanisms as a basis for crafting new theories. To that valuable suggestion, we add one: use our framework to support the systematic identification, comparison, and evaluation of alternative theories.

According to Weick (1989), theory cannot be improved without improvements in theorizing. The way to improve theorizing is by generating numerous "thought trials" and by progressively, through an evolutionary selection process, winnowing them down to the ones to be retained. The key, Weick argued, was to make these thought trials as heterogeneous as possible, so that "a broader range of possibilities" is tried (Weick, 1989, p. 522).[6] But Weick also noted how difficult it is to achieve that goal, because of the human tendency for "grooved" and redundant thought. To overcome that limitation, Weick recommended the use of strong theoretical classification systems, where a strong classification sys-

[6] For a detailed example of how to apply the framework to generate heterogeneous thought trials, see Markus and Rowe (2018).

tem is one in which the categories are mutually exclusive and collectively exhaustive (all observations assigned to one and only one category).

What Weick called thought trials, Yin (2009) called rival hypotheses. Although Yin discussed rival hypotheses in the context of realist case study research methods, the strategy of formulating rival hypotheses or explanations is also applicable to research methods adapted to positivist, interpretivist, and critical realist philosophies. More importantly, the strategy of formulating rival hypotheses or explanations is applicable to theory building, as well as to the methods of empirical investigation.

Rival explanations have traditionally been presented as mutually exclusive of each other, as in Weick's (1989) exposition. But a recent essay by Zaks (2017) cogently classified theoretical rivals into four categories. According to Zaks (2017), rival explanations can be:

1. Mutually exclusive. For example, X makes no difference, versus X does make a difference.
2. Coincident, that is, independent, such that evidence in favor of one explanation has nothing to say about the relevance of another explanation. For example, Markus and Mao (2004) proposed three different mechanisms by which participation in system development could lead to system success: buy-in, better design, and better relationships between users and developers. Evidence of better relationships and system success would not necessarily rule out the mechanisms of buy-in or better design.
3. Congruent, that is, different explanations that work together to produce a result. For instance, "ERP systems lead to business process improvement" and "good quality data, exception reporting capabilities, and trained users lead to business process improvement" are congruent hypotheses.
4. Inclusive, that is, a special case of congruent, in which one theory encompasses another. For example, "ERP systems in conjunction with good quality data, exception reporting capabilities, and trained users lead to business process improvement" is inclusive of "ERP systems lead to business process improvement."

Zaks' (2017) primary concern is the use of rival explanations in process tracing, an empirical research method. But as Weick (1989) has argued, the generation and winnowing of thought trials is also fundamental to theory building (see also Rivard (2014)). Our point is that Markus and Rowe's (2018) framework of causal structure is a theoretical classification system (Weick, 1989) that can be used to generate alternative explanations both for theory building purposes and for subsequent empirical investigation.

Because knowledge is inherently incomplete (Cilliers, 1998), theory building is a process of making choices and trade-offs among the myriad theoretical options offered by a phenomenon of interest. Our framework can be thought of as a broad "solution space" encompassing the variety of potential causal arguments that form the building blocks of theory. By arraying a range of theoretical options, the framework can be particularly valuable to scholars in two situations. The first is the situation in which one confronts what one believes to be a new sociotechnical phenomenon, not adequately covered by existing theories. Researchers who are reluctant to choose an existing theory may be able to bootstrap a new theory by choosing among, and then building upon, various positions in the framework. Second, for scholars whose primary contribution is empirical, that is, a rich "theory light" qualitative study (Avison & Malaurent, 2014), the framework may be able to suggest "theoretical implications" (Agerfalk, 2014). For instance, the framework might prompt a researcher to ask whether the data say more about how technology influences humans or about how humans shape technology to their own ends.

Research Design

A third way in which the Markus and Rowe (2018) framework can be used and useful is in choice of research design and in bridging the gap between theoretical aspirations and empirical achievements. Discussions of research epistemology often present configurations of theoretical assumptions and research techniques as if they are inextricable. If one believes that causality is a metaphysical concept for which statistical association is the best and only available proxy, then theory consists in statements like "the greater the organization's absorptive capacity, the greater the organization's IT business value," and the most appropriate method appears to be multivariate analysis. Similarly, if one believes that causation means how things happened, then "an incremental process of organizational learning punctuated by occasional grand insights explained the success of Company X's IT initiative" is an apt statement of theory, and qualitative process analysis seems to be the best research approach.

Most philosophers agree that no empirical evidence will ever prove a theory. At best, empirical evidence can strengthen or weaken our *beliefs* in a hypothesis. This suggests that high-quality relevant evidence achieved via any justified method should be acceptable as appropriate for a particular theory or research question, regardless of its epistemological inspiration. For example, survey evidence could support a qualitative interpretation; interview data could inspire or help justify a quantitative model. Evidence that a company learned from its IT investments over time could be used for any position on the Causal Ontology dimension, that is, to support a Directional Association-type theory of IT business value, a causal mechanism-type theory of ERP system success, or a process theory of ERP-occasioned changes in organizational practices. We believe there is no inextricable relation between what we learn and how we learned it on the one hand and what we make of it theoretically on the other.

Naturally, some methods will better match a theory or research question than others. But it is not always possible to select the method most appropriate to the nature of the research question: Relevant data might not be available; techniques might not be adequate to analyze the data according to the theoretically important questions; computational resources for analysis might be prohibitively expensive. Designing empirical research always involves making trade-offs among practical considerations like research access, data quality, cost, results generalizability, theoretical importance, practical relevance, funder requirements, and so on. There should always be room in our journals for research that contributes valuable knowledge even when theory-method fit is not ideal. And it should always be acceptable for qualitative researchers to employ quantitative evidence that fits their theories, and vice versa.

At the same time, when there is a choice of method, theory should be the decider, rather than the other way around. Moreover, the clarity and plausibility of theoretical statements should be of paramount importance in both research design and evaluating evidence. The significance of theoretical plausibility has been emphasized in different ways by leading interpretive and quantitative scholars. For example, Karl Weick (1989) proposed a set of criteria to be used for selecting good theoretical ideas and weeding out the bad ones from a set of heterogeneous thought trials. He asserted that empirical validation is not, and cannot be, the major criterion for theoretical selection. (See also Van Maanen, Sørensen, and Mitchell (2007).) "Plausibility," Weick argued, is the closest we can usually come to establishing the primacy of theoretical arguments. Weick's plausibility criteria are whether a theoretical conjecture is "interesting, obvious, connected, believable, beautiful, or real in the context of the problem [theorists] are trying to solve" (p. 524). But his discussion of these criteria should give every theorist pause, because they often have ironic twists. For instance, "the reaction *that's obvious* may be a clue to [theoretical] significance as well a clue to triviality" (p. 526, original emphasis). Weick noted that theorists should *welcome* disconfirmation of their ideas as an opportunity to learn something new. But he also observed that "once a theorist has a strong investment in a perspective .... then disconfirmation [will invoke] strong negative feelings .... [p. 526] ... It is a thin line from that's interesting to that's in my best interest" (p. 528). Weick warns against "self-serving translations of theoretical analysis into theoretical advocacy" (p. 528).

From the other end of the research spectrum, Pearl and Mackenzie (2018) emphasized the importance of theoretical plausibility in quantitative causal methods. Although these methods cannot be used to discover causal theory, they can be used to answer practically important causal questions, when combined with plausible theory and high-quality quantitative data:

> Many people still make [the] mistakes of thinking that the goal of causal analysis is to prove that X is a cause of Y or else to find the cause of Y from scratch. That is the problem of causal discovery, which was my ambitious dream when I first plunged into graphical modeling.... In contrast, the

> focus of … this book, is representing *plausible* causal knowledge, combining it with empirical data, and answering causal queries that are of practical value. (Pearl & Mackenzie, 2018, pp. 79–80, emphasis added)

This way of thinking makes the crafting of clear and plausible theory the most important step in research design, regardless of method. It is certainly not always possible to select the best method and evidence for one's causal claims—for reasons of data access, training and skill, personal preference, and so on. But this does not mean that one's *theorizing* needs to be as weak as the evidence inevitably is. Our framework can help scholars articulate their deep beliefs about causality clearly and plausibly, even when their methods and data do not provide strong enough support for their theoretical arguments.

Conclusion

Theoretical diversity is good for academic fields, particularly one like ours with a subject matter that continues to differentiate and evolve. But theoretical diversity also threatens incoherence and mutual unintelligibility. Strong theoretical classifications systems can aid in building new theories (Weick, 1989), synthesizing existing knowledge (Rowe, 2014) for varied stakeholder groups, and selecting plausible theories to guide the gathering and analysis of new evidence (Pearl & Mackenzie, 2018). As technology advances and knowledge grows, the theoretical classification framework of Markus and Rowe (2018) is unlikely to be the final word on a strong theoretical classification for IS and sociotechnical studies. However, we believe that the framework is good enough for now.

References

Agarwal, R., Animesh, A., & Prasad, K. (2009). Social interactions and the 'digital divide': Explaining variations in internet use. *Information Systems Research, 20*(2), 277–294.

Agerfalk, P. J. (2014). Insufficient theoretical contribution: A conclusive rationale for rejection? *European Journal of Information Systems, 23*(6), 593–599.

Alexander, J. C. (1985). *Neofunctionalism*. Beverly Hills, CA: Sage Publications.

Archer, M. S. (1995). *Realist social theory: The morphogenetic approach*. Cambridge, UK: Cambridge University Press.

Archer, M. S., Bhaskar, R., Collier, A., Lawson, T., & Norrie, A. (1998). *Critical realism: Essential readings*. London, UK: Routledge.

Arthur, W. B. (2009). *The nature of technology: What it is and how it evolves*. Simon and Schuster.

Avgerou, C. (2013a). Explaining trust in IT-mediated elections: A case study of e-voting in Brazil. *Journal of the Association for Information Systems, 14*(8), 420–451.

Avgerou, C. (2013b). Social mechanisms for causal explanation in social theory based IS research. *Journal of the Association for Information Systems, 14*(8), 399–419.

Avison, D., & Malaurent, J. (2014). Is theory king?: Questioning the theory Fetish in information systems. *Journal of Information Technology, 29*(4), 327–336.

Barley, S. R. (1998). What can we learn from the history of technology? *Journal of Engineering and Technology Management, 15*, 237–255.

Baum, J. A. C., & Singh, J. V. (1994). *Evolutionary dynamics of organizations*. Oxford, UK: Oxford University Press.

Beddington, J. (2010). *Foresight: The future of computer trading in financial markets*. Final project report, The Government Office for Science, London, UK.

Bhaskar, R. (1975). *A realistic theory of science*. Sussex, UK: Harvester Press.

Bhaskar, R. (1998). Philosophy and scientific realism. In M. S. Archer, R. Bhaskar, A. Collier, T. Lawson, & A. Norrie (Eds.), *Critical realism: Essential readings* (pp. 16–47). London, UK: Routledge.

Blue, J., & Andoh-Baidoo, F. K. (2010). Directive Decision Devices: Extending the reach of automation into the finance domain. *Expert Systems with Applications, 37*, 45–54.

Boudreau, M.-C., & Robey, D. (2005). Enacting integrated information technology: A human agency perspective. *Organization Science, 16*(1), 3–18.

Braverman, H. (1974). *Labor and monopoly capital: The degradation of work in the twentieth century*. New York: Monthly Review Press.

Bunge, M. (1996). *Finding philosophy in social science*. New Haven, CT: Yale University Press.

Bunge, M. (1999). *Social science under debate: A philosophical perspective*. Toronto: University of Toronto Press.

Burt, R. S. (1987). Social contagion and innovation: Cohesion versus structural equivalence. *American Journal of Sociology, 92*(6), 1287–1335.

Burton-Jones, A., & Gallivan, M. J. (2007). Toward a deeper understanding of system usage in organizations: A multilevel perspective. *MIS Quarterly, 31*(4), 657–679.

Callon, M. (1986). Some elements of a sociology of translation: Domestication of the scallops and the fishermen of St Brieux Bay. In J. Law (Ed.), *Power, action and belief: A new sociology of knowledge* (pp. 196–233). London: Routledge and Kegan Paul.

Choi, T., Dooley, K., & Rungtusanatham, M. (2001). Supply networks and complex adaptive systems: Control versus emergence. *Journal of Operations Management, 19*(3), 351–366.

Cilliers, P. (1998). *Complexity and post-modernism: Understanding complex systems*. London, UK: Routledge.

Coleman, J. (1986). Social theory, social research, and a theory of action. *American Journal of Sociology, 91*(6), 1309–1335.

Comte, A. (1830–1842). Cours de philosophie positive, volume 1, 16. Retrieved November 16, 2015, from http://www.egs.edu/library/auguste-comte/articles/cours-de-philosophie-positive-16/premiere-lecon/

Dahiyat, E. A. R. (2010). Intelligent agents and liability: Is it a doctrinal problen or merely a problem of explanation? *Artificial Intelligence and Law, 18*(1), 103–121.

De Pierris, G., & Friedman, M. (2013). "Kant and hume on causality," in *Stanford encyclopedia of philosophy*, E.N. Zalta (ed.). Palo Alto, CA: Stanford University. Retrieved July 6, 2015, from http://plato.stanford.edu/entries/kant-hume-causality/

Dilthey, W. (1989). *Introduction to the human sciences*. Princeton, NJ: Princeton University Press.

DiMaggio, P. J., & Powell, W. W. (1983). The iron cage revisited: Institutional isomorphism and collective rationality in organizational fields. *American Sociological Review, 48*(2), 147–160.

Durand, R., & Vaara, E. (2009). Causation, counterfactuals, and competitive advantage. *Strategic Management Journal, 30*, 1245–1264.

Durkheim, E. (1982). *New rules of sociological method*. New York, NY: The Free Press.

Ekbia, H., & Nardi, B. A. (2014). Heteromation and its (dis)contents: The invisible division of labor between humans and machines. *First Monday, 19*(6) Retrieved from http://firstmonday.org/ojs/index.php/fm/article/view/5331

Ellis, G. F. R., Murphy, N., & O'Connor, T. (Eds.). (2009). *Downward causation and the neurobiology of free will*. New York, NY: Springer.

Ellul, J. (1980). *The technological system* (trans. Joachim Neugroschel and originally published by Calmann-Lévy, 1977). New York: The Continuum Publishing Corporation.

Falcon, A. (2012). Aristotle on causality. In E. N. Zalta (Ed.), *Stanford encyclopedia of philosophy*. Palo Alto, CA: Metaphysics Research Lab, Center for the Study of Language and Information. Retrieved from http://plato.stanford.edu/entries/aristotle-causality/

Fan, M., Jan, S., & Andrew, W. (2004). The Internet and the future of financial markets. *Communications of the ACM, 43*(11), 82–88.

Fayard, A.-L., & DeSanctis, G. (2010). Enacting language games: The development of a sense of 'we-ness' in online forums. *Information Systems Journal, 20*(4), 383–416.

Garfinkel, H. (1967). *Studies in ethnomethodology*. New York, NY: Prentice-Hall.

George, A. L., & Bennett, A. (2005). *Case studies and theory development in the social sciences*. Cambridge, MA: MIT Press.

Giddens, A. (1979). *Central problems in social theory*. Berkeley, CA: University of California Press.

Gnassounou, B., & Kistler, M. (2007). Introduction. In M. Kistler & B. Gnassounou (Eds.), *Dispositions and causal powers* (pp. 1–10). Aldershot, UK: Ashgate.

Goh, J. M., Gao, G., & Agarwal, R. (2011). Evolving routines: Adaptive routinization of information technology in healthcare. *Information Systems Research, 22*(3), 565–585.

Goldthorpe, J. (1999). Causation, statistics, and sociology. In J. Goldthorpe (Ed.), *On sociology, numbers, narratives and the integration of research and theory*. Oxford: Oxford University Press.

Gregor, S. (2006). The nature of theory in information systems. *MIS Quarterly, 30*(3), 611–642.

Gregor, S., & Hovorka, D. (2011). *Causality: The elephant in the room in information systems epistemology*. European Conference on Information Systems (ECIS), Helsinki, FI.

Grodzinsky, F. S., Miller, K. W., & Wolf, M. J. (2008). The ethics of designing artificial agents. *Ethics and Information Technology, 10*(2–3), 115–121.

Gross, N. (2009). A pragmatist theory of social mechanisms. *American Sociological Review, 74*(3), 358–379.

Harman, G. (2010). *Towards speculative realism: Essays and lectures*. John Hunt Publishing.

Hayes, N., & Introna, L. (2005). Systems for the production of plagiarists? The implications arising from the use of plagiarism detection systems in UK universities for Asian learners. *Journal of Academic Ethics, 3*(1), 55–73.

Hedstrom, P., & Swedberg, R. (Eds.). (1998a). *Social mechanisms: An analytical approach to social theory*. Cambridge, UK: Cambridge University Press.

Hedstrom, P., & Swedberg, R. (1998b). Social mechanisms: An introductory essay. In *Social mechanisms: An analytical approach to social theory* (pp. 1–31). Cambridge, UK: Cambridge University Press.

Hedström, P., & Ylikoski, P. (2010). Causal mechanisms in the social sciences. *Annual Review of Sociology, 36*(1), 49–67.

Heeks, R., & Stanforth, C. (2007). Understanding e-government project trajectories from an actor-network perspective. *European Journal of Information Systems, 16*, 165–177.

Henfridsson, O., & Bygstad, B. (2013). The generative mechanism of digital infrastructure evolution. *MIS Quarterly, 37*(3), 907–931.

Hoekstra, R., & Breuker, J. (2007). Commonsense causal explanation in a legal domain. *Artificial Intelligence and Law, 15*(3), 281–299.

Hollis, M. (1994). *The philosophy of social science: An introduction*. Cambridge, UK: Cambridge University Press.

Hovorka, D., Germonprez, M., & Larsen, K. R. (2008). Explanation in information systems. *Information Systems Journal, 18*, 23–43.

Howcroft, D., Mitev, N., & Wilson, M. (2004). What we may learn from the social shaping of technology approach. In J. Mingers & L. Willcocks (Eds.), *Social theory and philosophy for IS* (pp. 329–371). London: John Wiley & Sons Ltd.

Hume, D. (1955). *An Inquiry concerning human understanding; with a supplement, An abstract of a treatise of human nature*. Indianapolis, IND: Bobbs-Merrill.

Introna, L., & Ilharco, F. (2004). Phenomenology, screens, and the world: A journey with Husserl and Heidegger into phenomenology. In J. Mingers & L. Willcocks (Eds.), *Social theory and philosophy for information systems* (pp. 56–102). Chichester, UK: Wiley.

Introna, L. D., & Hayes, N. (2011). On sociomaterial imbrications: What plagiarism detection systems reveal and why it matters. *Information and Organization, 21*(2), 107–122.

Jarvenpaa, S., & Leidner, D. (1998). Communication and trust in global virtual teams. *Journal of Computer-Mediated Communication, 3*(4). [Online].

Jasperson, J. S., Carte, T. A., Saunders, C. S., Butler, B. S., Croes, H., & Zheng, W. (2002). Review: Power and information technology research: A metatriangulation review. *MIS Quarterly, 26*(4), 397–459.

Juarrero, A. (2002). *Dynamics in action: Intentional behavior as a complex system.* Bradford Books.

Juarrero, A. (2011). Causality and explanation. In P. Allen, S. Maguire, & B. McKelvey (Eds.), *The Sage handbook of complexity and management* (pp. 155–163). London: SAGE.

Kaidesoja, T. (2013). Overcoming the biases of microfoundationalism: Social mechanisms and collective agents. *Philosophy of the Social Sciences, 43*(3), 301–322.

Kallinikos, J. (2011). *Governing through technology: Information artefacts and social practice.* Basingstoke, UK: Palgrave Macmillan.

Lapointe, L., & Rivard, S. (2005). A multilevel model of resistance to information technology implementation. *MIS Quarterly, 29*(3), 461–491.

Latour, B. (1996). *Aramis or the love of technology.* Cambridge, MA: Harvard University Press.

Latour, B. (2005). *Reassembling the social: An introduction to actor-network theory.* Oxford, UK: Oxford University Press.

Latour, B. (2007). Can we get our materialism back, please? *Isis, 98*(1), 138–142.

Law, J., & Callon, M. (1992). The life and death of an aircraft: A network analysis of technological change. In W. Bijker & J. Law (Eds.), *Shaping technology/building society: Studies in sociotechnical change* (pp. 21–52). Cambridge, MA: The MIT Press.

Lee, A. S., & Baskerville, R. L. (2003). Generalizing generalizability in information systems research. *Information Systems Research, 14*(2), 221–243.

Leonardi, P. (2011). When flexible routines meet flexible technologies: Affordance, constraint, and the imbrication of human and material agencies. *MIS Quarterly, 35*(1), 147–167.

Leonardi, P. (2012). Materiality, sociomateriality, and socio-technical systems: What do these terms mean? how are they different? do we need them? In P. Leonardi, B. A. Nardi, & J. Kallinikos (Eds.), *Materiality and organizing.* Oxford, UK: Oxford University Press.

Leonardi, P. M., Nardi, B. A., & Kallinikos, J. (2012). *Materiality and organizing: Social interaction in a technological world.* Oxford University Press.

Lincoln, Y. S., & Guba, E. G. (1985). *Naturalistic inquiry*. Thousand Oaks, CA: Sage Publications.

Lyytinen, K., Rowe, F., & McGuire, C. (2018). *Improving IS research relevance for practitioners: The role of knowledge networks*. International Conference on Information Systems, San Francisco.

Machamer, P., Darden, L., & Craver, C. F. (2000). Thinking about mechanisms. *Philosophy of Science, 67*(1), 1–25.

Majchrzak, A., & Markus, M. L. (2013). Technology affordances and constraints theory (of MIS). In E. H. Kessler (Ed.), *Encyclopedia of management theory* (pp. 832–835). Thousand Oaks, CA: Sage Publications.

Markus, M. L. (2005). The technology shaping effects of e-collaboration technologies—Bugs and features. *International Journal of e-Collaboration, 1*(1), 1–23.

Markus, M. L., & Mao, J. Y. (2004). Participation in development and implementation-updating an old, tired concept for today's IS contexts. *Journal of the Association for Information Systems, 5*(11), Article 14.

Markus, M. L., & Robey, D. (1988). Information technology and organizational change: Causal structure in theory and research. *Management Science, 34*(5), 583–598.

Markus, M. L., & Rowe, F. (2018). Is IT changing the world? Conceptions of causality for information systems theorizing. *MIS Quarterly, 42*(4), 1255–1280.

Markus, M. L., & Silver, M. S. (2008). A foundation for the study of IT effects: A new look at Desanctis and Poole's concepts of structural features and spirit. *Journal of the AIS, 9*(10/11), 609–632.

Marx, K. (1887). *Capital.* Retrieved February 6, 2015, from http://synagonism.net/book/economy/marx.1887-1867.capital-i.html

Mazmanian, M., Orlikowski, W., & Yates, J. (2013). The autonomy paradox: The implications of mobile email devices for knowledge professionals. *Organization Science, 24*(5), 1337–1357.

Meillassoux, Q. (2010). *After finitude: An essay on the necessity of contingency.* Bloomsbury Publishing.

Merali, Y. (2004). Complexity and information systems. In J. Mingers & L. Willcocks (Eds.), *Social theory and philosophy for information systems* (pp. 407–446). Chichester, UK: John Wiley & Sons.

Mingers, J. (2004). Re-establishing the real: Critical realism and information systems. In J. Mingers & L. Willcocks (Eds.), *Social theory and philosophy for information systems* (pp. 372–406). Chichester, UK: John Wiley & Sons, Ltd.

Mingers, J. (2014). *Systems thinking, critical realism and philosophy: A confluence of ideas*. London: Routledge.

Myers, M. D. (2004). Hermeneutics in information systems research. In J. Mingers & L. Willcocks (Eds.), *Social theory and philosophy for information systems* (pp. 103–128). Chichester, UK: Wiley.

Nan, N., & Lu, Y. (2014). Harnessing the power of self-organization in an online community during organizational crisis. *MIS Quarterly, 38*(4), 1135–1157.

Nan, N., & Tanriverdi, H. (2017). Unifying the role of IT in hyperturbulence and competitive advantage via a multilevel perspective of IS strategy. *MIS Quarterly, 41*(3), 937–958.

Ngwenyama, O., & Lee, A. (1997). Communication richness in electronic mail: Critical social theory and the contextuality of meaning. *MIS Quarterly, 21*(2), 145–167.

Orlikowski, W. (2002). Knowing in practice: Enacting a collective capability in distributed organizing. *Organization Science, 13*(2), 249–273.

Orlikowski, W., & Scott, S. (2008). Sociomateriality: Challenging the separation of technology, work and organization. *The Academy of Management Annals, 2*(1), 433–474.

Orlikowski, W. J. (2000). Using technology and constituting structures: A practice lens for studying technology in organizations. *Organization Science, 11*(4), 404–428.

Pawłowska, A. (2004). Failures in large systems projects in Poland: Mission [im]possible? *Information Polity, 9*(3/4), 167–180.

Pearl, J. (2009). *Causality: Models, reasoning, and inference* (2nd ed.). Cambridge, UK: Cambridge University Press.

Pearl, J., & Mackenzie, D. (2018). *The book of why: The new science of cause and effect*. Penguin Books.

Pollock, N., & Williams, R. (2009). *Beyond the ERP implementation study: A new approach to the study of packaged information systems: The biography of artifacts approach*. International Conference on Information Systems, Phoenix, AZ, Paper 6. Retrieved from http://aisel.aisnet.org/icis2009/2006

Poole, M. S., Van de Ven, A., Dooley, K., & Holmes, M. (2000). *Organizational Change and Innovation Processes: Theory and Method for Research*. New York, NY: Oxford University Press.

Ramiller, N., & Swanson, B. (2003). Organizing visions for information technology and the information systems executive response. *Journal of Management Information Systems, 20*(1), 13–50.

Ramiller, N. C., & Pentland, B. T. (2009). Management implications in information systems research: The untold story. *Journal of the Association for Information Systems, 10*(6), 474–494.

Rivard, S. (2014). Editor's commets: The ions of theory construction. *MIS Quarterly, 38*(2), iii–xiii.

Robertson, M., Swan, J., & Newell, S. (1996). The role of networks in the diffusion of technological innovation. *Journal of Management Studies, 33*(3), 333–359.

Rowe, F. (2014). Editorial: What literature review is not: Diversity, boundaries and recommendations. *European Journal of Information Systems, 23*(3), 241–255.

Rowe, F., & Markus, M. L. (2018). Taking on sacred cows: Openness, fair critique, and retaining value when revising classics. *European Journal of Information Systems, 27*(6), 623–628.

Salmon, W. C. (1998). *Causality and explanation*. New York, NY: Oxford University Press.

Sandis, C., & Tejedor, C. (2017). Wittgenstein on causation and induction. In H.-J. Glock & J. Hyman (Eds.), *A companion to Wittgenstein* (pp. 576–586). Chichester, UK: John Wiley & Sons.

Sayer, A. (2000). *Realism in social science*. London, UK: Sage Publications.

Schultze, U. (2014). Performing embodied identity in virtual worlds. *European Journal of Information Systems, 23*(1), 84–95.

Schwartz-Shea, P., & Yanow, D. (2013). *Interpretive research design: Concepts and processes*. Routledge.

Smith, M. R., & Marx, L. (1994). *Does technology drive history? The dilemma of technological determinism*. Cambridge, MA: The MIT Press.

Sorensen, A. B. (1998). Theoretical mechanisms and the empirical study of social processes. In *Social mechanisms: An analytical approach to social theory* (pp. 238–266). Cambridge, UK: Cambridge University Press.

Stahl, B. C. (2004). Information, ethics and computers: The problem of autonomous moral agents. *Minds and Machines, 14*(1), 67–83.

Stahl, B. C. (2006). Responsible computers? A case for ascribing quasi-responsibility to computers independent of personhood and agency. *Ethics and Information Technology, 8*(4), 205–213.

Stinchcombe, A. L. (2002). New sociological microfoundations for organizational theory: A postscript. In M. Lousbury & M. J. Ventresca (Eds.), *Social structure and organizations revisited* (pp. 415–433). Amsterdam: JAI.

Strong, D. M., & Volkoff, O. (2010). Understanding organization—Enterprise system fit: A path to theorizing the information technology artifact. *MIS Quarterly, 34*(4), 731–756.

Suchman, L. (1987). *Plans and situated actions: The problem of human machine communication*. Cambridge, UK: University of Cambridge Press.

Sutherland, J. W. (2008). Directive Decision Devices: Reversing the locus of authority in human–computer associations. *Technological Forecasting and Social Change, 75*(7), 1068–1089.

Tanriverdi, H., & Lim, S. Y. (2017). *How to survive and thrive in complex, hyper-competitive and distruptive ecosystems? The role of IS-enabled capabilities*. 38th International Conference on Information Systems, Seoul, KR.

Thornton, P. H., Ocasio, W., & Lounsbury, M. (2012). *The institutional logics perspective: A new approach to culture, structure, and process*. Oxford, UK: Oxford University Press.

Van de Ven, A., Polley, D., Garud, R., & Venkataraman, S. (1999). *The innovation journey*. New York, NY: Oxford University Press.

Van de Ven, A., & Poole, M. S. (1995). Explaining development and change in organizations. *Academy of Management Review, 20*(3), 510–540.

Van de Ven, A. H., & Poole, M. S. (2005). Alternative approaches for studying organizational change. *Organization Studies, 26*(9), 1377–1404.

Van Maanen, J., Sørensen, J. B., & Mitchell, T. R. (2007). The interplay between theory and method. *Academy of Management Review, 32*(4), 1145–1154.

Volkoff, O., & Strong, D. M. (2013). Critical realism and affordances: Theorizing IT-associated organizational change processes. *MIS Quarterly, 37*(3), 819–834.

Volkoff, O., Strong, D. M., & Elmes, M. B. (2007). Technological embeddedness and organizational change. *Organization Science, 18*(5), 832–848.

Von Wright, G. H. (1971). *Explanation and understanding*. Ithaca, NY: Cornell University Press.

Weber, M. (1968). *Economy and society*. Berkeley, CA: University of California Press.

Weick, K. E. (1989). Theory construction as disciplined imagination. *Academy of Management Review, 14*(4), 516–531.

Wilber, C., & Harrison, R. (1978). The methodological basis of institutional economics: Pattern models, storytelling, and holism. *Journal of Economic Issues, 12*(1), 61–89.

Winner, L. (1978). *Autonomous technology: Technics-out-of-control as a theme in political thought*. Cambridge, MA: MIT Press.

Winter, S., Berente, N., Howison, J., & Butler, B. S. (2014). Beyond the organizational "container": Conceptualizing 21st century sociotechnical work. *Information and Organization, 24*(4), 250–269.

Wittgenstein, L. (1958). *The blue and brown books: Preliminary studies for the "philosophical investigations"*. New York, NY: Harper & Row.

Wittgenstein, L. (1976). Cause and effect: Intuitive awareness. *Philosophia, 6*(3–4), 409–425.

Woodward, J. (2003). *Making things happen: A theory of causal explanation.* Oxford University Press.

Xiao, B., & Benbasat, I. (2007). E-Commerce product recommendation agents: Use, characteristics, and impact. *MIS Quarterly, 31*(1), 137–209.

Yin, R. K. (2009). *Case study research: Design and methods.* Thousand Oaks, CA: Sage.

Zaks, S. (2017). Relationships Among Rivals (RAR): A framework for analyzing contending hypotheses in process tracing. *Political Analysis, 25*(3), 344–362.

Zammuto, R. F., Griffith, T. L., Majchrzak, A., Dougherty, D., & Faraj, S. (2007). Information technology and the changing fabric of organizations. *Organization Science, 18*(5), 749–762.

Ziliak, S. T., & McCloskey, D. N. (2014). *The cult of statistical significance: How the standard error costs us jobs, justice, and lives.* Ann Arbor, MI: The University of Michigan Press.

Zuboff, S. (1988). *In the age of the smart machine: The future of work and power.* New York, NY: Basic Books.

4

Theory Building: Neither an Art nor a Science, But a Craft*

Suzanne Rivard

Introduction

The premise of this chapter is that many of us researchers hold a romantic view of theory and of theory building, a view that hinders our ability to effectively develop theory. The objective of this chapter is to present a more realistic view of theory and of its construction process. Accordingly, I propose a pragmatic model of the theory-building process. I developed the model inductively, by analyzing my experience as a theory builder and the experience I vicariously gained from supporting—as a senior editor—several authors in their theory development efforts and by guiding PhD students as they prepare a theory piece as part of the theory-building course I have been teaching for a decade. The model is enriched with

*An earlier version of this chapter appeared as Rivard, S. (2020) Theory building: Neither an art nor a Science. It is a craft. Journal of Information Technology. https://doi.org/10.1177/0268396220911938

S. Rivard (✉)
HEC Montreal, Montreal, QC, Canada
e-mail: suzanne.rivard@hec.ca

N. R. Hassan, L. P. Willcocks (eds.), *Advancing Information Systems Theories*, Technology, Work and Globalization, https://doi.org/10.1007/978-3-030-64884-8_4

methods, techniques, and heuristics that I have gleaned over the years and which I have found particularly useful or inspiring.

The romantic view of a theory portrays it as a complete, detailed, flawless, deep, and exhaustive explanation of a phenomenon, an object that Weick refers to as "Theory That Sweeps Away All Others" (Weick, 1995, p. 386). Such a conceptualization is intimidating, and most of us are likely to see reaching the level of explanatory power and generality it implies as unattainable. As an alternative to this romantic view, I suggest that it would be more productive if researchers and review teams alike agreed that "products of the theorizing process seldom emerge as full-blown theories, which means that most of what passes for theory in organizational studies consists of approximations [...and represents] interim struggles in which people intentionally inch toward stronger theories" (Weick, 1995, p. 385). This is indeed what we are developing: approximations. This notion is related to Hassan's (2014) products of theorizing "that only approximate theories but are critical to their development" (p. 1). This view is shared by several theorists. For instance, Bandura comments that "[t]heorists would have to be omniscient to provide an ultimate account of human behavior at the outset. They necessarily begin with an incomplete theory [...] Successive theoretical refinements bring one closer to understanding the phenomena of interest" (Bandura, 2005, p. 28). I suggest here that if, as authors, we come to terms with the idea that we are developing approximations and if, as reviewers and editors, we come to terms with the idea of publishing what may be approximations of theory, then we will learn to develop more refined approximations and become more proficient at theory building.

In addition, researchers sometimes hold one of two alternate romantic views of theory building. The first is that of theory building as an art. Under this view, the theorist has sparks of inspiration, and the theory emerges during trancelike periods of inspired writing. Once the theory is in writing, this object of art should not be "touched up." The alternate romantic view is that of theory building as a science, whereby there exists a series of activities—such as reviewing the literature to circumscribe the phenomenon of interest, analyzing the constructs and relationships from extant literature to identify gaps, and developing new theoretical propositions—that if dutifully followed will result in good theory. Unfortunately, neither of these two views is realistic, since few of us have sparks of inspiration that are powerful enough to result in a full-blown theory, and most of us need more than a well-defined series of steps to follow to produce compelling theories.

This chapter proposes that theory building resembles a craft more than either an art or a science. This does not imply that the theorist should not be creative and imaginative as the artist is, or does not need to be rigorous and methodical as a scientist is expected to be. Yet, in addition to possessing these qualities, the theorist as an artisan must also demonstrate care and ingenuity and work with patience and perseverance, being engaged in the work in and for itself (Sennet, 2008).

Model of the Theory-Building Process

Before outlining the model of theory building I propose, it is essential that I state my working assumptions. First, although I consider the outcome of the process an approximation, not an all-encompassing theory, I assume that the product that the theorist seeks to develop from the theorizing is an explanation—or explanation and prediction—of a phenomenon. This explanation will state the "relations among concepts within a boundary set of assumptions and constraints" (Bacharach, 1989, p. 496); it can be expressed as relational statements organized in a discursive essay or as a set of formal propositions (Bourgeois, 1979). Second, the process I present is better suited to developing theory under a deductive approach, one that commences "at the intersection of the theorist and the existing knowledge typically contained in the literature" (Shepherd & Sutcliffe, 2011, p. 361). My experience, however, suggests that the overall process and several of its activities are also relevant to inductive theory building. Third, the process that I present is intended as an explanation of how theory emerges from a set of activities. It is neither prescriptive nor predictive: it does not intend to tell theorists what they should do to develop a theory, nor does it pretend that faithfully following the proposed process will result in a good approximation of theory.

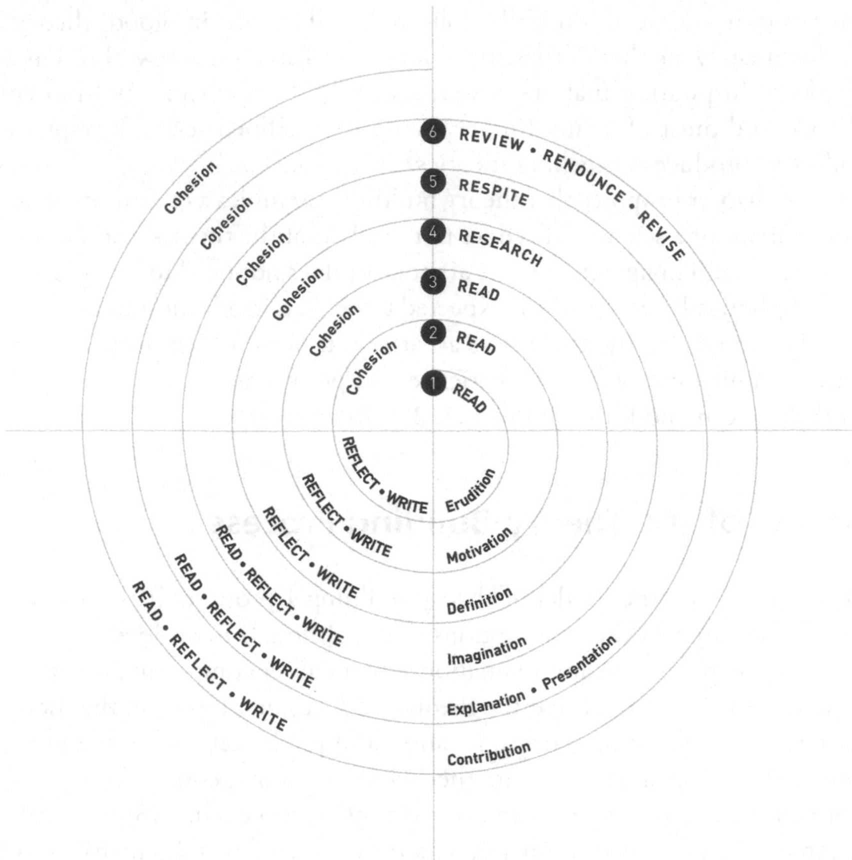

Fig. 4.1 A Spiral Model of Theory Building

Figure 4.1 presents my model, which is highly iterative. This spiral model[1] should be read as follows. The process comprises two main components: activities and their outcomes, which I call the "ions of theory construction" (Rivard, 2014). The term "ion" refers to small elemental particles that can be assembled into a compound—in the present case an approximation of theory—that has specific, unique, properties. It is not

[1] The reader familiar with the software risk management literature will have realized that my inspiration for this figure is Boehm's (1988) spiral model of the software process.

the presence of the elemental particles that gives value to the compound; it is how the theorist assembles them.

Three activities—read, reflect, and write—are part of every iteration, and are likely to take place almost concurrently. Such activities may take different forms depending on the iteration in which they are effectuated. For instance, at the beginning of the theory-building process, the theorist may only reflect, becoming introspective and contemplating her/his motivations in undertaking theory development. The reflect and read activities will also occur jointly, such as when the theorist consults the literature to understand it, map it, and make sense of it. Here, reading pertains to both scanning abstracts to select relevant sources and engaging with a source in greater depth, while reflecting takes the form of organizing, analyzing, and synthesizing. In the next iteration, the read and reflect activities will be motivated by a desire to identify trends or gaps in the literature. Later in the process, the reflect activity can occur conjointly with research, for instance when the theorist conducts thought experiments to enrich his/her theorization. In this case, reading would follow researching and reflecting, with the theorist going back to the literature to determine the extent to which the results of thought experiments are new and interesting. Reading often implies reading one's own work, and reflecting pertains to assessing the written piece. Writing can take the form of jotting down ideas, documenting findings or hunches, sketching a model, and evidently putting into words the actual theory piece. No matter what form it takes, writing is essential throughout the theory-building process to capture the explanation as it develops.

The model illustrated in Fig. 4.1 emphasizes that theorists are well advised to ensure that one specific ion—cohesion—is an ongoing outcome of their theory development process. Cohesion refers to unity within an ensemble and agreement among the parts of a whole. Checking for cohesion is essential throughout the process of theory building. Early on, cohesion must be established first between the phenomenon of interest and the literature that is reviewed, then between the gaps that the theorist observes and those he/she wishes to fill, then between the assumptions of the theory and the explanations it provides, and so on. Reflecting is a particularly important activity with respect to producing and ensuring cohesion, and this is one of the reasons why the reflect activity is present in each iteration.

Iteration 1: Activities—Read, Reflect, Write—Ion: Erudition

The first iteration situates the beginning of the deductive theory-building process at the intersection of the theorist and the existing knowledge represented by extant literature (Shepherd & Sutcliffe, 2011). This implies that one of its core activities is reading. The intensity of the review, in terms of its breadth and depth, will depend on the theorist's knowledge of the literature. A theorist with intimate knowledge of the extant research on a phenomenon will approach and read this literature differently from one with less extensive knowledge. In terms of methods, many excellent sources propose methods for reviewing the literature; the choice of a method will depend on the theorist's preferences, for instance for a top-down method or for the bottom-up, grounded theory method illustrated in Table 4.1.

The main outcome sought during the first iteration is erudition.[2] Defined as "profound knowledge," erudition refers to the breadth and depth of one's knowledge of a particular topic, "especially that based on learning and scholarship" (https://www.wordnik.com/words/erudition). Constructing a theory requires this kind of deep and broad knowledge. This implies an internalization of the literature's content in order to juggle concepts and organize them in a coherent whole that will provide a sound explanation of a phenomenon. To reach erudition, one must go through the steps proposed by literature review methodologists. One must identify the relevant literature; classify it according to various schemes, such as levels of analysis, antecedent variables, and foundational theories; construct descriptive tables that summarize this accumulated information; map the territory; and identify gaps. But these activities do not constitute erudition; they are the raw materials that serve as inputs to the ongoing reflect and write activities, which are essential if erudition is to develop as one reads the literature.

---

[2] The portions of the text that define the ions are in large part taken from Rivard (2014), with permission from the *MIS Quarterly*.

Table 4.1 Literature review methods

Top-down method of producing a literature review	Bottom-up method of producing a literature review
Stage 1: Searching for relevant sources Stage 2: Reading for the purpose of extracting materials based on themes Stage 3: Selecting extracts and writing notes on themes such as interpretations, concepts, data, theories, observations, methods, and assumptions Stage 4: Writing sections on the themes identified in Stage 3, with extracts from the sources identified in Stage 1. Adapted from Hart (2018, p. 4).	1. DEFINE 1.1 Define inclusion and exclusion criteria 1.2 Identify the fields of research 1.3 Determine the appropriate sources 1.4 Decide on the specific search terms 2. SEARCH 2.1 Search 3. SELECT 3.1 Refine the sample [of selected sources] 4. ANALYZE 4.1 Open coding 4.2 Axial coding 4.3 Selective coding 5. PRESENT 5.1 Represent and structure the content 5.2 Structure the article Reproduced from Wolswinkel et al. (2013, p. 47).

Iteration 2: Activities—Read, Reflect, Write—Ion: Motivation

Although literature review methodologists present their methods as a series of phases and stages, they emphasize the iterative nature of a literature review (e.g., Hart, 2018; Wolfswinkel et al., 2013). Similarly, the first four iterations of the model presented here take place when theorists engage with the literature, which is the raw material of deductive theory building. I propose that the main aim in the second iteration is to develop the motivation for the theory-building endeavor. In this context, read, reflect, and write will have a purpose quite different from that in the first iteration, although they will continue to support the development of the theorist's erudition. Here, the reading becomes more analytic and more critical. The objective is no longer to extract the essence of what is known on a phenomenon of interest, to make sense of what others have written, and to map the territory of extant research. Instead, the objective is to identify gaps in extant explanations or to relax the assumptions that undergird extant research, with the aim of expanding extant knowledge on the phenomenon of interest.

Being interested in a phenomenon is often sufficient as a *motivation* for many theorists to undertake theory development. Most readers, however, expect more convincing arguments if they are to devote time to reading a manuscript. One motivation often used is that of identifying gaps in extant research. This is shown in Sandberg and Alvesson's (2011) analysis of 52 articles published in premier organization studies journals, wherein gap-spotting was found to be the prevalent way to construct research questions. Sandberg and Alvesson further identified three modes of gap-spotting: confusion spotting, in which evidence from extant research is contradictory; neglect spotting, in which a phenomenon is deemed to be an overlooked area or an under-researched topic; and application spotting, in which extant literature needs to be extended or complemented. Gap-spotting is indeed a legitimate and often fruitful motivation for undertaking a theory development effort. Yet care must be exercised—it is not only because few people have studied a topic that the topic merits further research. It is quite possible that no one ever studied it because it is not interesting! Authors need to demonstrate that the gap they have spotted is important and that filling it will make a true contribution to the advancement of knowledge.

Another motivation is the desire to conceptually clarify a construct. This approach can lead to important contributions. One notable example (although not in the information systems (IS) field) is that of Ajzen and Fishbein, who endeavored to clarify the attitude construct and ended up developing the Theory of Planned Behavior (Ajzen & Fishbein, 1977; Jaccard & Jacoby, 2010). In the IS domain, some strong theoretical contributions have been made by authors spurred by this motivation. For instance, after observing confusion over the use of the terms "participation" and "involvement" in IS research, Barki and Hartwick (1989) argued for separation of the two constructs, with participation being a set of behaviors and involvement a psychological state. From this clarification, they developed a theoretical model explaining the relationships between involvement, participation, and system success (Hartwick & Barki, 1994).

A third and ostensibly powerful motivation for a theory development effort is problematization. Ranging from modifying the boundaries of a theory by relaxing assumptions (Whetten, 2002) to challenging the assumptions underlying existing theories (Alvesson & Sandberg, 2011), and even denying some assumptions (Davis, 1971), problematization is said to be a "central ingredient in the development of more interesting and influential theories" (Alvesson & Sandberg, 2011, 247). For instance, Simon (1972) proposed the theory of bounded rationality by modifying the assumptions of the classical theory of the firm, such as assuming that decision-makers have only incomplete information about their choice of alternatives, and altering the nature of decision-makers' goals, from maximizing to satisficing. In IS, Markus and Mao (2004) adopted an approach akin to problematization, arguing that the novel context of IS implementation—user populations that are much larger than when IS participation theory was first proposed, Enterprise Resource Planning (ERP) implementation, outsourcing—made it necessary to revisit some of the assumptions underlying participation research. For example, they challenged the assumption that the participants are the intended hands-on users. They relaxed this assumption and offered a number of theoretical propositions that encompassed a larger set of project stakeholders.

Being an "endeavour to know how and to what extent it might be possible to think differently, instead of what is already known" (Foucault, in Alvesson & Sandberg, 2011, 253), problematization "does not primarily question how well some constructs or relationships between constructs represent a particular subject matter […] Instead, it questions the necessary presuppositions researchers make about a subject matter in order to develop the specific theory about it" (Alvesson & Sandberg, 2011).

Iteration 3: Activities—Read, Reflect, Write—Ion: Definition

The theorist remains engaged with the literature during this iteration, but the purpose of this engagement switches from understanding and assessing extant knowledge to using it to set the foundations of the theory under development. The read, reflect, and write activities will focus on delineating the phenomenon that the theorist wishes to explain or explain and predict, stating the conceptual and contextual assumptions that set the boundaries of the theory under development, and developing clear conceptualizations of the constructs that will serve as the building blocks of the theory. Some of these defining elements will emerge from reading the literature anew and reflecting upon it to adopt extant definitions or develop new ones.

Therefore, *definition* is the key outcome of this iteration. Definitional clarity is indispensable if the authors are to convey the meaning of their theory. Although construct definition comes immediately to mind when one refers to definitional clarity, it is not the only element that requires the authors' careful attention. The phenomenon of interest, the type of theory that is proposed, and the boundary of the theory and its underlying assumptions all must be clearly defined.

Although the phenomenon under study may be obvious to the authors themselves, this is not always the case for the readers who, if they are confused, will either miss the explanation provided by the theory or misinterpret it. Consider the following example: the introduction to a manuscript announced that the authors' aim was to propose a model of information systems implementation. When reading this, an image formed in my mind that the paper would deal with the implementation—including initiation, adoption, adaptation, acceptance, routinization, and infusion, as per Kwon and Zmud (1987)—of any type of information system, be it a custom-developed system or configurable software such as ERP. Yet, several pages into the manuscript, I realized that the paper dealt with configurable software only and was solely interested in its configuration, not in the whole implementation process; the phenomenon under study was no longer the same. And it changed again later, in the model itself, which pertained essentially to user-analyst interactions during configuration.

Authors also ought to define the type of theory they propose. Although there is general agreement about what a theory is (e.g., "a statement of relations among concepts within a boundary set of assumptions and constraints" [Bacharach, 1989, 496]), a theory may have different purposes, as suggested by Gregor (2006). It is important to tell the reader whether one wants to explain a phenomenon or explain and predict some outcome. Also, because the phenomena we study may vary in terms of whether they are static or dynamic, authors should state whether they are proposing a variance or process type of theory (Poole, Van de Ven, Dooley, & Holmes, 2000) or a system or hybrid type (Burton-Jones, McLean, & Monod, 2015; Ortiz de Guinea & Webster, 2017).

It is also essential that authors specify the boundary of their theory, which is shaped by the theory's underlying assumptions (Bacharach, 1989). I will borrow from Whetten (2002), who defines two types of assumptions that help specify the boundary of a theory. The first type is that of conceptual assumptions, which "can be thought of as 'second order explanations'—the implicit whys underlying an explicit answer to a specific why question [and] are often articulated using the language of foundational theories" (Whetten, 2002, 58). For instance, when theorists adopt institutional theory as a foundational theory, they make—or espouse—the assumption that actors "accept and follow social norms unquestioningly, without any real reflection" (Tolbert & Zucker, 1996, 176) and that they seek legitimacy rather than efficiency. Similarly, adopting the theory of reasoned action (TRA) as a foundational theory implies the assumption that "most human social behavior is under volitional control and, hence, can be predicted from intentions alone" (Ajzen, 2002, 666). It is important that the manuscript makes such conceptual assumptions clear so that the reader is able to follow the authors' theoretical explanation.

The second type of assumption that helps specify the boundary of a theory is the set of contextual assumptions determining the conditions that circumscribe the explanation proposed by the theory (Whetten, 2002). According to Whetten, failing to specify the contextual boundaries reduces the "power" of explanations. On this issue, Whetten cites Sutton and Staw (1995, 376), who posit that "One indication that a strong theory has been proposed is that it is possible to discern conditions in which the major proposition or hypothesis is most and least likely to hold." Contextual assumptions pertain to when, where, and for whom the theory is assumed to hold (Whetten, 2002). For instance, Kappos and Rivard (2008) proposed a culture-based theoretical model of information systems development and use. They specified that the contextual boundary of their model comprises the processes of developing, using, and operating an information system, the information system itself, and the environments in which the system is developed, used, and operated.

Constructs are the essential components of a theory, and as such ought to be clearly defined (Barki, 2008; Weber, 2012). Failing to provide clear conceptual construct definitions can be particularly problematic because readers will themselves ascribe meanings to constructs, with the risk of ending up with as many meanings as there are readers. A good construct definition is said to be "a concise, clear verbal expression of a unique concept" (Wacker, 2004, 631). In contrast, a "bad" construct definition is "any verbal explanation that does not lead to a unique concept" (Wacker, 2004, 631). According to Suddaby (2010, 347), a construct definition should accomplish three main tasks. First, it ought to capture the essential properties and characteristics of the concept under consideration. Second, it should avoid tautology or circularity. Tautology refers to construct elements appearing in the definition, such as defining user resistance to IT implementation as "users who resist the implementation of a new IT." Circularity is present when the antecedents or outcome variables are part of the definition, such as defining IT strategic capability as "the capacity to use IT in a manner that leads to sustained competitive advantage." The third task in deriving a good construct definition involves being parsimonious. Indeed, it has been suggested that "[construct] definitions should have as few terms as possible in the conceptual definition to avoid violating the parsimony virtue of 'good' theory" (Wacker, 2004, 638).

Iteration 4: Activities—Research, Read, Reflect, Write—Ion: Imagination

Espousing Weick's (1989) idea that theory construction is a highly creative endeavor, the model portrays *imagination* as the ion to emerge from the next iteration. Although we all wish to find sudden inspiration, a eureka moment, as great minds such as Archimedes and Newton allegedly experienced (Falk, 2005), the creative process that most of us experience in our theory development efforts is somewhat more arduous. This is often noted by reviewers and editors who deplore that a manuscript's theoretical propositions are neither insightful nor meaningful and that they seem rather obvious, which makes the proposed theory simplistic. Furthermore, reviewers often deplore that not only the propositions, but also "the arguments supporting them seem to be too simplistic or underdeveloped and could have been justified more strongly. In many cases, the arguments supporting the propositions are based on what seem to be a simple translation or rephrasing of arguments drawn mainly from [the foundational theory or extant literature]." In a nutshell, reviewers appear to deplore a lack of imagination on the part of the theorist.

The spiral model of theory building described here proposes research as an activity that can stimulate the theorist's imagination and help develop more interesting and original theories. In the model, the term "research" refers to the effectuation of mental activities, such as role playing, analogizing, imaging, and conducting the thought experiments proposed in the theory-building literature. The following paragraphs present three types of such mental activities.

Generative Theorizing Practices. With the objective of informing the theorizing process and enriching the conversation on theorizing in the IS domain, Hassan, Mathiassen, and Lowry (2019) conceptualize theorizing by applying the Foucauldian notion of discursive formation. This leads them to propose four generative discursive practices that theorists can adopt to generate ideas and organize them to "make sense of the phenomenon" (p. 10) they wish to explain. As such, the four discursive practices are likely to be particularly effective in fostering a theorist's imagination; the practices are analogizing, metaphorizing, mythologizing, and modeling.

To the theorist, a major benefit of analogizing is to facilitate the development of an explanation for a new phenomenon or concept by analyzing the similarities between this new phenomenon and a well-known phenomenon, as one way to gain an understanding of the new phenomenon (Kuechler & Vaishnavi, 2012). Defining analogizing as "using a simplified or scaled-down reference to something familiar to explain or illustrate something more complex or less familiar" (p. 10), Hassan et al. provide several examples of how analogizing has helped theory development. These include project escalation (Keil & Robey, 1999), wherein an implicit analogy is made with a military scenario, or Goffman's (1959) presentation-of-self theory, which uses the analogy of theatrical performance in one's interactions with others.

Metaphorizing is similar to analogizing, as it calls for referring to known phenomena as a way to develop one's explanation of a new phenomenon. The two practices differ in that metaphorizing goes further than using the similarities between two phenomena; it also uses the behavior of the known phenomenon to explain that of the new. The mere suggestion of the metaphor suffices to evoke "a network of analogies" (Hassan et al., 2019, p. 11). One example is the metaphor of political systems that underlies Weill and Ross's (2004) typology of IT governance. Indeed, the mere mention of an archetype being a business monarchy, an IT monarchy or a feudal system has direct implications on the key characteristics of each archetype, such as the distribution of power within an organization, of the types of relationships among IT stakeholders, and even the types of conflicts that may emerge among stakeholders. It is important to note, however, that the metaphor does not constitute the theory (Bacharach, 1989). Theorists are encouraged to use metaphors as a tool for generating novel explanations for their object of study (Bacharach, 1989; Bauxenbaum & Rouleau, 2011).

Mythologizing refers to the use of "myths, mythologies, and hidden assumptions to provide or interrogate a means of explanation, as well as to study symbols of value, coherence, unity, social structure, conflict and contradictions" (Hassan et al., 2019, p. 11). The authors provide several examples of IS researchers who identified myths either in extant practice or in research, and developed explanations to "debunk" such myths, which include the myth of real-time systems that Dearden (1966) used to "expose several fallacies regarding the assumed capabilities of computers to support management functions" (p. 11).

The fourth generative practice proposed by Hassan et al. is modeling. They refer to a model as the "precise and economical statement of a set of relationships that are sufficient to produce the phenomenon in question" or the "actual biological, mechanical, or social system that embodies the relationships in an especially transparent way" (Schelling, 1978, p. 87 as cited by Hassan et al., 2019, p. 12). The authors suggest that models are "useful for building theories because they reveal the consequences of making certain assumptions and including or excluding certain elements in an economical way" (p. 12). Although a model is often used to represent the result of theorizing, the modeling process itself can serve as a valuable way to fire up one's imagination.

Modeling as Theorizing. While Hassan et al. view modeling as a generative practice, Whetten (2002) actually portrays it as theorizing and proposes a method to this end. The four-step method is based on Whetten's (1989) set of questions, the answers to which constitute theoretical contributions. The first step involves answering the question of "What" and identifying the key constructs that will constitute the theory. Although the main constructs will have been identified in a previous iteration, addressing the question anew while modeling may very well lead to the identification of new constructs. In the second step, the theorist will answer the question of "How" the constructs are related. The nature of the relationship will indeed vary depending on whether the theory under construction is a variance or a process theory. The third step involves determining "Why" the constructs are related. During this step, the conceptual assumptions the theorist has defined earlier are brought to the fore where, according to Whetten, they will play an important role in the development of the explanation. The fourth step is that of answering the questions of "When-where-who" and will help refine the contextual boundary of the theory being developed. In addition to comprising cognitive activities in terms of providing answers to key questions, Whetten's method implies physical activities, such as writing the names of constructs on Post-It® notes (PINs) and, starting with the focal construct as the center of the model, adding and moving constructs on the model, at times treating them as antecedents, consequences, mediators, or moderators. My experience is that physically rearranging PINs on a board or a desk is itself an exercise that fosters one's imagination.

Applying Heuristics. In their book on theory construction, Jaccard and Jacoby (2010) suggest heuristic techniques. Although they argue that "there is no simple strategy for generating good ideas or good explanations [and that] it is a creative process that is difficult to articulate, describe and teach" (p. 40), Jaccard and Jacoby nevertheless offer heuristics—from conducting thought experiments to shifting the unit of analysis—that can help generate ideas about the phenomenon one seeks to explain. I do not intend to review all 26 of the heuristics they propose.

Instead, I will introduce two of them as illustrations and invite authors to consult Jaccard and Jacoby's work when developing their theories.

Example 1: Alternate between abstractions and specific instances of explanations of the phenomenon under study. Here, Jaccard and Jacoby suggest that "[t]heory is developed by thinking about concrete instantiations of concepts and then abstracting upward to more general constructs that allow us to make theoretical propositions" (p. 56). In my view, this heuristic is readily applicable and useful, notwithstanding the theorist's preferred mode of thinking. For example, some minds work better at the level of concrete instantiations. Remaining at this level, however, may lead to propositions that seem obvious. When trying to explain why software project managers adopt risk management practices, for instance, one may identify the following antecedents from a literature review: project size, technological newness, lack of user support, project manager's training, and practices advocated by the project manager's professional association. Propositions developed directly from these antecedents would resemble the following: project size (or project manager's training) will have an effect on software project managers' choice of project risk management practices. If the theorist moves to a more abstract level, project size, technological newness, and lack of user support may become risk sources, while the project manager's training and practices as advocated by the project manager's professional association may become institutional norms. The theorist might then derive a more interesting proposition: software project managers' choice of project management practices will be influenced by project risk sources and institutional norms. It is when one purposefully makes the effort to think in terms of more general constructs that more interesting theoretical explanations seem to emerge.

In contrast, some people think at very abstract levels. According to Jaccard and Jacoby, remaining at abstract levels may "obscure important distinctions that should be made" (p. 56), distinctions that can be identified when one moves from abstractions to specific instances of the phenomenon under study. This would be the case of a theorist who theorizes that institutional pressures (Scott, 2008) explain software project managers' enactment of project risk management practices. Although this is an interesting proposition, it is too general. Moving to the concrete level might help the theorist draw a more finely grained portrait of institutional pressures, which might include practices advocated by project management professional associations, practices that are enforced by quasi-laws (such as the project risk management practices enforced by the Basel agreements), and practices that have developed within an organization and are now part of the organization's culture. Moving back to an abstract level would result in the proposition that the institutional pressures emerging from the three main institutional pillars—normative, regulative, and cultural-cognitive (Scott, 2008)—may influence project managers' enactment of risk management practices, hence providing a richer and more convincing explanation.

Example 2: Focus on processes or focus on variables. Most theoretical models developed in the IS domain are either variance or process models. As a way to foster the imagination of variance theorists, Jaccard and Jacoby suggest that they think about their phenomena of interest in terms of processes rather than variables. They suggest a strategy for invoking process perspectives by changing nouns into verbs. This would be the case, for instance, if someone started thinking about IT adoption in terms of adopting IT. Instead of theorizing on user, technology, or environment attributes that can lead to a certain degree of adoption, the theorist might reflect upon the stages in adopting a technology. Conversely, Jaccard and Jacoby suggest that for those who are accustomed to thinking in terms of processes, thinking in terms of variables might be more fruitful. They refer to Abbot and Alexander (2004), who proposed the heuristic of "'stopping the clock.' The idea is to 'freeze' the process at a given point in time and then describe the system in detail at the frozen moment" (p. 55).

Iteration 5: Activities—Respite, Read, Reflect, Write—Ions: Explanation and Presentation

The previous iteration, Iteration 4, was aimed at generating ideas about the relationships among the constructs of the theory being developed. Along with the other three iterations, Iteration 4 portrays theorists as constantly and purposefully working toward delineating, characterizing, describing, and understanding a phenomenon of interest. This ongoing mental activity is indeed fruitful, but from time to time it leads to a state of mental exhaustion, where the theorist feels that something is missing, preventing the emergence of a rich explanation, most often of why the constructs of the proposed theory are related. I contend that respite—defined as an interval of rest or relief[3]—is the activity of choice here. Not only should it relieve the theorist from some of the mental exhaustion experienced, but it can also be most fruitful for generating two essential ions: *explanation* and *presentation*.

In this chapter, I use the term "explanation" in reference to causal relationships (Sutton & Staw, 1995) espousing the view that a key goal of theorizing is to answer the question of why (Bacharach, 1989). There are several different views of causality that may be espoused by theorists seeking to explore the nature of causal relationships. For instance, based on Kim's (1999) discussion of causation, Gregor describes four approaches to the analysis of event causation: regularity analysis, counterfactual analysis, probabilistic causal analysis, and manipulation or teleological analysis (Gregor, 2006, 616). Similarly, Bacharach suggests that four types of causal linkages explain the substantive nature of the relationship between an antecedent and its consequence: recursive, teleological, dialectical, and reciprocal (Bacharach, 1989, 508).

A particularly illuminating discussion of the nature of causal relationships is offered by Poole et al. (2000), and Van de Ven and Poole (2005), who contrast and compare variance and process explanations in terms of their assumptions about causality. They first introduce Aristotle's four types of causes: material, formal, efficient, and final. They further explain that these four types of causes indicate, respectively, "that from which something was made (material cause); the pattern by which it is made (formal cause); that from which comes the immediate origin of movement or rest (efficient cause); and the end for which it is made (final cause)" (Van de Ven & Poole, 2005, 1396). In variance theories, explanations utilize efficient causality and variance theorists endeavor to identify the conditions, or antecedents, that are necessary and sufficient for the outcome (Poole et al., 2000). In process theories, explanations are based

[3] https://www.merriam-webster.com/dictionary/respite.

on necessary causality. In contrast to causality in variance explanations, in process theories, "[b]ecause causal influences come to bear 'eventwise'—through one or more events—rather than continuously, it is rare for a cause to be sufficient in narrative explanation. Only the entire set of forces that influence the developmental span, in the particular order and combinations in which they occur, are necessary and sufficient to explain a narrative" (Poole et al., 2000, 42). In addition to efficient causality, process explanations may involve final causality and formal causality. Final causality is at play when an end or a goal guides the unfolding of a phenomenon. Formal causality refers to "a pattern that informs change; the pattern must be applied to the developing entity somehow, either through plan or through some other governing mechanism" (Poole et al., 2000, 42).

Focusing on causal reasoning in IS research, Markus and Rowe (2018) propose a three-dimensional framework of causal structure. The first dimension, causal ontology, pertains to the theorist's "beliefs about the reality of causality" (p. 1258). The authors suggest that theorists will hold one of the following three beliefs: causality as a metaphor for a logical or metaphysical association; causality as a real mechanism; and causality as a "misnomer, because it incorrectly implies unidirectional, deterministic, external forces" (p. 1260). The second dimension, causal trajectory, refers to what is changed in causation. According to Markus and Row, there are three main trajectories of this type: across the boundaries of a stratified entity; within an undifferentiated entity; and "through the accretion (growth and complexification) over time of a heterogeneous entity" (p. 1260). The third dimension, causal autonomy, concerns the role played by human agents and/or by IT in a given change episode. Three arrangements of roles are proposed, as causality goes: from people to technology; from technology to people; and back-and-forth between people and technology. Positioning oneself and the theory under development within this framework can be most useful as a way to help the theorist develop a congruent and rich explanation.

An explanation that is clear in an author's mind does not automatically become so in the reader's mind. The *presentation* of the theory, which includes the structure of the manuscript, the syntax, the vocabulary, and sometimes a graphical representation are powerful tools that authors can use to convey their novel explanation. These elements must be assembled very carefully and fully exploited in order for demanding readers to be able to embrace the author's theoretical explanation.

For instance, referring to the structure of a paper, I once commented to the authors of a manuscript that it was only when I had read half of their paper's third section that I realized that the section was indeed where they had presented their theoretical development. Why did this happen? First, the authors had not announced that the third section presented their theoretical development. Second, while the previous section had introduced their foundational theory, the third section continued to introduce new concepts from that theory. Third, the boundary of the theory and its underlying assumptions were never defined. Fourth, the key elements of the process theory, which comprised events, triggers, and subprocesses, had not been introduced. Although the authors probably took my comments as petty details, as we all know, the devil is in the details!

In terms of presentation, what else might help better convey the results of one's theorizing efforts? Regarding another manuscript, an associate editor's report mentioned that reviewers felt that the paper's elements were disconnected because the authors did not provide an integrated visual depiction of their model. Although a figure does not constitute a theoretical explanation (Sutton & Staw, 1995), and is not always necessary for conveying an explanation, it often helps clarify it. Furthermore, as suggested above, the very activity of modeling—as in drawing a picture of one's model—can be an intrinsic part of the theorizing process (Hassan et al., 2019; Whetten, 2002).

Propositions play an important role in the formulation of a theory, as they state the relationships among the theory's constructs (Bacharach, 1989). Here, Bacharach portrays a theory as a system of constructs—bounded by the theorist's assumptions—that are related to each other by propositions. Because a key goal of theory is to provide an explanation, "[t]o the extent the propositions imply causality either implicitly or explicitly, the predictive and explanatory power of [the] theory is enhanced further" (Weber, 2012, 21). This implies that, when wording their propositions, authors should try to go beyond statements of correlation between constructs, and weave in explanation.

Should the formulation of a theory always be presented as propositions? Maybe not. Consider a theoretical model that I developed with a coauthor. We spent dozens of hours in "proposition formulation" to develop a crisp theoretical explanation of our phenomenon of interest. Nevertheless, the reviewers were somewhat unsatisfied with the explanation provided by these propositions, which they found limited. The associate editor made the following suggestion, which we applied: "It may be premature to lock into this form of theory before examining other more comprehensive forms. […] A better approach might be to consider this theoretical contribution not as a set of propositions but a set of insights and criteria that would set the stage for the development of a more comprehensive theory in the future. Over the long term, taking this approach may lead to a more significant contribution to theory development in this area."

In my experience and in that of others, the difficult and profound questions around explanation and the demanding issues that need to be addressed in order to arrive at a presentation that will draw and maintain readers' attention are better resolved when the mind of the theorist is not in a state of unrest. This is why respite is a key activity in the theory-building process.

> In his struggles with extremely complicated mathematics that led to the general theory of relativity of 1915, Einstein often turned for inspiration to the simple beauty of Mozart's music […] "Whenever he felt that he had come to the end of the road or into a difficult situation in his work, he would take refuge in music," recalled his older son, Hans Albert. "That would usually resolve all his difficulties." (Miller, 2006, http://www.nytimes.com/2006/01/31/science/31essa.html?_r=0)

Einstein turning to the music of Mozart when he met a stumbling block in his theory building was a form of combinatory play, which "describes the conscious and unconscious cognitive playful manipulation of two or more ideas, feelings, sensory experiences, images, sounds, words, or objects. In combinatory play, players experiment with hypotheses, they play with possible outcomes, and they adjust to unexpected results and even 'failures.' These players compare, contrast, synthesize, and break apart disparate elements or constructs in the service of reenvisioning a larger whole" (Stevens, 2014, p. 99). Although imagination is also nourished during combinatory play, it is my experience that periods of respite are particularly effective in yielding explanation and presentation.

Simply put, combinatory play pertains to situations where the theorist is not focused on theory building but rather lets her mind wander while being occupied by activities that are completely unrelated, such as bathing or washing the dishes (Stevens, 2014). During an interview on the process of planning the production of the opera Carmen, a stage director mentioned that decoupage and taking long walks were activities that he would perform when he had thorny issues to solve in staging the opera (Duchesne, 2019). While the research activity keeps the mind consciously focused on the phenomenon to be explained, during respite "the conscious mind remains busy with other tasks but the unconscious mind keeps working on the problem, combining or playing with ideas in ways rational thought might inhibit" (Stevens, 2014, p. 102). According to Stevens, this is when the "illumination" of the explanation can emerge. Although the outcome of combinatory play is not always an illumination in the sense of a "eureka" moment, it can nevertheless lead theorists to a more interesting explanation and a well-organized, clear, and interesting presentation.

Iteration 6: Activities—Review, Renounce, Revise, Read, Reflect, Write—Ion: Contribution

The review and revise activities require theorists to take some distance from their work and read it as if it was the work of another. Most of the time—except when the theorist is a perfectionist—when one thinks that a paper is finished, it is not. When reviewing our own work, we often identify issues that we just hope reviewers will not see. Most of the time they do, and the theorist would be well advised to complete a few iterations of self-imposed review and revision before considering the work complete. Obtaining friendly reviews and workshopping one's manuscript can also be profitable.

Another essential activity, one that is most difficult to accomplish, is to renounce. The term "renounce" here refers to giving up some part of the paper that we are writing or of the theory we are developing. Renouncing may imply to let go of one of the foundational theories we were using to develop our own theoretical explanation (e.g., because with hindsight we have one too many foundational theories), let go of a long literature review section that we painstakingly put together (e.g., when we realize that this literature review might have been too broad for the phenomenon we are studying), and let go of a set of propositions that we carefully crafted (e.g., when we realize that the large number of propositions we came up with challenges the parsimony of our theory). Oftentimes, we have worked so hard to develop each element of our theory piece that renouncing to any of them is unthinkable. Yet, it often happens that after reviewing our own work—and more often when our work has been reviewed by peers—that one of those elements suddenly appears superfluous and may even threaten the contribution we purport to make. Because of all the effort we have already invested in producing this work, we often resist renouncing. Yet, it may very well be the activity that is the trademark of true artisans, as they are engaged in the work in and for itself (Sennet, 2008).

In addition to checking for the presence of the ions of theory construction produced during the previous iterations, review, renounce, and revise activities should be focused on clarifying the *contribution* of the proposed theory. Having spent months, sometimes years, developing a theory—even an approximation of one—often comes with the conviction that a theoretical contribution is being made. Often the authors consider this so self-evident that they do not deem it necessary to spell out the contribution they are making. Nevertheless, one should not expect readers to see this contribution in the manuscript. Most often they do not, unless the authors explicitly state the following: how their theory is novel and different from extant explanations of the phenomenon of interest; what it can explain that other theoretical explanations have missed or ignored; how their theoretical explanation adds to extant knowledge about a phenomenon, to their field—and to other fields—of study; and how its use can change practice in the field of IS.

When authors have carefully reflected on their motivation and provided strong and convincing arguments for undertaking theory development, one way of presenting the proposed theory's contributions is to return to these motivating arguments and discuss how each gap was addressed, how each unclear construct was clarified, or how relaxing each assumption has helped move the field forward.

Did We Say Cohesion?

Cohesion is an outcome of each of the six iterations of the theory-building process I describe, and for a good reason. A theory is an artificial world that the theorist has created. This artificial world comprises constructs, assumptions, a boundary, and relationships among the constructs. When formulating a theory, authors ought to ensure that all these elements form a cohesive whole. Unfortunately, this does not always occur.

It sometimes happens that the model proposed in a manuscript does not correspond exactly to the phenomenon initially declared as the phenomenon of interest (see my earlier example about system implementation vs. software configuration). It also happens that the theoretical explanation does not remain within the boundaries initially set. This was the case of a manuscript for which I was senior editor. The author had announced that the proposed theory would explain an individual-level phenomenon. However, the theoretical development proposed market-level explanations of the phenomenon of interest, thus moving outside the boundary originally set. The issue was not that the market-level explanation was irrelevant or problematic. Rather, it was the lack of coherence between the stated boundary and the boundary-in-use. Since theory building is highly iterative, it may happen that the contextual assumptions specifying the boundary of the explanation will change, as in the example above. If this occurs, theorists ought to revise either their explanation or their contextual assumptions so that they fit the provided explanation. Authors should also ensure that their theoretical explanation is in line with the conceptual assumptions that help specify the boundary of their theory. For instance, if one develops a theory that draws on institutional theory as a foundational theory, one must remain faithful to its underlying assumption of actors seeking legitimacy.

The vocabulary also needs to belong to the theory and only to the theory. I once attended a presentation on theory building given by Ajay K. Kohli, former editor of the *Journal of Marketing*. During his presentation, Kohli referred to "the synonym as the enemy of the theorist." My reading of manuscripts and manuscript proposals strongly confirms this statement. What did Kohli mean? At the core of the artificial world that we build—our theory—stand our constructs, which we name and define. It often happens that authors, probably because they want to reduce repetitions, use a synonym to refer to a construct. For instance, one of my coauthors and I proposed a process theory of a phenomenon related to software development. Two important concepts in our theory were the goals and the means of a software development project. Reading the manuscript for the nth time, I realized that, because of the multiple iterations we went through developing our theory, we ended up using "objectives," "outcomes," and "ends" as synonyms for our concept of "goal"—all within a single page! As noted by a reviewer of a manuscript on IT adaptation by users, which had synonyms for "IT adaptation" including "IS change," "user modifications," and "alterations to system": this "slippage in terms" can be quite confusing for the reader and undermines the manuscript.

To make sure that cohesion exists, it is also a good idea to walk through the theory using different scenarios, as one would walk through an algorithm or program code, to make sure it works. Furthermore, authors should consider including such scenarios in the manuscript; they might help readers better understand the proposed theory, thus adding clarity to the presentation of their theory and power to its explanation.

Concluding Remarks

In this chapter I propose a pragmatic model of theory building, one that emphasizes the approximate nature of the theories we develop and the iterative nature of the theory-building process. The spiral model portrays theory building as an iterative process and suggests that each of six iterations produces an ion, each representing an element that is essential to good theory but whose presence is not sufficient to ensure that the theory under development has value. It is how the theorist assembles the ions that will provide value to the theory. As models do, this iterative representation is a simplification of reality. Indeed, if the emergence of a given ion is more salient as an outcome of a given iteration, in reality, all ions may emerge as an outcome of any iteration. It is for the theorist to be attentive to the emergence of each ion, to capture it, and reflect upon how it will be combined with the other ions into a theoretical approximation that will advance knowledge.

References

Abbot, A., & Alexander, J. C. (2004). *Methods of discovery: Heuristics for the social sciences*. New York: Norton.

Ajzen, I. (2002). Perceived behavioral control, self-efficacy, locus of control, and the theory of planned behavior. *Journal of Applied Social Psychology, 32*(4), 665–683.

Ajzen, I., & Fishbein, M. (1977). Attitude-behavior relations: A theoretical analysis and review of empirical research. *Psychological Bulletin, 84*(5), 888–918.

Alvesson, M., & Sandberg, J. (2011). Generating research questions through problematization. *Academy of Management Review, 36*(2), 247–271.

Bacharach, S. B. (1989). Organizational theories: Some criteria for evaluation. *Academy of Management Review, 14*(4), 496–515.

Bandura, A. (2005). The evolution of social cognition theory. In K. G. Smith & M. A. Hitt (Eds.), *Great minds in management: The process of theory development*. Oxford: Oxford University Press.

Barki, H. (2008). Thar's gold in them thar constructs. *The Database for Advances in Information Systems, 39*(3), 9–20.

Barki, H., & Hartwick, J. (1989). Rethinking the concept of user involvement. *MIS Quarterly, 13*(1), 53–63.

Bauxenbaum, E., & Rouleau, L. (2011). New knowledge products as bricolage: Metaphors and scripts in organizational theory. *Academy of Management Review, 36*(2), 272–296.

Boehm, B. B. (1988). A spiral model of software development and enhancement. *Computer, 21*, 61–72.

Bourgeois, L. J. I. I. I. (1979). Toward a method of middle-range theorizing. *The Academy of Management Review, 4*(3), 443–447.

Burton-Jones, A., McLean, E. R., & Monod, E. (2015). Theoretical perspectives in IS research: From variance and process to conceptual latitude and conceptual fit. *European Journal of Information Systems, 24*(6), 664–679.

Davis, M. S. (1971). That's interesting! Towards a phenomenology of sociology and a sociology of phenomenology. *Philosophy of the Social Sciences, 1*, 309–344.

Dearden, J. (1966). Myth of real-time management information systems. *Harvard Business Review*, 123–132.

Duchesne, A. (2019, May 4). Carmen: l'aboutissement de deux ans de travail. *La Presse* +. Retrieved May 15, 2019, from https://www.lapresse.ca/arts/spectacles/opera/201905/03/01-5224611-carmen-laboutissement-de-deux-ans-de-travail.php

Falk, D. (2005). Great Eureka moments in history. *U of T Magazine*. Retrieved May 25, 2019, from http://www.magazine.utoronto.ca/autumn-2005/great-eureka-moments-in-history-famous-inspirational-moments/

Goffman, E. (1959). *The Presentation of Self in Everyday Life*. Garden City, NY: Doubleday.

Gregor, S. (2006). The nature of theory in information systems. *MIS Quarterly, 30*(3), 611–642.

Hart, C. (2018). *Doing a literature review: Releasing the research imagination*. Thousand Oaks: Sage Publications.

Hartwick, J., & Barki, H. (1994). Explaining the role of user participation in information system use. *Management Science, 40*(4), 440–465.

Hassan, N. R. (2014). Useful products of theorizing for information systems. *Proceedings ICIS 2014*.

Hassan, N. R., Mathiassen, L., & Lowry, P. B. (2019). The process of information systems theorizing as a discursive practice. *Journal of Information Technology, 34*(3), 198–220.

Jaccard, J., & Jacoby, J. (2010). *Theory construction and model-building skills: A practical guide for the social scientist*. New York: The Guilford Press.

Kappos, A., & Rivard, S. (2008). A three-perspective model of culture, information systems and their development and use. *MIS Quarterly, 32*(3), 601–634.

Keil, M., & Robey, D. (1999). Turning around troubled software projects: An exploratory study of the de-escalation of commitment to failing courses of action. *Journal of Management Information Systems, 15*(4), 63–87.

Kim, J. (1999). Causation. In R. Audi (Ed.), *The Cambridge dictionary of philosophy* (pp. 125–127). Cambridge: Cambridge University Press.

Kuechler, W., & Vaishnavi, V. (2012). A framework for theory development in design science research: Multiple perspectives. *Journal of the Association for Information Systems, 13*(6), 395–423.

Kwon, T. H., & Zmud, R. W. (1987). Unifying the fragmented models of information systems implementation. In J. R. Boland & R. Hirshheim (Eds.), *Critical issues in information systems research* (pp. 227–251). New York: John Wiley.

Markus, M. L., & Mao, J. Y. (2004). Participation in development and implementation—Updating an old, tired concept for today's IS contexts. *Journal of the Association for Information Systems, 5*(11/12), 514–544.

Markus, M. L., & Rowe, F. (2018). Is IT changing the world? Conceptions of causality for information systems theorizing. *MIS Quarterly, 42*(4), 1255–1280.

Miller, A. J. (2006, January 31). A genius finds inspiration in the music of another. *The New York Times*. Retrieved May 24, 2019, from http://www.nytimes.com/2006/01/31/science/31essa.html?_r=0

Ortiz de Guinea, A., & Webster, J. (2017). Combining variance and process in information systems research: Hybrid approaches. *Information and Organization, 27*(3), 144–162.

Poole, M. S., Van de Ven, A. H., Dooley, K., & Holmes, M. E. (2000). *Organizational change and innovation processes*. New York: Oxford University Press.

Rivard, S. (2014). The ions of theory construction. *MIS Quarterly, 38*(2), iii–xiii.

Sandberg, J., & Alvesson, M. (2011). Ways of constructing research questions: Gap-spotting or problematization? *Organization, 18*(1), 23–44.

Schelling, T. C. (1978). Thermostats, lemons, and other families of models. In: Schelling, T. (ed.), *Micromotives and Macrobehavior*. New York: W. W. Norton, 81–134.

Scott, W. R. (2008). *Institutions and organizations: Ideas and interests*. Thousand Oaks: Sage Publications.

Sennet, R. (2008). *The craftsman*. New Haven: Yale University Press.

Shepherd, D. A., & Sutcliffe, K. M. (2011). Inductive Top-Down Theorizing: A Source of New Theories of Organizations. *Academy of Management Review, 36*(2), 361–380.

Simon, H. A. (1972). Theories of bounded rationality. In C. B. McGuire & R. Radner (Eds.), *Decision and organization* (pp. 161–176). North-Holland Publishing Company.

Stevens, V. (2014). To think without thinking. *American Journal of Play, 7*(1), 99–119.

Suddaby, R. (2010). Construct clarity in theories of management and organization. *Academy of Management Review, 35*(3), 346–357.

Sutton, R. I., & Staw, B. M. (1995). What theory is not. *Administrative Science Quarterly, 40*(3), 371–384.

Tolbert, P. S., & Zucker, L. G. (1996). The institutionalization of institutional theory. In S. Clegg, C. Hardy, & W. R. Nord (Eds.), *Handbook of organization studies* (pp. 175–190). Thousand Oaks: Sage Publications.

Van de Ven, A. H., & Poole, M. S. (2005). Alternative approaches for studying organizational change. *Organization Studies, 26*(9), 1377–1404.

Wacker, J. G. (2004). A theory of formal conceptual definitions: Developing theory-building measurement instruments. *Journal of Operations Management, 22*(6), 629–650.

Weber, R. (2012). Evaluating and developing theories in the information systems discipline. *Journal of the Association for Information Systems, 13*(1), 1–30.

Weick, K. E. (1989). Theory construction as disciplined imagination. *Academy of Management Review, 14*(4), 516–531.

Weick, K. E. (1995). What theory is not, theorizing is. *Administrative Science Quarterly, 40*(3), 385–390.

Weill, P., & Ross, J. (2004). *IT Governance: How Top Performers Manage IT Decision Rights for Superior Results*. Harvard Business Review Press.

Whetten, D. A. (1989). What constitutes a theoretical contribution? *Academy of Management Review, 14*(4), 490–495.

Whetten, D. A. (2002). Modelling-as-theorizing: A systematic methodology for theory development. In D. Partington (Ed.), *Essential skills for management research* (2nd ed., pp. 45–71). Thousand Oaks: Sage Publications.

Wolfswinkel, J. F., Furtmueller, E., & Wilderom, C. P. M. (2013). Using grounded theory as a method for rigorously reviewing literature. *European Journal of Information Systems, 22*(1), 45–55.

5

The Process of Information Systems Theorizing as a Discursive Practice*

Nik Rushdi Hassan, Lars Mathiassen, and Paul Lowry

Introduction

Compelling progress has been made in describing the nature of Information Systems (IS) theory (Gregor, 2006; Gregor & Jones, 2007) and in evaluating and refining existing theories (Grover, Lyytinen, Srinivasan, & Tan,

*Reprinted by permission, Hassan, N.R., Mathiassen, L. & Lowry, P.B. (2019) The process of information systems theorizing as a discursive practice. *Journal of Information Technology, 34*(3), pp. 198–220.

N. R. Hassan (✉)
Department of Management Studies, University of Minnesota Duluth, Duluth, MN, USA
e-mail: nhassan@d.umn.edu

L. Mathiassen
Georgia State University, Atlanta, GA, USA
e-mail: lars.mathiassen@ceprin.org

P. Lowry
Virginia Tech, Blacksburg, VA, USA
e-mail: pblowry@vt.edu

N. R. Hassan, L. P. Willcocks (eds.), *Advancing Information Systems Theories*, Technology, Work and Globalization, https://doi.org/10.1007/978-3-030-64884-8_5

2008; Weber, 2012). However, there is an intense debate regarding what constitutes IS theory and the role of theories in IS (Avison & Malaurent, 2014; Bichler et al., 2016; Gregor, 2014; Holmström & Truex, 2011; Lee, 2014; Markus, 2014) with disagreement concerning native theories in the IS field (Grover, Lyytinen, & Weber, 2012; Straub, 2012). Some IS scholars argue that a theoretical core is unnecessary and logically indefensible (King & Lyytinen, 2004; Lyytinen & King, 2004; Lyytinen & King, 2006), whereas others maintain that the IS field's legitimacy cannot be established without core theories (Orlikowski & Iacono, 2001; Weber, 1987, 2003, 2006). Much of this controversy can be traced to the general problem of defining what does or does not constitute theory:

> Theory belongs to the family of words that includes guess, speculation, supposition, conjecture, proposition, hypothesis, conception, explanation, [and] model, so if everything from a 'guess' to a general falsifiable explanation has a tinge of theory to it, then it becomes more difficult to separate what theory is from what isn't. (Runkel & Runkel (1984), as cited in Weick (1995b), p. 386)

Literature on theory and theory development in the human sciences are plentiful, but they focus on differing goals and issues, and they vary across different disciplines such as sociology (Blalock, 1969; Dubin, 1969; Jaccard & Jacoby, 2010; Kaplan, 1964; Merton, 1968; Stinchcombe, 1987), psychology (MacCorquodale & Meehl, 1948), management (Bacharach, 1989; Corley & Gioia, 2011; Corvellec, 2013; Eisenhardt, 1989; Eisenhardt & Graebner, 2007; Gioia & Pitre, 1990; Morgan, 1986; Weick, 1989), entrepreneurship (Reynolds, 1971), and nursing (Fawcett, 1998). Such disparate efforts have resulted in a landscape of theory that is complicated (Corvellec, 2013), and has long been described as nothing short of "incredible anarchy" (Freese, 1980, p. 189) with conflicting views of theory in management-related fields persisting to this day (Byron & Thatcher, 2016). In the context of IS research, Avison and Malaurent (2014) suggest that the desperate search for, and over-emphasis on IS theory, has produced uninteresting research, and Grover and Lyytinen (2015) claim that scripted research strategies that domesticate theories from other disciplines have led to lack of boldness and originality in IS research. Meanwhile, Markus (2014), in defence of theories, suggests that it is the narrow or conflicting notions of theory that lead to trivial and uninteresting findings.

Weick (1995b) anticipated these issues and argued that the problem lies not in the theories themselves, nor in arguing about whether research contributions constitute theories; rather, the problem and the solution lies in the *process of theorizing*. Instead of assuming the dichotomy between what theory is or is not, Weick (1995b) suggests viewing theory as taking the shape of a continuum that is often approximated. By their very nature, theories are incomplete, for no one theory can explain and include all phenomena, and thus, they can only be approximations. These approximations, which are essentially interim struggles in the process of theorizing (Runkel & Runkel, 1984), hold the key to building exciting theories by opening spaces for future thinking (H. L. Moore, 2004) and as critical steps towards developing better theories. Unfortunately, with the exception of several classical studies (e.g., Peirce (1893–1913/1931–1958)), and more recent studies concerning modes of logical reasoning such as deduction and induction (Adler & Rips, 2008; Ochara, 2013), most of the resources on theory development focus historically on articulating and testing hypotheses (Chamberlin, 1890) instead of what precedes these steps. In fact, the term "theorizing" has never been clearly defined and has consequently been ignored in the philosophy of science itself (Swedberg, 2012, 2014c). To wit, Weick (1989, p. 516) emphasizes:

> Theory cannot be improved until we improve the theorizing process, and we cannot improve the theorizing process until we describe it more explicitly, operate it more self-consciously, and decouple it from validation more deliberately.

As such, we agree with the scholars who emphasize the need for more theorizing and join a growing list of recent studies from our peers in the management and social science fields that are focusing more and more on the process of theorizing (Cornelissen & Durand, 2014; Ketokivi, Mantere, & Cornelissen, 2017; Mantere & Ketokivi, 2013; Swedberg, 2012, 2014c). Calls to focus on theorizing as a discursive and reflective practice have already been made by IS scholars (Burton-Jones, McLean, & Monod, 2014; Gregor, 2018; Truex, Holmström, & Keil, 2006), as well as by scholars from other disciplines, including education (Luke, 1995), organization science (Alvesson & Karreman, 2000), and nursing (Sargent, 2012). We add to this discussion by drawing on Foucault's (1972) notion of *discursive formation* to advance knowledge on the forgotten stage of research known as the *context of discovery* (Hanson, 1958; Kaplan, 1964) as a contribution to our understanding of IS theorizing and as a complement to studies on theory development in general. Foucauldian discourse analysis is not the only basis for informing the theorizing process. Many other philosophers and social theorists such as Giddens, Chomsky, Derrida, and Habermas have all contributed and even disagreed with Foucault on several topics. However, very few offer the kind of depth of analysis into discourses of theorizing as Foucault did, especially on how discourses are organized and how power and knowledge in discourse are mutually constructive. Our framework also includes supportive arguments from many other theorists including Reichenbach, Merton, Kaplan, Hesse, Weick, and Swedberg. We submit that a focus on the theorizing process within the context of discovery holds the key to building exciting IS theories.

The Context of Discovery

For most social scientists and IS researchers, the logic of discovery (Popper, 1959) implies the development and testing of hypotheses (Chen & Hirschheim, 2004; Orlikowski & Baroudi, 1991) as a process that requires strict adherence to rigorous rules in order to meet the requirements of research and science (Nickles, 1980; Schickore & Steinle, 2006). In this process, research starts with proposing hypotheses and then proceeds to the empirical stage during which data is collected and analysed to test those hypotheses. Reichenbach (1938) and Popper (1934) coined this process as the "context of justification" to prioritize it from what typically precedes it, which they call the "context of discovery." Thus, the *context of justification* is the stage of research in which the idealized logic of science, a reconstruction of the actual steps and thinking that took place, is presented in its perfected and refined form. Although many researchers begin this process with some kind of theoretical framework, theory is often added as an afterthought (Kaplan, 1964). By contrast, the *context of discovery* is the stage of research that represents the *actual* steps and thinking of the researcher, in which the practice of theorizing in the form of "disciplined imagination" (Weick, 1989) takes place and "intuitive leaps, false starts, mistakes, loose ends and happy accidents clutter up the inquiry" (Merton, 1967, p. 4). Despite this apparent messiness, it is this stage of research that exhibits the creativity and serendipity of discoveries. Table 5.1 depicts a summary of this stage of research consisting of various practices including problematizing the phenomena, leveraging paradigms, bridging discursive and non-discursive practices, analogizing, metaphorizing, modelling, and constructing the research framework, all taking place outside the context of justification.

Although the reconstructed logic (Kaplan, 1964) of the context of justification is cleaner, easier for reviewers and editors to understand, and facilitates publication, the deductive logic that underpins it cannot infer anything beyond the data provided by its premises (Gauch, 2003), which, in turn, limits the possibilities for new discoveries that makes for interesting research (Orlitzky, 2012; Schwab, Abrahamson, Starbuck, & Fidler, 2011). This distinction is noteworthy to IS researchers because the research approach characterized by the context of justification, which is "concerned with the hypothetic-deductive testability of theories" (Chen & Hirschheim, 2004, p. 201), remains the dominant approach within IS research not only in North America but also in Europe (Liu & Myers, 2011). As Reichenbach (1938) notes, the researcher's subjective thinking processes, and the discursive activities that follow, which represent the context of discovery, are more valuable than the same researcher's "rational reconstructions" (p. 5), which take place in the context of justification. The creativity of the researcher is most strongly pronounced within the context of discovery, and foregrounding this stage of theorizing allows us to understand the researchers' creative strategies that led them to realize their goals (Swedberg, 2014b).

We are not suggesting an abandonment of the context of justification and its related logic and methods. The associated rigor that constitutes the context of justification provides the scientific enterprise its credibility and authority. We suggest, however, that the preceding stage of research characterized by a logic-in-use—the modus operandi of great scientists—has been largely ignored within the IS field and allied disciplines. The likes of Emile Durkheim, Max Weber, Karl Marx, and Bronisław Malinowski did not begin their research with a scripted research approach or a theory domesticated (Grover & Lyytinen, 2015) from their reference disciplines. Instead, they imagined and theorized the core concerns of their phenomenon of interest (Rappaport, 1987), including the occurrence of suicide, the growth of capitalism, and the question of class conflict and universal culture—while not ignoring the fruits of serendipity.

Our juxtaposition of the context of discovery and the context of justification does not imply that the context of discovery is applicable only to hypothetico-deductive research. Other approaches such as interpretive and critical research naturally place a focus on the context of discovery, as seen for example, in the grounded theory method. Unlike hypothetico-deductive research, the theorizing process in grounded theory (Glaser & Strauss, 1967) is documented in detail through the various steps such as comparative analysis, conceptual clarification and developing theoretical sensitivity. This study enriches those approaches by describing the theorizing process at a deeper level for researchers. Foucault (1972, p. 64) describes this process of theorizing as the formation of strategies in the human sciences, giving

> rise to certain organizations of concepts, certain regroupings of objects, certain types of enunciation, which form, according to their degree of coherence, rigor, and stability, themes or theories.

For example, even though the computer as an object of study in the IS field is the same as in computer science, the IS field formulates its propositions surrounding that object using a strategy that is different from the one based on symbol-processing rules in computer science (Denning, 1999; Newell & Simon, 1976). Because each field of study follows different rules of forming its discourse, and strategizes in different ways, each field builds different theories concerning their phenomenon of interest. Thus, each discipline lays claims to their own unique theories.

Viewing theorizing as strategizing is not unlike witnessing how a good chess player strategizes his or her game. A chess player who follows the rules of the game is not guaranteed a win, but it would be wise for that player to follow the rules if the player seeks to win (Kaplan, 1964). Beyond following the rules of the game—which metaphorically represent how elements of theorizing can be applied to the process within each discipline—the chess player strategizes each move to win the match. Similarly, in the context of discovery, strategizing requires intuitively and imaginatively working with the elements of theorizing. Although there may not be a prescribed set of rules for how that can be accomplished, theorizing can be learned and taught (Swedberg, 2014c) in the same way that Rivard's (2014) "Ions of Theory Construction" can be marshalled for crafting new theories. Accordingly, our goal is to advance knowledge on how theorizing, and, most importantly, theorizing in the context of discovery, can be undertaken by IS scholars with key examples from the field.

We define theorizing as making certain claims in the form of statements that reflexively apply specific rules of formation to constitute a discourse within a field, thus creating "a group of statements in so far as they belong to the same discursive formation" (Foucault, 1972, p. 117). Thus, when the claim that "user involvement in systems development enhances the likelihood of system success" was examined by Ives and Olson (1984) early in the history of IS, they engaged in theorizing using a set of rules pertaining specifically to IS, and not, say, to computer science. This set of rules, or discursive formation, governs how additional statements are enunciated by that field, and those additional statements constitute the field of study itself. As Ives and Olson (1984) examined the various concepts and constructs surrounding user involvement, user roles, system type, and the expected outcomes in system quality and level of acceptance, the discourse of IS was simultaneously constructed, and, from their practice of theorizing, ironically discovered a lack of theory in earlier frameworks, thereby raising doubts about the claimed benefits of user involvement.

Theorizing within a field of study implies that the statements enunciated by that field of study should differ from the statements enunciated by another field, thus distinguishing economic discourse from legal discourse, medical discourse from biological discourse, and computer science discourse from IS discourse. At the same time, each discourse is comprised of different sub-discourses. For example, the economic discourse developed mercantilist, physiocratic, classical, Keynesian (Foucault, 1970), and monetarist discourses throughout its history, each making different claims based on different rules of discourse concerning how value and prices are determined and how human economic needs and wants could be satisfied.

Even with these differences, a sense of unity in discourse allows a community of scholars to say that they are talking about "the same things," "at the same level," or "applying the same or different principles" with their colleagues. This practice of theorizing is what Foucault (1972) calls *discursive practices* (pp. 46, 48–49), in which certain relations among a heterogeneous group of concepts, claims, and other discursive practices are built. Discursive theorizing practices include, but are not limited to, formulating ideas, creating imagery, and engaging in deductive or inductive reasoning or logical inferencing. These discursive practices consist of "a body of anonymous, historical rules, always determined in the time and space that have defined a given period, and for a given social, economic, geographical, or linguistic area," which define the conditions for the formation of concepts and claims (Foucault, 1972, p. 117). This formation of the discourse from claims and statements is not unlike how discursive practices enact identities in social media, business consulting, and market categories (Vaast, Davidson, & Mattson, 2013). The outcome of these discursive practices—the discourse—often develops into a field of study that is stable enough to be given a name (e.g., how specific discursive practices became to be known as *Information Systems*) but at the same time is always in a constant state of renewal, subject to ongoing discoveries, criticisms, and corrections.

A Framework for Foundational and Generative Discursive Practices

Following these ideas of the context of discovery as discursive practice, Fig. 5.1 depicts IS theorizing as a set of foundational and generative discursive practices that deploy different strategies to produce specific theory components. Hence, the ultimate goal of foundational and generative theorizing practices is to produce the components of a theory, such as concepts, claims, and theory boundaries (Bacharach, 1989) that name

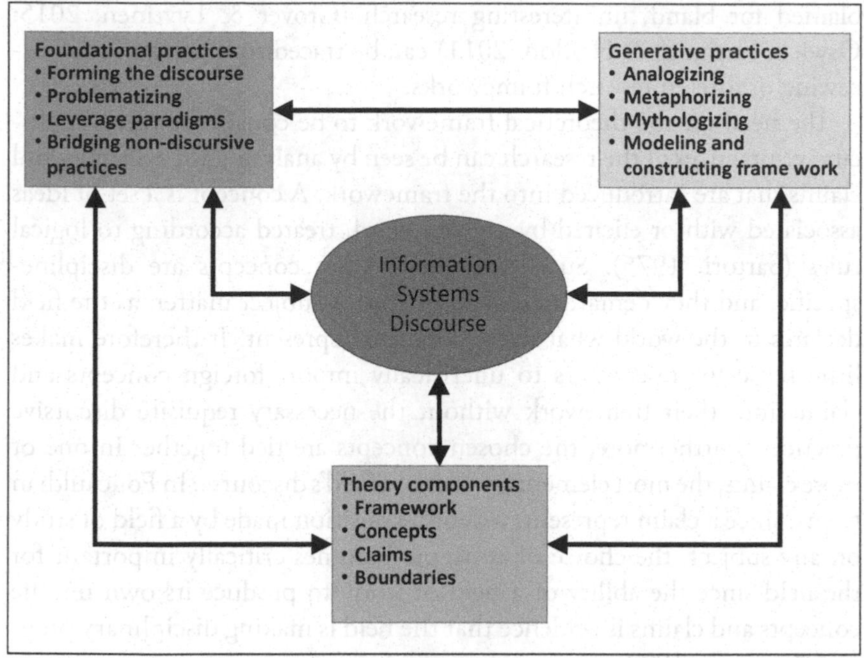

Fig. 5.1 IS theorizing as discursive practices

and describe the phenomenon of interest, all of which are organized into a research framework. Theorizing as discursive practices requires that the building of such a framework be organically connected to the foundational and generative practices.

Often students are told, after crafting their research or dissertation proposal, that they lack a theoretical framework (Ågerfalk, 2014). A frequent reaction to this critique is to force an ill-fitting theoretical scaffold over that existing effort, which brings with it additional problems that distract the research from its goals. Others may start by uncritically importing a theoretical framework without the requisite theorizing practices. By distinguishing between the "research" framework and the "theoretical" framework, Ravitch and Riggan (2012) highlight the need for requisite theorizing practices to take place in building the research framework before importing them. The domestication of theory that is often blamed for bland, uninteresting research (Grover & Lyytinen, 2015; Oswick, Fleming, & Hanlon, 2011) can be traced to this wholesale borrowing of foreign research frameworks.

The need for the theoretical framework to be consistent with the discursive practices of the research can be seen by analysing the concepts and claims that are introduced into the framework. A concept is a set of ideas associated with or elicited by a given word, treated according to logical rules (Sartori, 1975). Such rules imply that concepts are discipline-specific, and they demarcate a field of study's subject matter, as the field declares to the world what those concepts represent. It therefore makes little sense for researchers to uncritically import foreign concepts and claims into their framework without the necessary requisite discursive practices. Furthermore, the chosen concepts are tied together in one or more claims, the most elementary unit of a field's discourse. In Foucauldian terms, since a claim represents a definite position made by a field of study on any subject, the choice of concepts becomes critically important for the field since the ability of a field of study to produce its own unique concepts and claims is evidence that the field is making disciplinary progress (Foucault, 1972).

As Schön (1963) emphasizes, producing new concepts and claims is the raison d'etre of all disciplines and has been a mystery since antiquity. The same goes with the IS field in its efforts to invent its own concepts (Markus & Saunders, 2007), in particular surrounding its core concerns—information and systems. Peter Keen (1980) once counselled:

> Until we have a coherent definition of "information" we have nothing to measure. Surrogates for improved information, such as user satisfaction or terminal hours of usage, will continue to mislead us. (p. 9)

Only recently has the IS field seriously engaged in what "information" means (Boell, 2017; McKinney & Yoos, 2010; J. C. Mingers, 1995, 1996) and what "systems" entail when they are coupled with information (Lee, 2010; Lee, Thomas, & Baskerville, 2015). The theorization of these terms should result in concepts and claims that belong to the IS field (Markus & Saunders, 2007) because it is through such meaningful and precise terms that the IS field declares its subject matter to others (Kaplan, 1964).

According to Foucault (1972), a theory does not just define relationships between several concepts as is often stated (Bacharach, 1989; Whetten, 1989), but also acts as a strategic choice of addressing its phenomenon of interest, arranging different forms of enunciations, manipulating concepts and giving them rules for their use, and placing those concepts into a constellation that could create new discourses. This definition fulfils the goals of theory not just to describe, analyse, explain, and predict (Gregor, 2006), but also to uncover, to excite, to inspire, and to be productive. It is no surprise that the expression "crafting strategy" alludes to the same essential activity as the expression "crafting theory," an activity that produces rules but in itself does not have a set of explicit rules (Swedberg, 2014b). Therefore, based on the nature of theories as strategic choices, theories can also be defined as regulated ways of practising the possibilities of discourse.

The nature of theory as a productive force that produces new discourses is lost when it is viewed merely as means for explanation or prediction, as is commonly understood within IS circles. This generative nature of theory is especially obscured when theory is added as an afterthought to dress up the research for publication. The work of Ferdinand de Saussure (1916/1966) illustrates this productive nature of theory when he proposed a theory that distinguished two concepts that were previously assumed to be inseparable: *langue* (language) and *parole* (speech). Saussure's strategic choice made possible a new discourse of historically studying languages which, because of the assumed inseparability of language and speech, was hitherto impossible. Doing so allowed Saussure to formalize a new theory of the root word, concluding that all Indo-European languages were derived from one original language. The particular rules of his discourse, which is known as structuralism, were applied to other fields beyond the study of languages. Lévi-Strauss (1955) applied the discourse of structuralism to study myths and founded the new school of structural anthropology that opened more possibilities for discourse in a new field of study. This process of theories spawning not just other theories but whole disciplines shows that theory is not just the product of intellectual activity within a field of study. Rather, theory becomes the formative element of that field of study and becomes the inspiration for other fields of study. As such, our theory of IS theorizing as discursive practices uncovers the creative activities wielded within the context of discovery.

Foundational Theorizing Practices

Foundational practices are discursive theorizing practices that involve high-level concepts such as its discursive formation, disciplinary questions, and paradigms shared by the community that characterize the phenomenon of interest. Foundational practices that consist of (1) forming the discourse, (2) problematizing the IS phenomena, (3) leveraging paradigms, and (4) bridging discursive and non-discursive practices, assist members of the field in recognizing opportunities for crafting theories, applying their expertise, and helping others understand the distinguishing features of the discourse.

More importantly, foundational practices bound our thinking processes to that which concerns our phenomenon of interest. Counter-intuitively, it is this bounding process that actually engenders our creativity. As Kant's formulation of the phenomena and the noumena implies, the very factors that make us finite (being subjected to space, time, context, and history) are also conditions for the possibility of knowledge. Foucault (1970, p. 340) calls this need to bound theorizing the "analytic of finitude." By establishing boundaries, one is forced to carve out knowledge within those boundaries. For example, to the untrained eye, snow is snow. Yet, anthropologists and linguists have found that Eskimo and Sami tribes, whose lives are bounded by their frigid and hostile environments, have developed dozens (Krupnik & Müller-Wille, 2010), and, in some cases, hundreds, of names and classifications of snow and ice (Magga, 2006). The limitations of their environment made possible knowledge about snow that others outside their environment could not have fathomed. This extraordinary knowledge is made possible by foundational theorizing practices. The following foundational theorizing practices illustrate this process of delineating our phenomenon of interest.

Forming the Discourse

In the same way the language of the Eskimo and Sami tribes theorizes the numerous descriptions for snow, a field of study carves out its knowledge by following a set of rules that governs the formation of concepts and claims concerning its phenomena of interest. This set of rules which Foucault (1972) calls the discursive formation, defines the basis of the unity surrounding different phenomena associated with that field of study. As instruments of power, discourses identity the field of study, establish its disciplinary authority, and limit what can or cannot be said and what is or is not acceptable. Disciplinary history demonstrates how rules of discourse delineates the boundaries of disciplines. When Auguste Comte (1830–42) envisioned the then new field of sociology, he framed it as "social physics" to describe how order can be maintained in society by applying the mechanistic rules of physics and other natural sciences. Borrowing from these natural sciences, Comte constructed new rules of discourse (i.e., formed a new discourse) that became known as sociology, which addressed social phenomena. New concepts and theories "native" to the new discipline are developed with the help of the new rules of discourse. Forming the discourse involves all the rest of the foundational theorizing practices and all the generative theorizing practices to define the set of rules that govern the formation of concepts and claims concerning the IS phenomenon.

Thus, when an IS researcher applies economic theory, or studies the use of computers using rules concerning value, prices, costs, and trade-offs, which are part of the discursive formation of economics, the power of the economic discourse shapes and colours that research. One question that can arise in using this discourse is whether the research is primarily about economics, IS or whether the research is about IS in economics. The choice of applying a specific discursive formation has wide-ranging implications, not just for the direction of a study, but also for the direction of the entire IS field, especially if a similar discursive formation is ubiquitously applied in that field of study. The same object of study—information—can be researched in as many different ways as there are different rules of how claims about information can be construed. Because the IS discourse is yet to be clearly articulated (Nik R. Hassan, 2011; Nik R. Hassan & Will, 2006), IS researchers need to be sensitive as to which discursive formation is primarily in operation in their research. For instance, both the IS and the information and library sciences fields study information. However, how the study of information contributes to knowledge in IS differs from how information contributes to knowledge in library sciences because the two fields apply different rules of formation concerning information (Ellis, Allen, & Wilson, 1999).

The wide-ranging implications of choosing a specific discursive formation can be seen in the IS field in the case of technology adoption with its long and varied historical tradition that included implementation, user involvement, user satisfaction, studies of IS failures, innovation, assimilation and media richness. Despite these diverse traditions, it was the discourse of social psychology—in the form of the technology acceptance model TAM and its variants—that overwhelmed other discourses in theorizing technology adoption. The genesis of TAM can be traced to Davis' (1986, p. 7) adaptation of Ajzen and Fishbein's extensive research on human behaviour (Ajzen & Fishbein, 1972, 1973; Fishbein & Ajzen, 1974, 1975) to "provide a theoretical basis for a practical 'user acceptance testing' methodology … prior to implementation." Fishbein and Ajzen's discourse, subsequently instantiated as the Theory of Reasoned Action (TRA) (Ajzen & Fishbein, 1980) and the Theory of Planned Behaviour (TPB) (Ajzen, 1991), was critical of the widespread assumption that beliefs directly impacted behaviour. Since people with the same beliefs or attitudes do not necessarily display similar overt behaviour, by distinguishing beliefs, attitudes, and intentions as different concepts in predicting behaviour, Fishbein and Ajzen were able to explain the historically poor and conflicting results in existing attitude research. By omitting the evaluation of beliefs, social norms, and intention, TAM led adoption research back to the conflicting research results and discourse that Fishbein and Ajzen sought to remedy. As Benbasat and Barki (2007) suggest, subsequent versions of TAM merely brought the IS field back full circle to the original Fishbein and Ajzen models that TAM has dismissed at the beginning.

An analysis of TAM's discourse uncovers several other more fecund alternatives for studying adoption. For example, Rogers' (1983) diffusion of innovation discourse (G. C. Moore & Benbasat, 1991; Rogers, 1983) revolves around how new ideas are diffused over time among members of a social system, whereas, TAM's discourse is about how perceived features of the technology motivate use; two very different discourses that could theorize adoption. As such, the Rogers' diffusion model is a more comprehensive model that includes pre-adoption stages, the innovation-decision process, continued adoption and discontinuance. The extensive scope of this discourse enables prediction of the rate of adoption (i.e., early adopters and laggards), not possible with other discourses. More broadly, studying the discursive formation of reference theories opens IS researchers to alternative, perhaps more parsimonious, discourses for adoption studies, including those from network science (Katz & Shapiro, 1985, 1986), philosophy of technology (Ellul, 1973; Feenberg, 1991; Heidegger, 1977), and social constructionist discourses (Bijker, 1995; Bijker, Hughes, & Pinch, 1987).

Problematizing the IS Phenomenon

The notion of problematizing as a theorizing practice has been extensively addressed, especially within the organization sciences (Alvesson & Sandberg, 2013; Locke & Golden-Biddle, 1997; Sandberg & Alvesson, 2011). Generally, it is defined as identifying and challenging the assumptions underlying existing theories and research. In relation to IS, this definition can be extended to mean asking questions other disciplines are not asking or are incapable of asking. Unfortunately, current research approaches, including those applied by the IS field, are increasingly neglecting the important role of questions in research (Meyer, 1995). Intuitively, researchers often focus on providing and constructing answers to questions, but Alvesson and Sandberg (2013) consider question construction to be the more crucial aspect of research because questions encourage intellectual reflection, whereas answers tend to encourage closure and inactivity. It is much easier to follow a scripted method of research (Grover & Lyytinen, 2015) that outlines how to provide and construct answers to existing questions, even though it is the construction of the questions that helps the researcher venture into new territory and become more reflective and intellectually productive. For example, in the case of the IS field, after decades of asking questions that were limited to the concerns of IS management within the organization (Brancheau & Wetherbe, 1987; Kappelman, McLean, Luftman, & Johnson, 2013), IS scholars are expanding their list of questions beyond organizational management to global challenges such as environmental sustainability, poverty, cyber-attacks, diseases, and global conflict (Becker, vom Brocke, Heddier, & Seidel, 2015; Gholami, Watson, Hasan, Molla, & Bjørn-Andersen, 2016; Winter & Butler, 2011).

The key to problematizing is to focus on the *disciplinary question* instead of just the research question. Every field of study has its own unique set of questions. Since questions need to pertain to the discipline, not all research questions can be admitted into that discipline (Bal, 2002; Bromberger, 1992; Meyer, 1995). Asking the wrong questions wastes valuable research resources, as the results often do not address the research problem and the entire research programme could proceed in a less productive or unintended direction, which in turn prevents the field from demonstrating its value (Agarwal & Lucas Jr., 2005; Nik R. Hassan, 2014b). A disciplinary question is one that addresses the phenomenon of interest as a problem requiring a solution based on the field's rules of discourse. Thus, when Durkheim (1951/1897, p. 324) posed the problem of suicide and asked the question of why a definite proportion of people commit suicide in any given period in every society, he was not focusing on the state of mind (e.g., despair, neurosis, depression, or any psychological state of individual members of the society), as one would expect in the case of suicide; rather, he was linking suicide, the object of study, to the newly emerging discipline of sociology. His questions essentially defined his discipline.

Given the significance of disciplinary questions, it is not surprising that the IS field emerged as a result of addressing questions its reference disciplines had not satisfactorily addressed. Mason and Mitroff's (1973) early framework for IS and Davis' (1989) TAM asked questions that did not fit exclusively into either management, computer science, or psychology. In media-richness studies, the triggering question concerned managers' activities. If managers spent 80% of their time communicating, as studies had found, then what kind of communication media do managers use, and are some of these media more effective than others? By asking these questions, Daft and Lengel (1983) and Lengel and Daft (1984) modified the rules of their discourse, which originated in communication (Bodensteiner, 1970) and management studies, towards the IS discourse. Following from the possibility that certain structures and IS artefacts translate organizational messages at different levels of richness, Daft and Lengel asked if the richness of media is related to the translation richness of information, which directly impacts information-processing needs. By the time this study was published in *Management Science* (Daft & Lengel, 1986), the first sentence in the article no longer asked a communication-related question; rather, it asked the IS questions: "Why do organizations process information?" (p. 554) and, relating to IS artefacts, "How do organizations process information?" (p. 568).

The right questions not only make the research relevant to the IS field, but they also embody curiosity and inquisitiveness and spawn other interesting questions. What kinds of information-processing mechanisms and IT artefacts are most helpful to organizations? How can such IT artefacts be evaluated? Do different environments and problems require different kinds of mechanisms and IT artefacts? Can these different mechanisms be integrated? Despite the availability of new advanced communications, why do managers still prefer face-to-face meetings (Daft, Lengel, & Trevino, 1987)? Accordingly, problematizing the IS phenomenon of interest implies asking questions that are not being asked by other disciplines or asking questions that other disciplines are incapable of answering.

Leverage Paradigms

Partly as a result of criticisms of Kuhn's (1970) paradigm concept and its varied interpretations, the role of the paradigm in IS theorizing has been largely neglected and misunderstood (Hassan, 2014a; Hassan & Mingers, 2018). Although there are several notable exceptions (Chen & Hirschheim, 2004; Goles & Hirschheim, 2000; Iivari, Hirschheim, & Klein, 1998; Khazanchi & Munkvold, 2003; J. Mingers, 2004; Moody, Iacob, & Amrit, 2010; Richardson & Robinson, 2007), the IS field is discouraged from actively engaging paradigms in theorizing (e.g., Adam & Fitzgerald, 2000; Avison, 1997; Banville & Landry, 1989; Cushing, 1990; Jones, 1997; Khazanchi & Munkvold, 2000). This tendency can be traced back to the earliest phases of the IS field's development, when attempts to theorize using paradigms were met with resistance because of the "disrepute into which this word [had] fallen" (Ein-Dor & Segev, 1981, p. vii).

This state of affairs is unfortunate. Kuhnian *paradigms*, defined as shared exemplars for research that provide concrete problem-solutions, have historically been widely accepted as useful objects for theorizing. They can take the form of research achievements, widely agreed-upon political bases and legal precedents, and standard classical textbooks and illustrations (T. Kuhn, 1970), and leveraging them involves using the organizing principles, recognized scientific achievements, heuristic illustrations and other concrete scientific problem-solutions on existing research. The paradigm concept has been applied successfully in many other fields of study. The software development subfield of computer science relied on the engineering paradigm to develop its discourse, as is evident from the rallying cry of software engineering in the 1960s to establish it as a professional discipline (Naur & Randell, 1969; Shaw, 1990). Historian David H. Fischer (1970, p. 161) asserts that "historians in every field have much to learn" from Kuhn and the historian David Hollinger (1980) explains how the Kuhnian paradigm helps neutralize the biases of social and anthropological theories without excluding them from developing historical theories.

Minsky (1975), a pioneer of artificial intelligence, admitted his debt to Kuhn for his frame theory: "The basic frame idea itself is not particularly original—it is in the tradition of the 'schema' of Bartlett and the 'paradigms' of Kuhn" (p. 113). In the social sciences, Berger and Luckmann (1966) credited Kuhn for their understanding of the social construction of reality, and Ritzer's (1980) *Sociology: A Multiple Paradigm Science* was based on the Kuhnian paradigm. The influence of Kuhn's paradigms is particularly evident in science and technology studies, in which Kuhnian concepts of normal science, worldviews, and scientific revolutions forever changed the understanding of progress. The field of social construction of technology, which is often cited by IS researchers, is based on the Kuhnian paradigm. Explaining the basis of Kuhn's concept of the "technological frame," Bijker (1995) noted: "The analogy with Kuhn's 'paradigm,' among other concepts, is obvious" (p. 123). Bijker went on to claim that the "technological frame is evidently one of the many children of Kuhn's (1970) disciplinary matrix" (p. 126). Other such children include Collins and Pinch's (1982) "frame of meaning," Constant's (1980) "technological tradition," Rosenberg's (1976) "focusing devices," Gutting's (1980) "technological paradigm," and Jenkins' (1975) "technological mind-set." All of these cognate terms reflect how fundamental paradigms are to theoriing.

As illustrated in the case of problematizing in IS, media richness theory (MRT) was inspired by communication studies, which applied a linguistic paradigm thus suggesting that managers preferred natural languages to formal mathematical languages. Sensing the limitations of this paradigm for their work, Daft and Lengel (1983) integrated two other paradigms into MRT (Daft & Lengel, 1986) to describe two complementary dimensions that explain why organizations process information: to reduce task uncertainty, drawing from Galbraith (1973, 1977), and to reduce equivocality, drawing from Weick (1979, 1995a). Galbraith's information-processing paradigm offered a concrete problem-solution that links information-processing and the notion of uncertainty to organization design. Weick's sensemaking paradigm provided a means of explaining media richness using the concept of equivocality. In this example, paradigms play a double role of limiting the discourse to what the researcher is already familiar with while allowing the researcher to "see his problem as like a problem he has already encountered" (T. Kuhn, 1970, p. 189), thus making possible new discourses that can be constructed to describe the phenomena being researched.

Bridging Non-Discursive Practices

Foucault (1972) argues that every field of study has both discursive and non-discursive practices. *Non-discursive practices* are material relations that enunciate the same discursive formation and items of knowledge as their corresponding discursive practices, but take the shape of repeatable materiality in things unsaid in the form of routines, processes, and events in social, legal, economic, and political institutions (Bacchi & Bonham, 2014; Foucault, 1972). In contrast to the received view of the IS field as being "applied" (Keen, 1980; Robey, 2003; Taylor, Dillon, & Van Wingen, 2010; Vessey, Ramesh, & Glass, 2002), the notion of the non-discursive practice implies that the IS field is an applied field that is inseparably connected to its discursive side. Consequently, the idea of a non-discursive practice closes the oft-repeated gap between "basic" and "applied" sciences and between theory and practice. Indeed, non-discursive practices provide the horizon, background, and justification for any discursive strategy to be intelligible (Dreyfus & Rabinow, 1983) and bridging with non-discursive practices imply providing the tacit knowledge, fore-meaning (the horizon), context and tradition (the background), and warrants (the justification) from practice for the theorizing process.

The recent rise of "applied mathematics" or "applied statistics" in the form of non-discursive practices of data analytics and Big Data (Davenport & Patil, 2012) for the "pure science" of mathematics illustrates the inseparability of discursive and non-discursive practices. These non-discursive practices contribute to the revival of a discursive practice within IS known as business intelligence. Therefore, certain discursive practices within the IS field are shaped, appropriated, or even abandoned as a result of the non-discursive practices of these data scientists and statisticians. As these examples illustrate, theory cannot be separated from practice.

This inseparability is in part why non-discursive practices are considered foundational: when they are articulated, these practices form the basis for a discipline. The history of the IS field itself speaks of such a phenomenon (Caminer, 1997; Ferry, 2003; R. Hirschheim & Klein, 2012), as the non-discursive practices of the J. Lyons and Company in the United Kingdom, and of the military implementations of ENIAC in the United States, inspired other companies to set up their own non-discursive practices in the form of the earliest MIS departments, which, in time, led to several discursive practices (textbook publications) in the late 1950s and early 1960s (Gregory & Van Horn, 1960; Langefors, 1966), and eventually the first graduate MIS programme at the University of Minnesota in 1968. As Foucault (1972) explains, a "whole non-discursive field of practices, appropriation, interests, and desires" (p. 69) with other discursive practices and discourses external to the discourse itself, together define the discipline. Therefore, theories cannot be crafted separately from their relationship with the non-discursive practices that surround discursive practices, and provide the authority to take one or another strategic choice in theorizing (Dreyfus & Rabinow, 1983).

Some aspects of the non-discursive practice are difficult to articulate. The illusive "gut feel" and almost automatic decision-making processes that are associated with practitioner activities are often opaque to the prying eyes of the researcher. These aspects of non-discursive practices are known by different names and descriptions. Aristotle (1934) calls it phronesis; Polanyi (1958) calls it tacit knowledge; Bourdieu (1977) calls it habitus; Ryle (1949) calls it "knowing how"; and Knorr-Cetina (2014) calls it intuitionist theorizing. Scholars deal with frameworks, concepts, claims, and theory boundaries (Fig. 5.1), while practitioners typically deal with more pragmatic concerns. However, both scholars and practitioners within a field are bounded by the same rules of discourse.

To illustrate, the rules of discourse for the decision support system (DSS) area in the 1970s were concerned with the different characteristics of information (source, scope, time horizon, and frequency) and the different levels of decision-making in the organization based on the spectrum of programmability between structured and unstructured tasks (Gorry & Scott, 1971). By the time Keen and Scott (1978) and Alter (1977, 1980) formalized the discourse in the form of the DSS, considerations on how DSS might be designed, proposals for different types of DSS, and ideas for how they could be deployed in the organization were added to the discourse. Although these contributions added new elements to the discourse, the rules of that discourse did not change substantially. Even when Rockart and colleagues (1979, 1988, 1982) extended the discourse by proposing a similar support system for executives (i.e., ESS), or when the researchers associated with the Minnesota Experiments (Dickson, Senn, & Chervany, 1977; Watson, DeSanctis, & Poole, 1988) and the Center for the Management of Information at the University of Arizona (Applegate, Konsynski, & Nunamaker, 1986; Dennis, George, Jessup, Nunamaker Jr, & Vogel, 1988; Nunamaker Jr, Applegate, & Konsynski, 1987) began investigating how such systems (called GDSS) could be used in a group environment to enhance collaboration, the same rules of discourse applied, albeit with a few additional rules (e.g., anonymity in the case of GDSS). Foucault (1972) describes this process of bridging between theory and practice as the "procedures of intervention" and "rewriting" (pp. 58–59), during which certain claims are transferred from one domain or context to another, without losing their enunciative homogeneity. This process allows scholars and practitioners to recognize the same phenomena in their disciplines, albeit in different contexts.

Generative Theorizing Practices

As shown in Fig. 5.1, foundational discursive practices define the information systems discourse and impact generative practices. Generative theorizing practices support or modify the development of the discourse once it is founded or once its nature is clearly delineated. These generative practices wield "the power of putting our finite resources to virtually infinite use" (Leary, 1995, p. 267) to name and describe the phenomenon of interest using components of theory, such as concepts and claims, and to construct frameworks to organize such components into a theory. Generative practices are not to be perceived as "exploratory research," as merely the under-labourer to the "real research" task of developing and testing hypotheses and investigating propositions. Instead, these practices play an ineluctable role in research by offering the "magnified tendency to call up ideas" (Peirce, 1992, p. 182) and organizing those ideas to make sense of the phenomenon. The following generative theorizing practices involve analogizing, metaphorizing, mythologizing, and modelling (Fig. 5.1), discursive practices that T. S. Kuhn (1987, p. 20) states are "the most obvious and most consequential" to scientific progress.

Analogizing

Among the most powerful of the generative practices is *analogizing*, which played a highly constructive role in the development of Western knowledge up to the Age of Enlightenment (Foucault, 1970). William

James (1890, p. 530), as cited in Leary (1995, pp. author-year), views analogizing as "the leading fact of genius of every order." An *analogy*—from the Latin *analogia*—refers to ratio or proportion and the practice of analogizing involves using a simplified or scaled-down reference to something familiar to explain or illustrate something more complex or less familiar (Bagnall, 2012; M. Hesse, 1967). Tsoukas (1993) argues that analogies are not merely literary devices; rather, they supply the raw materials for theorizing and, if suitably handled, yield theories. M. B. Hesse (1966) goes even further to insist that analogies do not just yield theories but are an ineradicable part of them. For example, the analogy of the flow of electrons in an electrical circuit as the flow of people in the subway is what helped theorize the flow of electricity, and is, at the same time, part of the theory itself (Gentner, 1983, 1989).

Within the context of discovery, analogies allow for demonstrative inferences that are difficult or impossible to achieve in purely positivist schemes of explication and justification. Darwin (1859) drew an analogy between artificial selection (i.e., the breeding of domesticated animals) and natural selection to argue for the plausibility of the latter and, as a result, distinguished his discourse on biological evolution from that of earlier natural history. In the management field, Beer (1972, 1979), drawing an analogy between the human body and the enterprise, theorized that only five major subsystems are required to coordinate and control any organization. Accordingly, N. R. Campbell (1920) highlighted the ineradicable nature of analogy in theorizing:

> The value of the theory is derived largely, not from the formal constitution, but from an analogy displayed by the hypothesis. This analogy is essential to and inseparable from the theory and is not merely an aid to its formulation. (p. 119)

Although analogizing in the IS field has produced many research programmes, most of them are undertaken implicitly rather than explicitly. For example, when Keil and colleagues (1995, 1999, 2000) applied the term "project escalation" in the context of software project management, they used an analogy originally applied in a military scenario (Kahn, 1965), which draws on the similarity between intensifying conflict and climbing higher up the rungs of a ladder. When MRT researchers propose that "rich" information processing mechanisms are necessary to successfully address complex and ambiguous environments, they are drawing an analogy between managerial work and complex human biological systems. Other IS scholars that study "punctuated equilibrium" or "systemic change" (Street & Denford, 2012), or discover a "contagion" (Angst, Agarwal, Sambamurthy, & Kelley, 2010) of system adoption, are implicitly applying analogies from other disciplines, such as geology or biology, to inform and explain IS phenomena. Explicit analogical reasoning has only recently captured the attention of IS researchers (e.g., Kuechler & Vaishnavi, 2012).

Explicit analogizing harnesses the potential of analogies for explaining how the world and societies work to their fullest extent. In sociology, Erving Goffman's ethnographic studies are prime examples of explicit analogizing (Vaughan, 2014). In *The Presentation of Self in Everyday Life* (Goffman, 1959), he draws an analogy between the face-to-face interaction that everyone has with others and theatrical performances. In this work, he theorizes that when an individual comes into contact with another person, that individual will control or guide the impression that the other person forms by altering appearance and manner, much like actors in movies and theatre. For IS, the usefulness of this generative theorizing practice should be obvious if we consider the offline and online lives and activities of Internet users. For example, a study of the relationship between identity verification and knowledge contribution in online communities (Ma & Agarwal, 2007) finds that IT features that support persistent labelling, self-presentation, and deep profiling, all of which enhance identity verification, promote satisfaction and knowledge contribution in those online communities.

Whereas IT project failure research that applies analogies from escalation literature assumes the presence of negative information, failures often occur in the absence of negative information (e.g., the Obamacare website crash; see T. Cohen, 2013). In such cases, the escalation literature might not be appropriate. Other forms of failure research, such as disaster ethnography (Vaughan, 1996), safety science (Le Coze, 2008), and disaster prevention and mitigation (Weichselgartner, 2001), could offer better analogies for those kinds of failures. These alternative genres of research offer what existing IT project failure lacks—identifying counter-intuitive causes of failure, spotting red flags of impending disasters, providing post-disaster management, and identifying the normalization of deviance, all of which IT project managers can apply to better prevent and manage failures.

Metaphorizing

Metaphorizing involves extending the goal of finding similarities in analogizing by selecting often familiar physical or linguistic objects to not only carry the meanings of analogies but also to elegantly clarify and impress on those meanings using the characteristics of those familiar objects (Ortony, 1979). Isaiah Berlin (1999) states:

> To think of one phenomenon or cluster of phenomena is to think in terms of its resemblances and differences with others ... All language and thought are, in this sense, necessarily metaphorical. (p. 158)

Thus, *metaphors* are essentially linguistic forms of analogies and have been used in discourse since Aristotle's time (Ricoeur, 1977; Schön, 1963). Whereas analogies are abstractions of similarities, metaphors select a term or sets of terms that carry the meanings of those similarities (Geary, 2009). In this way, metaphors represent powerful generative practices. In *Poetics 21*, Aristotle defined *metaphora* as a "carrying over" from one thing to another, with *phor* meaning "carrying" and *meta* meaning "beyond" (Kirby, 1997). Whereas an analogy finds similarities between two different things, a metaphor "consists in giving the things a name that belongs to something else" (Aristotle, cited by McKeon (1941, p. 1476)).

Metaphors are valuable to theorizing for their ability not only to transfer meaning but also to highlight, clarify, enrich, and enlighten (Ortony, 1979). Therefore, the origin of the metaphor is usually elegant, beautiful, and impressive (Kirby, 1997). The metaphor harnesses an entire network of analogies to accomplish its task. For example, when computer scientists use the metaphor of the brain to describe the computer's central processing unit (CPU), they quickly transfer well-known functions of the brain to explain something often unfamiliar to the public—computer processing. Aristotle suggests that the more dissimilar the objects are where analogies are found, the more powerful is the metaphor:

> The observation of likeness (homoiou theoria) is useful with a view both to inductive arguments and to hypothetical deductions, and also with a view to the production of definitions. (Aristotle, translation cited in Kirby, 1997, p. 536)

Only a handful of IS studies demonstrate extensive metaphorizing that leads to inductive arguments, hypothetical deductions, or production of definitions. R. M. Mason (1991) proposed organismic, sports team, and city-state metaphors for IS strategic planning, offering alternatives to the war metaphor that dominated strategic thinking at the time. The area of IS development attracted most of the work that applied metaphors. Some studies used the metaphor of magic, as it is applied to generally accepted practices in IS development (R. A. Hirschheim & Newman, 1991; Kaarst-Brown & Robey, 1999), to theorize about the social nature of IS development and how it impacts a project's probability of success, while others described how useful metaphors can be when communicating with users during the systems development life cycle (Kendall & Kendall, 1993) or in persuading users to support the integration of two different systems (Oates & Fitzgerald, 2007).

Metaphors are not only useful for drawing similarities, but also for distinguishing differences and highlighting incompatibilities. Carr (2003) applies the metaphor of household utilities to argue that because IT has become for all intents and purposes as common as electricity and plumbing, it can no longer support the goals of achieving competitive advantage. Brynjolfsson, Hofmann, and Jordan (2010) apply the same metaphor of utilities to arrive at the opposite conclusion. They argue that IT, unlike utilities, is scalable and incorporates digital innovations and complementary services; that electricity- and plumbing-like utilities do not offer these services; and thus, IT supports efforts towards achieving competitive advantage. In theorizing, metaphors cut through difficult and complex issues and enable the researcher to view those issues in a more familiar light.

Mythologizing

Mythologizing involves using myths, mythologies, and hidden assumptions to provide or interrogate a means of explanation, as well as to study symbols of value, coherence, unity, social structure, conflict, and contradictions (P. S. Cohen, 1969; R. A. Hirschheim & Newman, 1991; Mousavidin & Goel, 2007). A *myth* is:

> A dramatic narrative of imagined events, usually used to explain origins or transformations of something ... an unquestioned belief about the practical benefits of certain techniques and behaviors that is not supported by demonstrated facts. (Trice & Beyer, 1984, p. 655)

Although myths are frequently referred to as mistaken beliefs or popular misconceptions, they can address unquestioned assumptions within existing belief systems and theories. Lévi-Strauss (1963, 1966) viewed myths as parallels to science, especially in the science of relations, while Cassirer developed a theory of symbolic forms inspired by his study of myths (Bidney, 1955; Cassirer & Verene, 1979). As such, myths provide a means of explanation; a language for studying symbols of value, solidarity, and social structure; and a means of managing contradictions (P. S. Cohen, 1969).

To illustrate the use of myths Daft and Lengel's (1983)and Lengel's (1983) MRT studies can be traced back to certain myths concerning what managers do and how management was assumed to consist of the essential activities of planning, organizing, coordinating, and controlling. Mintzberg (1972, 1973) debunked this myth and found instead that managers rarely plan; rather, they spend more than 70% of their time in verbal communication and act spontaneously on trigger information. Daft and Lengel's MRT studies began as a result of interrogating this myth. The notion of the "total information system" propagated in the 1960s, which was thought to enable planning, organizing, coordinating, and controlling, leveraged such a myth. The earliest critics of MIS invoked the "myth of real-time systems" (Dearden, 1966) to expose several fallacies regarding the assumed capabilities of computers to support management functions. Mintzberg (1972) observed that because managers rely on informal communication channels—which often carry gossip, hearsay, and speculative information—the information provided by formal IS will be at odds with a manager's information requirements.

Because myths are often viewed pejoratively, this negative view of myths occupies most of early theorizing of myths in IS. Boland (1987) described five universally claimed myths, which he called "fantasies," about information that might distort the progress of research in IS. R. A. Hirschheim and Newman (1991) identified six common myths in IS development, such as the overriding advantage of user involvement, the need to ameliorate user resistance, and the necessity of system integration. In artificial intelligence, myths make up much of its hype and Roszak (1994) meticulously uncovers the layers of fabulous myths and claims in support of the "information age" or "information economy" to enhance quality of life, when in reality, these claims carry with them an equal, if not a disproportionate, weight of threats to human and societal well-being.

What is yet to be developed in the IS field is viewing myths in their positive sense, which Lévi-Strauss (1966) calls "mythical reflection" (p. 17). This form of theorizing on the intellectual plane is similar to *bricolage* on the technical plane. In its positive sense, mythologizing involves the *bricoleur* going beyond standard tools, methods, or data to using "devious means" from "whatever is at hand" (p. 16) to take advantage of the heterogeneous repertoire that is available. As Lévi-Strauss (1955, 1966) argues, myths and their derivatives, rituals, are universally found to be extremely organized, ordered, and precise, thus suggesting that they contain scientific parallels that are yet to be discovered, and it is these scientific parallels that the process of bricolage targets. The scientist as *bricoleur* is constantly on the lookout for images and signs from their phenomena of interest.

In the IS field, Claudio Ciborra was among the few who highlighted the significance of bricolage as a theorizing practice (Avgerou, Lanzara, & Willcocks, 2009; Lanzara, 2009). Realizing that established IS strategies were becoming increasingly ineffective for studying rapidly changing technological environments, Ciborra argued that much of the IT innovation taking place in current volatile marketplaces did not come from methodically evaluating the industry by evaluating threats, opportunities, threats, and weaknesses; identifying key success factors and distinctive competencies; and selecting the optimal strategy. Rather, innovation came from opportunistically adapting to highly unpredictable environments and continuously learning from direct experience. Ciborra (1992) refers to what essentially is bricolage as "tinkering," which involves extracting solutions "embedded in everyday experience and local knowledge" (pp. 301–302).

Going beyond experimenting and improvising, bricolage involves IS researchers essentially becoming mythologists, who, like Lévi-Strauss, are able to encode the structures of IS phenomena based merely on symbols and signs from their various myths. Robey and Markus' (1984) classic on rituals of IS design exemplifies this kind of work, as they uncover the political and symbolic activities that routinely take place as stakeholders of systems with conflicting interests compete for dominance within what might appear to be a rational process of systems design.

Modelling and Constructing the Framework

Modelling is a generative practice that builds different forms of models, including mechanical, mathematical, computational, graphical, and narrative models. A model is often confused with framework and theory. Also referred to as *analogues* (M. B. Hesse, 1966), models apply analogies to build precise and economical representations of selected elements and relationships to produce and examine the phenomenon of interest. Emphasizing the importance of models for theorizing, Suppe (2000, p. S110) stated that "models are the heart of scientific experimentation, observation, instrumentation, and experimental design." Using notions of positive analogies (common properties between two different objects), negative analogies (properties that differ between objects), and neutral analogies (uncertain as to whether positive or negative analogies exist) (M. B. Hesse, 1966), a *model* can be defined as an imperfect copy of the phenomenon of interest, consisting of positive and neutral analogies. Nobel laureate Thomas Schelling (1978a, p. 87) defines the model as the "precise and economical statement of a set of relationships that are sufficient to produce the phenomenon in question" or the "actual biological, mechanical, or social system that embodies the relationships in an especially transparent way."

Models are useful for building theories because they reveal the consequences of making certain assumptions and including or excluding certain elements in an economical way (Swedberg, 2014b). For example, William Gilbert (1893/1600) applied the model of the earth as a magnet with the poles as the ends of that magnet to explain why compasses point north. In economics, Schelling (1978b) applied the model of the thermostat to explain the reasons and the social mechanisms behind the return of measles to the United States after its elimination in the 1960s. As these examples show, models are not theories but simplifications of the phenomenon of interest that offer limited explanation and may serve as part of a theory. Also, models are not frameworks; models may become part of a framework that represents a map of the elements of the research process and helps researchers to assess and refine goals, develop questions, select appropriate methods and models, and identify potential validity threats.

M. B. Hesse (1966) categorizes the process of modelling into two approaches: Continental modelling, which is the more abstract, logical, and systematic approach, and English modelling, which is the more visual, imaginative, and intuitive approach. Researchers in IS are most familiar with the former, which often takes the shape of box-arrow diagrams that depict causal or associative relationships. In IS and other fields, these models are called "conceptual models" and are mentioned synonymously with "conceptual frameworks." Jaccard and Jacoby (2010) include various models as part of their discussion of theory construction. In nursing theory, for example, models are treated as part of a theory's a priori frame of reference that defines what questions will be asked, guides the generation of new theories, focuses the researcher on specific problems, and facilitates the selection of methods for the discovery of new theories (Fawcett, 1995, 1998). For more mathematically inclined disciplines, the ultimate goal of this systematic and logical form of generative practice is a mathematical system with a deductive structure that succinctly explains the phenomenon of interest during theorizing (M. B. Hesse, 1966). Notwithstanding its systematicity, formal box-arrow diagram modelling can be counterproductive because it (1) bypasses many other forms such as mechanical, computational, narrative, and alternative graphical models that could bring insights into the research, and (2) encourages arbitrary extensions in which "splitting of concepts and their endless rearrangement becomes the central endeavor" (Mills, 1959, p. 23).

The IS field is replete with models that use Continental modelling, and these models are often loosely referred to as theories, obscuring the theorizing process. The technology acceptance model (TAM) and the DeLone and McLean IS success model are claimed to be the two most applied IS theories (Moody et al., 2010; Straub, 2012) even though both are labelled and depicted as models. Problems also emerge when theories are uncritically imported from another field into the IS field because these outside theories may be based on different models. For example, two popular theories in the social sciences, the diffusion of innovations theory (DIT) (Rogers, 1983) and the theory of reasoned action (TRA) (Fishbein & Ajzen, 1977), are among the two most applied theories in assessing the influence of IT on individuals (Lim, Saldanha, Malladi, & Melville, 2009). These theories describe two different models of innovation. The DIT originates in the communications field and models innovation in terms of the flow of information. Consequently, flow-related analogies, such as channels that carry information, the time taken for the rate of adoption, stages of adoption, and the social system engaging in the flow, provide a rich set of concepts and constructs to be researched. The TRA is a theory of behaviour predicated on the individual's behavioural intention, which, in turn, is affected by the individual's attitude. Because the DIT includes a time element, it can describe the logistic curve of innovation, which is not possible when using the TRA. Conversely, the TRA's focus on attitude is only tangentially addressed by the DIT. Being aware of the model underlying the research is critical during IS theorizing and models are often ceremoniously introduced into research before enough modelling is undertaken.

Following a formal modelling practice, TAM researchers compare eight conceptual models to assess their relative utility for theorizing adoption. Based on this comparison, they suggested the UTAUT version of TAM incorporating ten constructs consisting of four predictors, four moderators, one mediator, and one dependent variable chosen from eight models from social psychology: the original Fishbein's behavioural intention model (1975), Ajzen's (1991) planned behavioural model, the intrinsic-extrinsic motivation model (Vallerand, 1997), Triandis' (1971) attitude change model, Rogers' (1983) innovation diffusion theory model, Bandura's (1982) social cognitive model, the gender differences model (Bem, 1981; Helmreich, Spence, & Wilhelm, 1981), and the age differences model (Hall & Mansfield, 1975). Unpacking this complex web of models makes it difficult to explain the results from applying the unified model.

The more intuitive English modelling practice rejects the view that models are mere aids to theorizing that can be disposed of when the theory is formulated. This approach presupposes the mutability of theories, which, as they are extended and modified to account for new phenomena, are not divorced from the analogies that were originally used to build them. Instead of being mere aids to theorizing, analogies are an essential part of theories without which the theories would lose their value. While the formal modelling practice allows any model to be attached to a working theory, the intuitive model depends on the analogy that builds the theory. In his classic text, *Micromotives and Macrobehavior*, Nobel laureate Thomas Schelling (1978b) emphasized the usefulness of intuitive modelling following the English approach in reproducing the essential features of complex behavioural systems. Using the home thermostat as analogy, Schelling explained in detail how the spread of disease follows upswings and downswings in a cyclical way similar to how a thermostat mechanism reaches a tipping point and the temperature keeps rising even when someone attempts to lower the temperature. This process of modelling affords the researcher several advantages during the context of discovery. The model helps visualize social epidemiology as a cyclical process that involves a tipping point, an overlapping phenomenon that a box-arrow diagram may not be able to elucidate. As such, the researcher can identify several key concepts that are critical to the process of containing the epidemic. There are few English modelling practices within the IS field. Kirsch's (1996, 1997) work on controlling and managing complex systems development processes includes merging the Ouchi's control model (another example of a English model consisting of behavioural and outcome measures) with agency theory to build an intuitive model that comprises of behaviour, outcome, clan, and self-control methods.

Towards General Principles for Theorizing

In addition to discussing the use of specific discursive practices, and how these practices organize different forms of enunciations to produce and manipulate concepts and claims about the phenomena of interest, we offer general principles that apply to all discursive practices in the context of theorizing. These general principles start with where theorizing begins, how theories can serve as inspiration, how to establish relationships with disciplines outside the IS field, and how to evaluate the success of theorizing. Most of these general principles follow from Peirce's (1893–1913/1931–1958) view of how to theorize (Swedberg, 2014a).

Starting in the Context of Discovery

Theorizing does not begin when hypotheses or propositions are considered or when they are tested or validated. As the superset to reasoning, theorizing is as natural to human beings as thinking, and it is this rich natural capability endowed in all human beings that characterizes the activities within the context of discovery well before any claims are considered. Unfortunately, graduate research training may have preconditioned many researchers to not theorize as freely as we should when we encounter an interesting phenomenon. Instead of focusing on the context of discovery by taking advantage of all plausible avenues to explain the phenomenon, research practice tends to limit thinking to the sanitized, rationalized reconstructions found in published works. In addition to digging deeper into the insights and creative thinking that characterize the context of discovery, researchers must use the natural human capabilities they are endowed with and have a certain level of willingness to question and even forget previous thinking to engage in original research. As Whitehead (1917, p. 115) observes: "A science which hesitates to forget its founders is lost."

Deriving Inspiration from Other Theories

Theories can be inspired, borrowed, or adapted from other disciplines. Because a theory is a strategic choice taken within a discourse, theories are tethered to the discourse and ultimately to the associated discipline. Thus, when theories are inspired, borrowed, or adapted from other disciplines, they carry with them the same rules of discourse by which they were constituted. Theories are bounded to the discursive formation of their discipline. These rules of discourse are consistent with what Truex et al. (2006) describe as the "underlying notions" and "methodological implications" (p. 798) of those theories, as well as the need to "be inculcated into the internal logic and intellectual tradition associated with the theory" (p. 801). Theories in IS, they argue, cannot simply be uncritically borrowed from other disciplines, and any borrowing and adapting, they advise, must be undertaken in a more "reflexive manner" (p. 799). By applying the various foundational and generative theorizing practices, theorizing naturally takes place in a reflexive way because doing so requires a deeper examination of the rules of discourse, analogies, and other elements underlying those theories that serve as inspiration for theorizing.

The case of structuration theory in IS that was borrowed from Giddens (1976, 1984) exemplifies this need. For example, despite efforts in the form of the adaptive structuration theory (AST) (DeSanctis & Poole, 1994) to address issues that structuration theory presented (Jones, 1999; Rose, Jones, & Truex, 2005), the rules of discourse that operated in structuration theory remained in operation in the proposed theory. This resulted in a disproportionate application of structuration theory itself and a lack of overall coherence and cumulative development in studies that apply structuration (Jones & Karsten, 2008; Orlikowski, 2000). A close study of the rules of the discourse of those theories provides a more consistent approach towards theorizing.

Connecting to Other Discourses

Within a discipline, theorizing does not take place in isolation. Theories are developed within the larger discursive constellation in which the discipline belongs. Concepts and claims that are invented within the field should not be inconsistent with well-known concepts and claims outside the field. Theories in IS therefore cannot be developed or operate independently of other disciplines; rather, they must demonstrate a coherent relationship with others. In addition to establishing a certain level of coherency with other disciplines, this close connection establishes the IS field's relevance to the stock of knowledge of the world and helps make the field intellectually influential.

For example, when the IS field theorizes about "technology" or about "artefacts," any inconsistencies with or divergence from what is well-known about technology or artefacts within the larger discourse in allied fields, or generally accepted definitions, need to be clearly justified and made clear to all. When Weick (1979) redefined the concept of organizations, which are typically interpreted as static, bounded entities, to the concept of organizing, he not only switched a noun into a verb, his work built on and connected with other discourses in the larger constellation of discourses. His theorizing released management theory from the boundaries of the static organization and connected to other discourses including among others, the discourse of ethnomethodology from Garfinkel (1967), Merton's (1948) self-fulfilling prophecy, Campbell's (1965a, 1965b) sociocultural evolution, and Simon's (1957) and Allport's (1962) studies in social psychology.

Evaluating Success in Theorizing

The criterion for evaluating success in theorizing lies in the value of the concepts and claims that the theorizing process produces. In theorizing, concepts and claims in existing theories are not merely reorganized but are often reconstituted and given a new meaning, and, if theorizing is especially productive, new concepts are invented. Theories might not last the test of time, but often concepts and claims are what spawn new creative endeavours. For instance, although older biological theories are replaced by more recent ones, the concept of "organic structure," originally coined from biology (Cuvier, 1800–1805), remains useful as it was redefined in the context of social psychology (Spencer, 1897) and was later made famous by management studies to explain how organizations innovate (Burns & Stalker, 1961). This process of inventing new concepts and crafting new theories is what qualifies a field of study for becoming a discipline, following what Foucault (1972, pp. 186–188) describes as the threshold of positivity and the threshold of epistemology. The threshold of positivity occurs when the field of study starts applying its own set of rules for producing original mutually exclusive concepts, while the threshold of epistemology is reached when the field of study becomes coherent, ordered, is accepted as legitimate by others, and ultimately produces theories that exert an influence on the stock of knowledge in the world (Nik R. Hassan, 2011). The success of theorizing is reflected in how it either carves out its own space within existing knowledge or builds something novel over and above that knowledge. "A theory must somehow fit God's world, but in an important sense it creates a world of its own" (Kaplan, 1964, pp. 308–309).

We summarize the discursive theorizing practices in Table 5.1 showing how various elements of theorizing are marshalled within the context of discovery. The activity of forming the IS discourse comprises the entire list of discursive practices, the related major elements of theorizing, as well as key principles associated with practising each element. Although the focus is on the context of discovery, each of the discursive practices have implications for the context of justification and it is very likely that activities in the context of justification may inform theorizing practices in the context of discovery. Similarly, the discursive practices typically interact during IS theorization. For example, the construction of the framework may raise further questions related to the phenomenon and suggest further disciplinary questions. Each theorizing practice may lead to any number of other theorizing practices.

Table 5.1 Summary of discursive theorizing practices

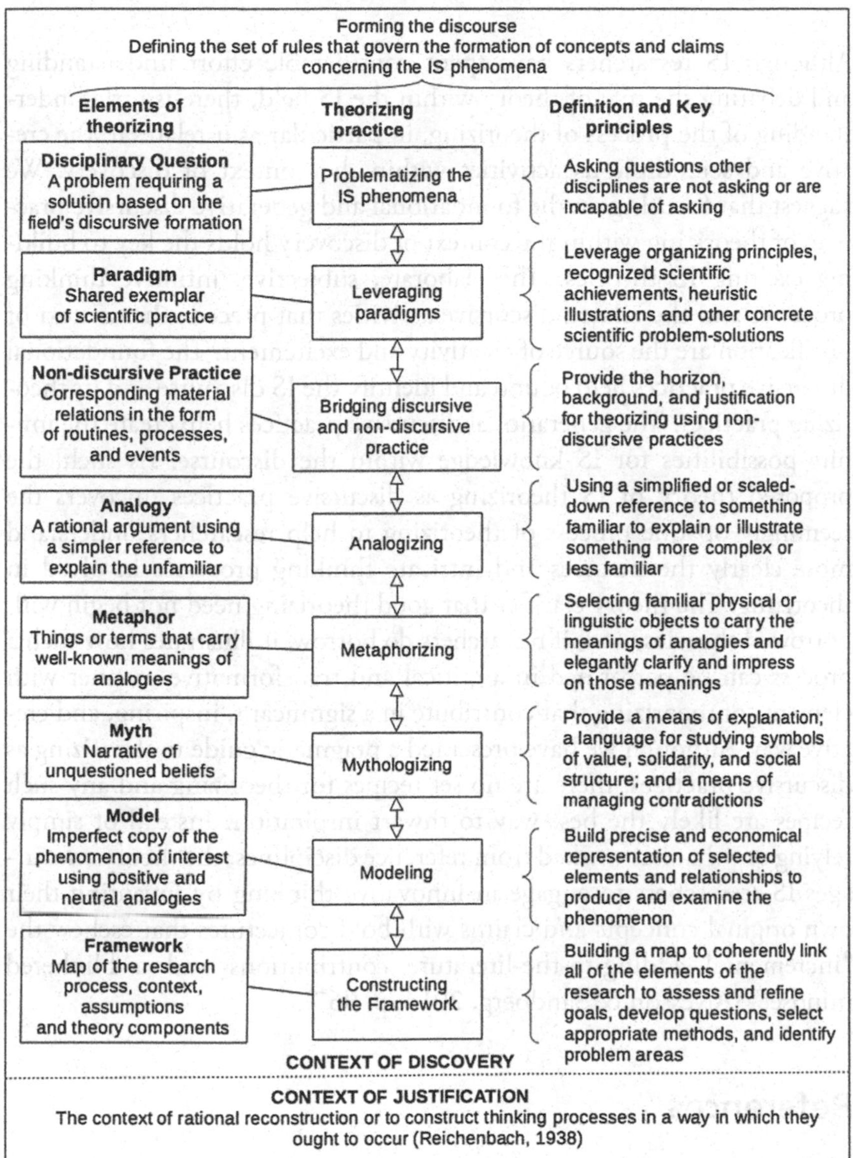

Forming the discourse
Defining the set of rules that govern the formation of concepts and claims concerning the IS phenomena

Elements of theorizing	Theorizing practice	Definition and Key principles
Disciplinary Question A problem requiring a solution based on the field's discursive formation	Problematizing the IS phenomena	Asking questions other disciplines are not asking or are incapable of asking
Paradigm Shared exemplar of scientific practice	Leveraging paradigms	Leverage organizing principles, recognized scientific achievements, heuristic illustrations and other concrete scientific problem-solutions
Non-discursive Practice Corresponding material relations in the form of routines, processes, and events	Bridging discursive and non-discursive practice	Provide the horizon, background, and justification for theorizing using non-discursive practices
Analogy A rational argument using a simpler reference to explain the unfamiliar	Analogizing	Using a simplified or scaled-down reference to something familiar to explain or illustrate something more complex or less familiar
Metaphor Things or terms that carry well-known meanings of analogies	Metaphorizing	Selecting familiar physical or linguistic objects to carry the meanings of analogies and elegantly clarify and impress on those meanings
Myth Narrative of unquestioned beliefs	Mythologizing	Provide a means of explanation; a language for studying symbols of value, solidarity, and social structure; and a means of managing contradictions
Model Imperfect copy of the phenomenon of interest using positive and neutral analogies	Modeling	Build precise and economical representation of selected elements and relationships to produce and examine the phenomenon
Framework Map of the research process, context, assumptions and theory components	Constructing the Framework	Building a map to coherently link all of the elements of the research to assess and refine goals, develop questions, select appropriate methods, and identify problem areas

CONTEXT OF DISCOVERY

CONTEXT OF JUSTIFICATION
The context of rational reconstruction or to construct thinking processes in a way in which they ought to occur (Reichenbach, 1938)

Conclusion

Although IS researchers have spent considerable effort understanding and debating the role of theory within the IS field, there is little understanding of the process of theorizing, in particular as it relates to the creative and serendipitous activities within the context of discovery. We suggest that focusing on the foundational and generative discursive practices of theorizing within the context of discovery holds the key to building exciting IS theories. The elaborate, subjective, intuitive thinking processes and the related discursive activities that precede the context of justification are the source of creativity and excitement. The foundational discursive practices help bound and identify the IS discourse and its theorizing practices. The generational discursive practices help create the infinite possibilities for IS knowledge within the discourse. As such, the proposed theory of IS theorizing as discursive practices uncovers the seemingly opaque process of theorizing to help researchers understand more clearly the nuances and intricate thinking processes involved in theorizing. The theory clarifies that good theorizing need not begin with borrowed theories; and, if researchers do borrow, it illustrates how such a process can be performed in a critical and transformative manner with elements of theorizing that contribute in a significant, inspiring, and creative way. Although we have presented a pragmatic guide to theorizing as discursive practices, there are no set recipes for theorizing and any such recipes are likely the best way to thwart inspiration. Instead of simply relying on what has worked from reference disciplines, the theory encourages IS researchers to engage in innovative thinking by inventing their own original concepts and claims with bold conjectures that eschew the "incremental adding-to-the-literature contributions and a blinkered mindset" (Alvesson & Sandberg, 2014, p. 967).

References

Adam, F., & Fitzgerald, B. (2000). The status of the IS field: Historical perspective and practical orientation. *Information Research, 5*(4), 5–14.

Adler, J. E., & Rips, L. J. (Eds.). (2008). *Reasoning: Studies of human inference and its foundations*. Cambridge, UK: Cambridge University Press.

Agarwal, R., & Lucas Jr., H. C. (2005). The information systems identity crisis: focusing on high-visibility and high-impact research. *MIS Quarterly, 29*(3), 381–398.

Ågerfalk, P. J. (2014). Insufficient theoretical contribution: A conclusive rationale for rejection? *European Journal of Information Systems, 23*(6), 593–599.

Ajzen, I. (1991). The theory of planned behavior. *Organizational Behavior and Human Decision Processes, 50*(2), 179–211.

Ajzen, I., & Fishbein, M. (1972). The prediction of behavior from attitudinal and normative variables. *Journal of Experimental Social Psychology, 6*(4), 466–487.

Ajzen, I., & Fishbein, M. (1973). Attitudinal and normative variables as predictors of specific behaviors. *Journal of Personality and Social Psychology, 27*(1), 41–57.

Ajzen, I., & Fishbein, M. (1980). *Understanding attitudes and predicting social behavior*. Englewood Cliffs, NJ: Prentice-Hall.

Allport, F. H. (1962). A structuronomic conception of behavior: Individual and collective: I. Structural theory and the master problem of social psychology. *The Journal of Abnormal and Social Psychology, 64*(1), 3–30.

Alter, S. L. (1977). A taxonomy of decision support systems. *Sloan Management Review, 19*(1), 39–56.

Alter, S. L. (1980). *Decision support systems: Current practice and continuing challenge*. Reading, MA: Addison-Wesley.

Alvesson, M., & Karreman, D. (2000). Varieties of discourse: On the study of organizations through discourse analysis. *Human Relations, 53*(9), 1125–1149.

Alvesson, M., & Sandberg, J. (2013). *Constructing research questions: Doing interesting research*. Thousand Oaks, CA: SAGE Publications.

Alvesson, M., & Sandberg, J. (2014). Habitat and Habitus: Boxed-in versus Box-breaking research. *Organization Studies, 35*(7), 967–987.

Angst, C. M., Agarwal, R., Sambamurthy, V., & Kelley, K. (2010). Social contagion and information technology diffusion: The adoption of electronic medical records in US hospitals. *Management Science, 56*(8), 1219–1241.

Applegate, L. M., Konsynski, B. R., & Nunamaker, J. F. (1986). *A group decision support system for idea generation and issue analysis in organization planning*.

Paper presented at the Proceedings of the 1986 ACM Conference on Computer-supported Cooperative Work.

Aristotle. (1934). *Nicomachean Ethics* (H. Rackham, Trans. Vol. 19). London: William Heinemann Ltd.

Avgerou, C., Lanzara, G. F., & Willcocks, L. P. (Eds.). (2009). *Bricolage, care and information: Claudio Ciborra's legacy in information systems research*. Basingstoke, UK: Palgrave Macmillan.

Avison, D. (1997). The 'Discipline' of information systems: Teaching, research and practice. In J. Mingers & F. Stowell (Eds.), *Information systems: An emerging discipline?* (pp. 113–139). London: McGraw-Hill.

Avison, D., & Malaurent, J. (2014). Is theory king?: Questioning the theory fetish in information systems. *Journal of Information Technology, 29*(4), 327–336.

Bacchi, C., & Bonham, J. (2014). Reclaiming discursive practices as an analytic focus: Political implications. *Foucault Studies, 17*, 179–192.

Bacharach, S. B. (1989). Organizational theories: Some criteria for evaluation. *Academy of Management Review, 14*(4), 496–515.

Bagnall, J. (2012). What's the difference between analogy, metaphor and simile. Retrieved from https://www.quora.com/Whats-the-difference-between-analogy-metaphor-and-simile

Bal, M. (2002). *Travelling concepts in the humanities: A rough guide*. Toronto, CA: University of Toronto Press.

Bandura, A. (1982). Self-efficacy mechanism in human agency. *American Psychologist, 37*(2), 122–147.

Banville, C., & Landry, M. (1989). Can the field of MIS be disciplined? *Communications of the ACM, 32*(1), 48–60.

Becker, J., vom Brocke, J., Heddier, M., & Seidel, S. (2015). In search of information systems (grand) challenges: A community of inquirers perspective. *Business Information Systems Engineering, 57*(6), 377–390.

Beer, S. (1972). *Brain of the firm*. London, UK: The Penguin Press.

Beer, S. (1979). *The heart of enterprise*. Chichester, UK: Wiley and Sons.

Bem, S. (1981). Gender schema theory: A cognitive account of sex typing. *Psychological Review, 88*(4), 354–364.

Benbasat, I., & Barki, H. (2007). Quo vadis TAM? *Journal of the Association for Information Systems, 8*(4), 211–218.

Berger, P. L., & Luckmann, T. (1966). *The social construction of reality*. New York: Anchor Books.

Berlin, I. (1999). *Concepts and categories*. London: Pimlico.

Bichler, M., Frank, U., Avison, D., Malaurent, J., Fettke, P., Hovorka, D., et al. (2016). Theories in business and information systems engineering. *Business Information Systems Engineering, 58*(4), 291–319.

Bidney, D. (1955). Myth, symbolism, and truth. *Journal of American Folklore, 68*(270), 379–392.

Bijker, W. E. (1995). *Of bicycles, Bakelites, and Bulbs: Toward a theory of socio-technical change*. Cambridge, MA: MIT Press.

Bijker, W. E., Hughes, T. P., & Pinch, T. J. (1987). *The social construction of technological systems*. Cambridge, MA: MIT Press.

Blalock, H. M. (1969). *Theory construction: From verbal to mathematical formulations*. Englewood Cliffs, NJ: Prentice-Hall.

Bodensteiner, W. D. (1970). *Information channel utilization under varying research and development project conditions: An aspect of inter-organizational communication channel usages*. PhD Dissertation, The University of Texas, Austin, Austin, TX.

Boell, S. K. (2017). Information: Fundamental positions and their implications for information systems research, education and practice. *Information and Organization, 27*(1), 1–17.

Boland, R. J. (1987). The in-formation of information systems. In R. J. Boland & R. A. Hirschheim (Eds.), *Critical issues in information systems research* (pp. 363–379). Chichester: John Wiley & Sons.

Bourdieu, P. (1977). *Outline of a theory of practice*. Cambridge, UK: Cambridge University Press.

Brancheau, J. C., & Wetherbe, J. C. (1987). Key issues in information systems management. *MIS Quarterly, 11*, 23–45.

Bromberger, S. (1992). *On what we know we don't know: Explanation, theory, linguistics, and how questions shape them*. Chicago: University of Chicago Press.

Brynjolfsson, E., Hofmann, P., & Jordan, J. (2010). Cloud computing and electricity: Beyond the utility model. *Communications of the ACM, 53*(5), 32–34.

Burns, T., & Stalker, G. M. (1961). *The management of innovation*. London: Tavistock Institute.

Burton-Jones, A., McLean, E. R., & Monod, E. (2014). Theoretical perspectives in IS research: From variance and process to conceptual latitude and conceptual fit. *European Journal of Information Systems, 24*(6), 664–679.

Byron, K., & Thatcher, S. M. B. (2016). Editors' comments: "What I know now that I wish I knew then" – Teaching theory and theory building. *Academy of Management Review, 41*(1), 1–8.

Caminer, D. T. (Ed.). (1997). *LEO: The incredible story of the world's first business computer*. New York, NY: McGraw-Hill.

Campbell, D. T. (1965a). Ethnocentric and other altruistic motives. In D. Levine (Ed.), *Nebraska symposium on motivation* (pp. 283–311). Lincoln, NE: University of Nebraska Press.

Campbell, D. T. (1965b). Variation and selective retention in socio-cultural evolution. In H. R. Barringer, G. I. Blanksten, & R. Mack (Eds.), *Social change in developing areas* (pp. 19–49). Cambridge, MA: Schenkman.

Campbell, N. R. (1920). *Foundations of science: The philosophy of theory and experiment*. New York, NY: Dover Publications, Inc..

Carr, N. G. (2003). IT doesn't matter. *Harvard Business Review, 81*(5), 41–49.

Cassirer, E., & Verene, D. P. (1979). *Symbol, myth, and culture: Essays and lectures of Ernst Cassirer, 1935–1945*. New Haven: Yale University Press.

Chamberlin, T. C. (1890). The method of multiple working hypotheses. *Science, 15*(366), 92–96.

Chen, W., & Hirschheim, R. (2004). A paradigmatic and methodological examination of information systems research from 1991 to 2001. *Information Systems Journal, 14*(3), 197–235.

Ciborra, C. (1992). From thinking to tinkering: the grassroots of strategic information systems. *Information Society, 8*(4), 297–309.

Cohen, P. S. (1969). Theories of myth. *Man, 4*(3), 337–353.

Cohen, T. (2013). Rough Obamacare rollout: 4 reasons why. Retrieved from http://www.cnn.com/2013/10/22/politics/obamacare-website-four-reasons/

Collins, H. M., & Pinch, T. J. (1982). *Frames of meaning: The social construction of extraordinary science*. London: Routledge and Kegan Paul.

Comte, A. M. (1830–42). *The positive philosophy of Auguste Comte translated by Harriet Martineau*. Chicago: Belford, Clarke & Co.

Constant, E. W. (1980). *The origins of the Turbojet revolution*. Baltimore: Johns Hopkins University.

Corley, K. G., & Gioia, D. A. (2011). Building theory about theory building: What constitutes a theoretical contribution. *Academy of Management Review, 36*(1), 12–32.

Cornelissen, J. P., & Durand, R. (2014). Moving forward: Developing theoretical contributions in management studies. *Journal of Management Studies, 51*(6), 845–1023.

Corvellec, H. (2013). Why ask what theory is? In H. Corvellec (Ed.), *What is theory? Answers from the social and cultural sciences* (pp. 9–24). Copenhagen: Liber CBS Press.

Cushing, B. E. (1990). Frameworks, paradigms, and scientific research in management information systems. *Journal of Information Systems, 5*(1), 38–59.

Cuvier, G. (1800–1805). *Leçons d'anatomie comparée (Lessons of comparative anatomy)*. Paris: Baudouin.

Daft, R. L., & Lengel, R. H. (1983). *Information richness: A new approach to managerial behavior and organization design* (TR-ONR-DG-02). Retrieved from www.dtic.mil/dtic/tr/fulltext/u2/a128980.pdf

Daft, R. L., & Lengel, R. H. (1986). Organizational information requirements, media richness and structural design. *Management Science, 32*(5), 554–571.

Daft, R. L., Lengel, R. H., & Trevino, L. K. (1987). Message equivocality, media selection, and manager performance: Implications for Information Systems. *MIS Quarterly, 11*(3), 355–366.

Darwin, C. (1859). *On the origin of species*. London: John Murray.

Davenport, T. H., & Patil, D. J. (2012). Data scientist: The sexiest job of the 21st century. *Harvard Business Review, 90*(10), 70–76.

Davis, F. D. (1989). Perceived usefulness, perceived ease of use, and user acceptance of information technology. *MIS Quarterly, 13*(3), 318–340.

Davis, F. D., Jr. (1986). *A technology acceptance model for empirically testing new end-user information systems: theory and results*. PhD Dissertation, Massachusetts Institute of Technology, Cambridge, MA.

Dearden, J. (1966). Myth of real-time management information. *Harvard Business Review, 44*(3), 123–132.

Denning, P. J. (1999). *Computing the profession*. Paper presented at the Proceedings of the 30th Technical Symposium on Computer Science Education (SIGCSE), New Orleans, LA, March 24–28.

Dennis, A. R., George, J. F., Jessup, L. M., Nunamaker Jr., J. F., & Vogel, D. R. (1988). Information technology to support electronic meetings. *MIS Quarterly, 12*(4), 591–624.

DeSanctis, G., & Poole, M. S. (1994). Capturing the complexity in advanced technology use – Adaptive structuration theory. *Organization Science, 5*(2), 121–147.

Dickson, G. W., Senn, J. A., & Chervany, N. L. (1977). Research in management information systems: The Minnesota experiments. *Management Science, 23*(9), 913–923.

Dreyfus, H. L., & Rabinow, P. (1983). *Michel Foucault: Beyond structuralism and hermeneutics*. Chicago, IL: University of Chicago Press.

Dubin, R. (1969). *Building theory*. New York: The Free Press.

Durkheim, É. (1951/1897). *On suicide: A study in sociology*. New York, NY: Free Press.

Ein-Dor, P., & Segev, E. (1981). *A paradigm for management information systems*. New York: Praeger Publishers.

Eisenhardt, K. M. (1989). Building theories from case study research. *Academy of Management Review, 14*(4), 532–551.

Eisenhardt, K. M., & Graebner, M. E. (2007). Theory building from cases: Opportunities and challenges. *Academy of Management Journal, 50*(1), 25–32.

Ellis, D., Allen, D., & Wilson, T. (1999). Information science and information systems: Conjunct subjects disjunct disciplines. *Journal of the American Society for Information Science, 50*(12), 1095–1107.

Ellul, J. (1973). *The technological society*. New York: Alfred A. Knopf.

Fawcett, J. (1995). *Analysis and evaluation of conceptual models of nursing* (3rd ed.). Philadelphia, PA: F. A. Davis Company.

Fawcett, J. (1998). *The relationship of theory and research*. Philadelphia: F. A. Davis Company.

Feenberg, A. (1991). *The critical theory of technology*. New York: Oxford University Press.

Ferry, G. (2003). *A computer called LEO: Lyons teashops and the world's first office computer*. London, UK: Fourth Estate.

Fischer, D. H. (1970). *Historian's Fallacies: Toward a logic of historical thought*. New York: Harper Perennial.

Fishbein, M., & Ajzen, I. (1974). Attitudes towards objects as predictors of single and multiple behavioral criteria. *Psychological Review, 81*(1), 59–74.

Fishbein, M., & Ajzen, I. (1975). *Belief, attitude, intention and behavior*. Reading, MA: Addison-Wesley.

Fishbein, M., & Ajzen, I. (1977). *Belief, attitude, intention and behavior*. Reading, MA: Addison-Wesley.

Foucault, M. (1970). *The order of things: An archeology of the human sciences*. New York: Pantheon Books.

Foucault, M. (1972). *The archaeology of knowledge and the discourse on language*. (A. M. S. Smith, Trans. New York: Pantheon Books.

Freese, L. (1980). Formal theorizing. *Annual Review of Sociology, 6*, 187–212.

Galbraith, J. R. (1973). *Designing complex organizations*. Reading, MA: Addison-Wesley.

Galbraith, J. R. (1977). *Organization design*. Reading, MA: Addison-Wesley.

Garfinkel, H. (1967). *Studies in ethnomethodology*. New York: Prentice-Hall.

Gauch, H. G. (2003). *Scientific method in practice*. Cambridge, UK: Cambridge University Press.

Geary, J. (2009). Metaphorically speaking. Retrieved from https://www.ted.com/talks/james_geary_metaphorically_speaking?language=en

Gentner, D. (1983). Structure-mapping: A theoretical framework for analogy. *Cognitive Science, 1*, 155–170.

Gentner, D. (1989). Mechanisms of analogical reasoning. In S. Vosniadou & A. Ortony (Eds.), *Similarity and analogical reasoning* (pp. 199–241). New York: Cambridge University Press.

Gholami, R., Watson, R. T., Hasan, H., Molla, A., & Bjørn-Andersen, N. (2016). Information systems solutions for environmental sustainability: How can we do more? *Journal of the Association for Information Systems, 17*(8), 521–536.

Giddens, A. (1976). *New rules of sociological method: A positive critique of interpretive sociologies* (2nd ed.). New York: Basic Books.

Giddens, A. (1984). *The constitution of society: Outline of the theory of structuration*. Berkeley: University of California Press.

Gilbert, W. (1893/1600). *On the loadstone and magnetic bodies and on the great magnet the earth*. New York: John Wiley & Sons.

Gioia, D. A., & Pitre, E. (1990). Multiparadigm perspectives on theory building. *The Academy of Management Review, 15*(4), 584–602.

Glaser, B. G., & Strauss, A. L. (1967). *The discovery of grounded theory: Strategies for qualitative research*. New York, NY: Aldine de Gruyter.

Goffman, E. (1959). *The presentation of self in everyday life*. Garden City, NY: Doubleday.

Goles, T., & Hirschheim, R. (2000). The paradigm is dead, the paradigm is dead…long live the paradigm: the legacy of Burrell and Morgan. *Omega, 28*(3), 249–268.

Gorry, G. A., & Scott, M. S. (1971). A framework for management information systems. *Sloan Management Review, 13*(1), 55–70.

Gregor, S. (2006). The nature of theory in information systems. *MIS Quarterly, 30*(3), 611–642.

Gregor, S. (2014). Theory – Still king but needing a revolution! *Journal of Information Technology, 29*(4), 337–340.

Gregor, S. (2018). On theory. In R. Galliers & M.-K. Stein (Eds.), *The Routledge companion to management information systems* (pp. 57–72). New York: Routledge.

Gregor, S., & Jones, D. (2007). The anatomy of a design theory. *Journal of the Association for Information Systems, 8*(5), 312–335.

Gregory, R. H., & Van Horn, R. L. (1960). *Automatic data-processing systems: Principles and procedures*. San Francisco: Wadsworth Publishing Co., Inc.

Grover, V., & Lyytinen, K. (2015). New state of play in information systems research: The push to the edges. *MIS Quarterly, 39*(2), 271–296.

Grover, V., Lyytinen, K., Srinivasan, A., & Tan, B. C. Y. (2008). Contributing to rigorous and forward thinking explanatory theory. *Journal of the Association for Information Systems, 9*(2), 40–47.

Grover, V., Lyytinen, K., & Weber, R. (2012). *Panel on native IS theories.* Paper presented at the Special Interest Group on Philosophy and Epistemology in IS (SIGPHIL) Workshop on IS Theory: State of the Art, Orlando, FL, Dec 16–19.

Gutting, G. (Ed.). (1980). *Paradigms and revolutions: Applications and appraisals of Thomas Kuhn's philosophy of science.* Notre Dame, IN: University of Notre Dame Press.

Hall, D. T., & Mansfield, R. (1975). Relationships of age and seniority with career variables of engineers and scientists. *Journal of Applied Psychology, 60*(2), 201–210.

Hanson, N. R. (1958). *Patterns of discovery: An inquiry into the conceptual foundations of science.* London: Cambridge University Press.

Hassan, N. R. (2011). Is information systems a discipline? Foucauldian and Toulminian insights. *European Journal of Information Systems, 20*(4), 456–476.

Hassan, N. R. (2014a). Paradigm lost … paradigm gained: A hermeneutical rejoinder to Banville and Landry's 'Can the Field of MIS be Disciplined?'. *European Journal of Information Systems, 23*(6), 600–615.

Hassan, N. R. (2014b). Value of IS research: Is there a crisis? *Communications of the Association for Information Systems, 34*(Art 41), 801–816.

Hassan, N. R., & Mingers, J. C. (2018). Reinterpreting the Kuhnian paradigm in information systems. *Journal of the Association for Information Systems, 19*(7), 568–599.

Hassan, N. R., & Will, H. J. (2006). Synthesizing diversity and pluralism in information systems: Forging a unique disciplinary subject matter for the information systems field. *Communications of the Association for Information Systems, 17*(7), 152–180.

Heidegger, M. (1977). *The question concerning technology, and other essays.* New York: Harper & Row.

Helmreich, R. L., Spence, J. T., & Wilhelm, J. A. (1981). A psychometric analysis of the personal attributes questionnaire. *Sex Roles, 7*(11), 1097–1108.

Hesse, M. (1967). Models and analogy in science. In P. Edwards (Ed.), *The Encyclopedia of philosophy* (Vol. 5, pp. 354–359). New York: The Macmillan Co. & The Free Press.

Hesse, M. B. (1966). *Models and analogies in science*. Notre Dame, IN: University of Notre Dame Press.

Hirschheim, R., & Klein, H. K. (2012). A glorious and not-so-short history of the information systems field. *Journal of the Association for Information Systems, 13*(4), 188–235.

Hirschheim, R. A., & Newman, M. (1991). Symbolism and information systems development: Myth, metaphor and magic. *Information Systems Research, 2*(1), 29–62.

Hollinger, D. A. (1980). T. S. Kuhn's theory of science and its implications for history. In G. Gutting (Ed.), *Paradigms and revolutions: Applications and appraisals of Thomas Kuhn's philosophy of science* (pp. 195–222). Notre Dame, IN: University of Notre Dame Press.

Holmström, J., & Truex, D. (2011). Dropping your tools: Exploring when and how theories can serve as blinders in IS research. *Communications of the Association for Information Systems, 28*(1), 283–294. Article 219.

Iivari, J., Hirschheim, R., & Klein, H. K. (1998). A paradigmatic analysis contrasting information systems development approaches and methodologies. *Information Systems Research, 9*(2), 164–193.

Ives, B., & Olson, M. H. (1984). User involvement and MIS success. *Management Science, 30*(5), 586–603.

Jaccard, J., & Jacoby, J. (2010). *Theory construction and model-building skills: A practical guide for social scientists*. New York, NY: The Guilford Press.

James, W. (1890). *The principles of psychology*. New York: Henry Holt and Company.

Jenkins, R. V. (1975). *Images and enterprise: Technology and the American photographic industry, 1839 to 1925*. Baltimore: Johns Hopkins University Press.

Jones, M. (1997). It all depends what you mean by discipline. In J. Mingers & F. Stowell (Eds.), *Information systems: An emerging discipline?* (pp. 97–112). London: McGraw-Hill.

Jones, M. (1999). Structuration theory. In W. Currie & B. Galliers (Eds.), *Rethinking management information systems* (pp. 103–135). Oxford: Oxford University Press.

Jones, M., & Karsten, H. (2008). Giddens's structuration theory and information systems research. *MIS Quarterly, 32*(1), 127–157.

Kaarst-Brown, M. L., & Robey, D. (1999). More on myth, magic and metaphor: Cultural insights into the management of information technology in organizations. *Information Technology & People, 12*(2), 192–217.

Kahn, H. (1965). *On escalation: Metaphors and scenarios*. New York, NY: Praeger.

Kaplan, A. (1964). *The conduct of inquiry: Methodology for behavioral science*. San Francisco: Chandler Pub. Co.

Kappelman, L., McLean, E., Luftman, J., & Johnson, V. (2013). Key issues of IT organizations and their leadership: The 2013 SIM IT trends study. *MIS Quarterly Executive, 12*(4), 227–240.

Katz, M. L., & Shapiro, C. (1985). Network externalities, competition, and compatibility. *American Economic Review, 75*(3), 424–440.

Katz, M. L., & Shapiro, C. (1986). Technology adoption in the presence of network externalities. *Journal of Political Economy, 94*(4), 822–841.

Keen, P. G. W. (1980). *MIS research: Reference disciplines and a cumulative tradition*. Paper presented at the International Conference on Information Systems (ICIS), Philadelphia, PA.

Keen, P. G. W., & Scott, M. S. (1978). *Decision support systems: An organizational perspective*. Reading, MA: Addison-Wesley Pub. Co.

Keil, M. (1995). Pulling the plug: Software project management and the problem of project escalation. *MIS Quarterly, 19*(4), 421–447.

Keil, M., & Robey, D. (1999). Turning around troubled software projects: An exploratory study of the deescalation of commitment to failing courses of action. *Journal of Management Information Systems, 15*(4), 63–87.

Keil, M., Tan, B. C. Y., Wei, K.-K., Saarinen, T., Tuunainen, V., & Wassenaar, A. (2000). A cross-cultural study on escalation of commitment behavior in software projects. *MIS Quarterly, 24*(2), 299–325.

Kendall, J. E., & Kendall, K. E. (1993). Metaphors and methodologies: Living beyond the systems machine. *MIS Quarterly, 17*, 149–171.

Ketokivi, M., Mantere, S., & Cornelissen, J. (2017). Reasoning by analogy and the progress of theory. *Academy of Management Review, 42*(4), 637–658.

Khazanchi, D., & Munkvold, B. E. (2000). Is information systems a science? An inquiry into the nature of the information systems discipline. *Database for Advances in Information Systems, 31*(3), 24–42.

Khazanchi, D., & Munkvold, B. E. (2003). *On the rhetoric and relevance of IS research paradigms: A conceptual framework and some propositions*. Paper presented at the Hawaii International Conference on System Sciences (HICSS-36), Hawaii, January 6–9.

King, J. L., & Lyytinen, K. (2004). Reach and grasp. *MIS Quarterly, 28*(4), 539–552.

Kirby, J. T. (1997). Aristotle on metaphor. *American Journal of Philology, 118*(4), 517–554.

Kirsch, L. J. (1996). The management of complex tasks in organizations: Controlling the systems development process. *Organization Science, 7*(1), 1–22.

Kirsch, L. J. (1997). Portfolios of control modes and IS project management. *Information Systems Research, 8*(3), 215–239.

Knorr-Cetina, K. D. (2014). Intuitionist theorizing. In R. Swedberg (Ed.), *Theorizing in social science: The context of discovery* (pp. 29–60). Stanford, CA: Stanford University Press.

Krupnik, I., & Müller-Wille, L. (2010). Franz Boas and inuktitut terminology for ice and snow: From the emergence of the field to the "Great Eskimo Vocabulary Hoax". In I. Krupnik, C. Aporta, S. Gearheard, G. J. Laidler, & L. K. Holm (Eds.), *SIKU: Knowing our ice. Documenting Inuit Sea Ice knowledge and use* (pp. 377–400). New York: Springer.

Kuechler, W., & Vaishnavi, V. (2012). A framework for theory development in design science research: Multiple perspectives. *Journal of the Association for Information Systems, 13*(6), 395–423.

Kuhn, T. (1970). *The structure of scientific revolutions* (2nd ed.). Chicago: University of Chicago Press.

Kuhn, T. S. (1987). What are scientific revolutions? In L. Kruger, L. J. Daston, & M. Heidelberger (Eds.), *The probabilistic revolution* (pp. 7–22). New York: Cambridge University Press.

Langefors, B. (1966). *Theoretical analysis of information systems*. Lund: Studentlitteratur.

Lanzara, G. F. (2009). Introduction: Information systems and the quest for meaning – An account of Claudio Ciborra's intellectual journey. In C. Avgerou, G. F. Lanzara, & L. P. Willcocks (Eds.), *Bricolage, care and information: Claudio Ciborra's legacy in information systems research* (pp. 1–27). Basingstoke, UK: Palgrave Macmillan.

Le Coze, J.-C. (2008). Disasters and organisations: From lessons learnt to theorising. *Safe Science, 46*(1), 132–149.

Leary, D. E. (1995). Naming and knowing: Giving forms to things unknown. *Social Research, 62*(2), 267–298.

Lee, A. S. (2010). Retrospect and prospect: Information systems research in the last and next 25 years. *Journal of Information Technology, 25*(4), 336–348.

Lee, A. S. (2014). Theory is king? But first, what is theory? *Journal of Information Technology, 29*(4), 350–352.

Lee, A. S., Thomas, M., & Baskerville, R. L. (2015). Going back to basics in design science: From the information technology artifact to the information systems artifact. *Information Systems Journal, 25*(1), 1–65.

Lengel, R. H. (1983). *Managerial information processing and communication media source selection behavior*. PhD dissertation, Texas A&M University, College Station, TX.

Lengel, R. H., & Daft, R. L. (1984). *An exploratory analysis of the relationship between media richness and managerial information processing* (TR-ONR-DG-08). Retrieved from www.dtic.mil/dtic/tr/fulltext/u2/a143503.pdf

Lévi-Strauss, C. (1955). The structural study of myth. *Journal of American Folklore, 68*(270), 428–444.

Lévi-Strauss, C. (1963). *Structural anthropology*. New York: Basic Books.

Lévi-Strauss, C. (1966). *The savage mind*. London: Weidenfeld & Nicolson.

Lim, S., Saldanha, T., Malladi, S., & Melville, N. P. (2009). *Theories used in information systems research: Identifying theory networks in leading IS journals*. Paper presented at the International Conference on Information Systems, December 15–18 Phoenix AZ.

Liu, F., & Myers, M. D. (2011). An analysis of the AIS basket of top journals. *Journal of Systems and Information Technology, 13*(1), 5–24.

Locke, K., & Golden-Biddle, K. (1997). Constructing opportunities for contribution: Structuring intertextual coherence and 'problematizing' in organizational studies. *Academy of Management Journal, 40*(5), 1023–1062.

Luke, A. (1995). Text and discourse in education: An introduction to critical discourse analysis. *Review of Research in Education, 21*, 3–48.

Lyytinen, K., & King, J. L. (2004). Nothing at the center? Academic legitimacy in the information systems field. *Journal of the Association for Information Systems, 5*(6), 220–246.

Lyytinen, K., & King, J. L. (2006). The theoretical core and academic legitimacy: A response to professor Weber. *Journal of the Association for Information Systems, 7*(11), 714–721.

Ma, M., & Agarwal, R. (2007). Through a glass darkly: Information technology design, identity verification, and knowledge contribution in online communities. *Information Systems Research, 18*(1), 42–67.

MacCorquodale, K., & Meehl, P. E. (1948). On a distinction between hypothetical constructs and intervening variables. *Psychological Review, 55*(2), 95–107.

Magga, O. H. (2006). Diversity in Saami terminology for reindeer, snow, and ice. *International Social Science Journal, 58*(187), 25–34.

Mantere, S., & Ketokivi, M. (2013). Reasoning in organization science. *Academy of Management Review, 38*(1), 70–89.

Markus, M. L. (2014). Maybe not the king, but an invaluable subordinate: A commentary on Avison and Malaurent's advocacy of 'theory light' IS research. *Journal of Information Technology, 29*(4), 341–345.

Markus, M. L., & Saunders, C. S. (2007). Editorial comments: Looking for a few good concepts…and theories…for the information systems field. *MIS Quarterly, 31*(1), iii–vi.

Mason, R. M. (1991). *Metaphors and strategic information systems planning.* Paper presented at the Proceedings of the 24th Hawaii International Conference on System Sciences, Kauai, HI.

Mason, R. O., & Mitroff, I. I. (1973). A program for research on management information systems. *Management Science, 19*(5), 475–487.

McKeon, R. P. (1941). *The basic works of Aristotle.* New York, NY: Random House.

McKinney Jr., E. H., & Yoos, C. J. (2010). Information about information: A taxonomy of views. *MIS Quarterly, 34*(2), 329–344.

Merton, R. K. (1948). The self-fulfilling prophecy. *The Antioch Review, 8*(2), 193–210.

Merton, R. K. (1967). *On theoretical sociology.* New York, NY: Free Press.

Merton, R. K. (1968). Sociological theories of the middle range. In *Social theory and social structure* (pp. 39–72). New York: Free Press.

Meyer, M. (1995). *Of problematology: Philosophy, science, and language.* Chicago: University of Chicago Press.

Mills, C. W. (1959). *The sociological imagination.* New York: Oxford University Press.

Mingers, J. (2004). Paradigm wars: Ceasefire announced who will set up the new administration. *Journal of Information Technology, 19*(3), 165–171.

Mingers, J. C. (1995). Information and meaning – Foundations for an intersubjective account. *Information Systems Journal, 5*(4), 285–306.

Mingers, J. C. (1996). An evaluation of theories of information with regard to the semantic and pragmatic aspects of information systems. *Systems Practice, 9*(3), 187–209.

Minsky, M. (1975). A framework for representing knowledge. In J. Haugeland (Ed.), *Mind Design II* (pp. 111–142). Cambridge MA: MIT Press.

Mintzberg, H. (1972). The myths of MIS. *California Management Review, 15*(1), 92–97.

Mintzberg, H. (1973). *The nature of managerial work*. New York, NY: Harper and Row.

Moody, D., Iacob, M.-E., & Amrit, C. (2010). *In search of paradigms: Identifying the theoretical foundations of the IS field*. Paper presented at the European Conference on Information Systems, June 6–9, Pretoria, South Africa.

Moore, G. C., & Benbasat, I. (1991). Development of an instrument to measure the perceptions of adopting an information technology innovation. *Information Systems Research, 2*(3), 192–222.

Moore, H. L. (2004). Global anxieties: Concept-metaphors and pre-theoretical commitments in anthropology. *Anthropological Theory, 4*(1), 71–88.

Morgan, G. (1986). *Images of organization*. Beverly Hills Sage Publications.

Mousavidin, E., & Goel, L. (2007). *Seeking Dragons in IS Research*. Paper presented at the Americas Conference on Information Systems (AMCIS 2007), August 9–12, Keystone, CO.

Naur, P., & Randell, B. (Eds.). (1969). *Software engineering: Report on a conference sponsored by the NATO committee, Garmisch, Germany, 7–11th Oct. 1968*. Brussels: Scientific Affairs Division, North American Treaty Organization (NATO).

Newell, A., & Simon, H. A. (1976). Computer science as empirical inquiry: Symbols and search. *Communications of the ACM, 19*(3), 113–126.

Nickles, T. (Ed.). (1980). *Scientific discovery, Logic, and rationality*. Dordrecht, Holland: Springer.

Nunamaker Jr., J. F., Applegate, L. M., & Konsynski, B. R. (1987). Facilitating group creativity: Experience with a group decision support system. *Journal of Management Information Systems, 3*(4), 5–19.

Oates, B. J., & Fitzgerald, B. (2007). Multi-metaphor method: Organizational metaphors in information systems development. *Information Systems Journal, 17*(4), 421–449.

Ochara, N. M. (2013). *Linking Reasoning to Theoretical Argument in Information Systems Research*. Paper presented at the Americas Conference on Information Systems (AMCIS 2013), Aug 15–17, Chicago, IL.

Orlikowski, W. J. (2000). Using technology and constituting structures: A practice lens for studying technology in organizations. *Organization Science, 11*(4), 404–428.

Orlikowski, W. J., & Baroudi, J. J. (1991). Studying information technology in organizations: Research approaches and assumptions. *Information Systems Research, 2*(1), 1–28.

Orlikowski, W. J., & Iacono, C. S. (2001). Research commentary: Desperately seeking the 'IT' in IT research – A call to theorizing the IT artifact. *Information Systems Research, 12*(2), 121–134.

Orlitzky, M. (2012). How can significance tests be deinstitutionalized? *Organizational Research Methods, 15*(2), 199–228.

Ortony, A. (Ed.). (1979). *Metaphor and thought*. Cambridge: Cambridge University Press.

Oswick, C., Fleming, P., & Hanlon, G. (2011). From borrowing to blending: Rethinking the processes of organizational theory-building. *Academy of Management Review, 36*(2), 318–337.

Peirce, C. S. (1893–1913/1931–1958). *Collected writings* (8 Vols.). Cambridge, MA: Harvard University Press.

Peirce, C. S. (1992). *Reasoning and the logic of things*. Cambridge, MA: Harvard University Press.

Polanyi, M. (1958). *Personal knowledge: Towards a post-critical philosophy*. Chicago: University of Chicago Press.

Popper, K. R. (1934). *Logik Der Forschung (The logic of scientific discovery)*. Tübingen: Mohr.

Popper, K. R. (1959). *The logic of scientific discovery*. New York: Basic Books.

Rappaport, J. (1987). Terms of empowerment/exemplars of prevention: Toward a theory of community psychology. *American Journal of Community Psychology, 15*(2), 121–148.

Ravitch, S. M., & Riggan, M. (2012). *Reason and rigor: How conceptual framework guides research*. Thousand Oaks: Sage.

Reichenbach, H. (1938). *Experience and prediction: An analysis of the foundations and the structure of knowledge*. Chicago, IL: University of Chicago Press.

Reynolds, P. D. (1971). *A primer in theory construction*. Boston, MA: Allyn & Bacon.

Richardson, H., & Robinson, B. (2007). The mysterious case of the missing paradigm: A review of critical information systems research 1991–2001. *Information Systems Journal, 17*(3), 251–270.

Ricoeur, P. (1977). *The rule of metaphor: The creation of meaning in language*. (R. Czerny, Trans. Toronto, CA: University of Toronto Press.

Ritzer, G. (1980). *Sociology: A multiple paradigm science*. Boston, MA: Allyn and Bacon, Inc.

Rivard, S. (2014). Editor's comments: The ions of theory construction. *MIS Quarterly, 38*(2), iii–xiii.

Robey, D. (2003). Identity, legitimacy and the dominant research paradigm: An alternative prescription for the IS discipline! A response to Benbasat and Zmud's call for returning to the IT artifact. *Journal of the Association for Information Systems, 4*(7), 352–359.

Robey, D., & Markus, M. L. (1984). Rituals in information system design. *MIS Quarterly, 8*(1), 5–14.

Rockart, J. F. (1979). Chief executives define their own data needs. *Harvard Business Review, 52*(2), 81–113.

Rockart, J. F., & DeLong, D. W. (1988). *Executive support systems*. Homewood, IL: Dow Jones-Irwin.

Rockart, J. F., & Treacy, M. E. (1982). The CEO goes on-line. *Harvard Business Review, 60*(1), 82–88.

Rogers, E. M. (1983). *Diffusion of innovations* (3rd ed.). New York: The Free Press.

Rose, J., Jones, M., & Truex, D. (2005). Socio-theoretic accounts of IS: The problem of agency. *Scandinavian Journal of Information Systems, 17*(1), 133–152.

Rosenberg, N. (1976). *Perspectives on technology*. Cambridge: Cambridge University Press.

Roszak, T. (1994). *The cult of information: A neo-luddite treatise on high tech, artificial intelligence, and the true art of thinking*. Berkeley, CA: University of California Press.

Runkel, P. J., & Runkel, M. (1984). *A guide to usage for writers and students in the social sciences*. Totowa, NJ: Rowman & Allanheld.

Ryle, G. (1949). *The concept of mind*. London, UK: Hutcheson.

Sandberg, J., & Alvesson, M. (2011). Ways of constructing research questions: Gap-spotting or problematization? *Organization, 18*(23), 23–44.

Sargent, A. (2012). Reframing caring as discursive practice: A critical review of conceptual analyses of caring in nursing. *Nursing Inquiry, 19*(2), 134–143.

Sartori, G. (1975). The Tower of Babel. In G. Sartori, F. W. Riggs, & H. Teune (Eds.), *Tower of Babel: On the definition and analysis of concepts in the social sciences* (Vol. 6, pp. 7–37). Pittsburgh, PA: International Studies Association.

Saussure, F. D. (1916/1966). *Course in general linguistics* (W. Baskin, Trans.). New York: McGraw-Hill.

Schelling, T. C. (1978a). *Micromotives and macrobehavior*. New York, NY: Norton.

Schelling, T. C. (1978b). Thermostats, lemons, and other families of models. In *Micromotives and macrobehavior* (pp. 81–134). New York, NY: Norton.

Schickore, J., & Steinle, F. (Eds.). (2006). *Revisiting discovery and justification: Historical and philosophical perspectives on the context distinction*. Dordrecht, Holland: Springer.

Schön, D. A. (1963). *The displacement of concepts*. London: Tavistock Publications.

Schwab, A., Abrahamson, E., Starbuck, W. H., & Fidler, F. (2011). Researchers should make thoughtful assessments instead of null-hypothesis significance tests. *Organization Science, 22*(44), 1105–1120.

Shaw, M. (1990). Prospects for an engineering discipline of software. *IEEE Software, 7*(6), 15–24.

Simon, H. (1957). *Administrative behavior*. New York: Free Press.

Spencer, H. (1897). *The principles of sociology*. New York: D. Appleton.

Stinchcombe, A. L. (1987). *Constructing social theories*. Chicago, IL: University of Chicago Press.

Straub, D. (2012). Editorial: Does MIS have native theories. *MIS Quarterly, 36*(2), iii–xii.

Street, C. T., & Denford, J. S. (2012). Punctuated equilibrium theory in IS research. In Y. K. Dwivedi, M. R. Wade, & S. L. Schneberger (Eds.), *Information systems theory: Explaining and predicting our digital society, Volume 1* (pp. 335–354). New York, NY: Springer.

Suppe, F. (2000). Understanding scientific theories: An assessment of developments, 1969–1998. *Philosophy of Science, 67*(Supplement), S102–S115.

Swedberg, R. (2012). Theorizing in sociology and social science: Turning to the context of discovery. *Theory and Society, 41*(1), 1–40.

Swedberg, R. (2014a). *The art of social theory*. Princeton, NJ: Princeton University Press.

Swedberg, R. (2014b). From theory to theorizing. In R. Swedberg (Ed.), *Theorizing in social science: The context of discovery* (pp. 1–28). Stanford, CA: Stanford University Press.

Swedberg, R. (Ed.). (2014c). *Theorizing in social science: The context of discovery*. Stanford, CA: Stanford University Press.

Taylor, H., Dillon, S., & Van Wingen, M. (2010). Focus and diversity in information systems research: Meeting the dual demands of a healthy applied discipline. *MIS Quarterly, 34*(4), 647–667.

Triandis, H. C. (1971). *Attitude and attitude change*. New York, NY: John Wiley and Sons.

Trice, H. M., & Beyer, J. M. (1984). Studying organizational cultures through rites and ceremonials. *Academy of Management Review, 9*(4), 653–669.

Truex, D. P., Holmström, J., & Keil, M. (2006). Theorizing in information systems research: A reflexive analysis of the adaptation of theory in information systems research. *Journal of the Association for Information Systems, 7*(12), 797–821.

Tsoukas, H. (1993). Analogical reasoning and knowledge generation in organization theory. *Organization Studies, 14*(3), 323–346.

Vaast, E., Davidson, E. J., & Mattson, T. (2013). Talking about technology: The emergence of a new actor category through new media. *MIS Quarterly, 37*(4), 1069–1092.

Vallerand, R. J. (1997). Toward a hierarchical model of intrinsic and extrinsic motivation. In M. P. Zanna (Ed.), *Advances in Experimental Social Psychology* (pp. 271–360). New York, NY: Academic Press, Inc.

Vaughan, D. (1996). *The challenger launch decision: Risky technology, culture, and deviance at NASA*. Chicago, IL: University of Chicago Press.

Vaughan, D. (2014). Analogy, cases, and comparative social organization. In R. Swedberg (Ed.), *Theorizing in social science: The context of discovery* (pp. 61–84). Stanford, CA: Stanford University Press.

Vessey, I., Ramesh, V., & Glass, R. L. (2002). Research in information systems: An empirical study of diversity in the discipline and its journals. *Journal of Management Information Systems, 19*(2), 129–174.

Watson, R. T., DeSanctis, G., & Poole, M. S. (1988). Using a GDSS to facilitate group consensus: Some intended and unintended consequences. *MIS Quarterly, 12*(3), 463–478.

Weber, R. (1987). Toward a theory of artifacts: A paradigmatic base for information systems research. *Journal of Information Systems, 1*(2), 3–19.

Weber, R. (2003). Editor's comments: Theoretically speaking. *MIS Quarterly, 27*(3), iii–xii.

Weber, R. (2006). Reach and grasp in the debate over the IS core: An empty hand? *Journal of the Association for Information Systems, 7*(10), 703–713.

Weber, R. (2012). Evaluating and developing theories in the information systems discipline. *Journal of the Association for Information Systems, 13*(1), 1–30.

Weichselgartner, J. (2001). Disaster mitigation: The concept of vulnerability revisited. *Disaster Prevention and Management: An International Journal, 10*(2), 85–95.

Weick, K. E. (1979). *The social psychology of organizing* (2nd ed.). New York, NY: Random House.

Weick, K. E. (1989). Theory construction as disciplined imagination. *Academy of Management Review, 14*(4), 516–531.

Weick, K. E. (1995a). *Sensemaking in organizations*. Thousand Oaks, CA: Sage Publications.

Weick, K. E. (1995b). What theory is not, theorizing Is. *Administrative Science Quarterly, 40*(3), 385–390.

Whetten, D. A. (1989). What constitutes a theoretical contribution? *Academy of Management Review, 14*(4), 490–495.

Whitehead, A. N. (1917). *The organization of thought*. London: Williams and Norgate.

Winter, S. J., & Butler, B. S. (2011). Creating bigger problems: Grand challenges as boundary objects and the legitimacy of the information systems field. *Journal of Information Technology, 26*(2), 99–108.

6

Theorizing Digital Experience: Four Aspects of the Infomaterial

David Kreps

Introduction

Theorizing in the information systems field (IS) seems largely stuck in the late twentieth century, according to three quite different assessments by thought leaders in the field (Gregor, 2006; Grover & Lyytinen, 2015; Mingers, 2015). Although the issues raised vary—the ontological character of IS, Business School positivism, a 'mid-range' cul-de-sac—the prescribed solutions are in fact reasonably similar: a bolder, fresh approach to the nature of theory in, and thus the foundations of, the field. It is this project—this fresh approach—upon which I embarked some years ago (Kreps, 2015, 2018a, 2018b; Kreps & Kimppa, 2015; Kreps, Rowe, & Muirhead, 2020). Specifically, I am interested in finding new philosophical foundations, upon which indigenous theory can be built, within IS. This chapter presents an introduction to this ongoing work. Firstly, I

D. Kreps (✉)
National University of Ireland, Galway, UK
e-mail: david.kreps@nuigalway.ie

N. R. Hassan, L. P. Willcocks (eds.), *Advancing Information Systems Theories*,
Technology, Work and Globalization, https://doi.org/10.1007/978-3-030-64884-8_6

introduce some of the philosophical underpinnings of the new approach and then give a very brief outline of the notion of 'infomateriality.' This leads into an embrace of Varun Grover's four 'fundamental aspects' of 'the digital' that distinguish it from IT and how these four aspects can be evidenced in the data gathered in a research project I undertook over the course of 2018, using diary studies to record individual experience of the digital world. The chapter concludes that the 'infomaterial' could offer a philosophical grounding upon which Grover's 'four fundamental aspects' can rest and that together they represent a powerful philosophical and theoretical tool for understanding the digital world.

As Allen Lee put it in his 2001 Editorial in *MISQ*, "research in the information systems field examines more than just the technological system, or just the social system, or even the two side by side; in addition, it investigates the phenomena that emerge when the two interact" (Lee, 2001, p. iii). Ray Paul echoed this in his 2007 Editorial in *EJIS*: "The IS is what emerges from the usage and adaptation of the IT and the formal and informal processes by all of its users" (Paul, 2007, p. 193). This emergent 'socio-technical' interaction is the space where I believe some philosophical examination might help reveal new means of understanding such phenomena. It is, moreover, (a) more pressing that such new philosophical foundations be sought, because older, 'tried and trusted' understandings don't seem to be getting us very far (Grover & Lyytinen, 2015), and (b) more fitting—because the nature of that interaction has changed quite fundamentally in recent years. As Grover depicts, the 'digital' is different from the old world of 'IT' and needs a different approach (Grover, 2018). The relatively recent shift away from a screen-focused experience of the internet, toward one mediated by myriad internet connected devices, some with screens and some without, constitutes a fundamental change in the nature of the relationship between the social and technological systems our field is devoted to understanding. The theoretical paradigms and approaches aligned to Web 1.0—the read-only web, and to Web 2.0—the read-write web, must be radically different to understand Web 3.0—the newly *mangled* 'internet of things and people' (Kreps & Kimppa, 2015). But the implications of this 'mangle' (Pickering, 1995) run very deep. Orlikowski's ideas for the socio-technical (2002, 2005, 2006), and the relationship of her 'sociomateriality' with the sociological approach of Anthony Giddens (1984) and the ideas of Karen Barad (2007), whilst a good start, in fact merely reveal the great philosophical vistas ready to be explored at the intersection of technological and social systems (Kreps, 2018b). Much more, in short, must be chanced, if indigenous theory in IS is to evolve and have impact not just within but beyond our field.

My own project to discover and develop a fresh approach turns directly to the philosophical underpinnings of the relationship at the heart of information systems—between people and technology—and finds there some key and fundamental questions about the nature of existence. Now, ontology—stressed by Gregor in her 2006 essay on theory in IS—in its original form of metaphysics, hinges, inevitably, upon the question of *our* existence. This is because *we* are the ones who are asking, and so in the end, this is where we must inevitably start. Therefore, it is in the nature of *experience* that we must begin. Such a view, indeed, goes back as far as Greek philosopher Epicurus (Furley, 1967; Oates, 1940), whose unique formulation of the previously weaker notion of 'atomism' rendered it stronger against the criticisms of Aristotle—whose scholasticism was nonetheless to envelop Medieval Europe in circular reasoning for more than a thousand years. Not until Descartes, and perhaps even more fundamentally, Spinoza—one of the great rationalists of seventeenth-century philosophy—did the Epicurean appreciation for the experience of the senses return to the centre of scholarly enquiry, spawning the scientific revolution (Israel, 2002; Spinoza, 2003). Yet, as the eighteenth and nineteenth centuries of scientific advance unfolded, one question we might usefully ask is, 'Did this return to the testimony of the senses go too far?'

As Gregor relates, describing the early twentieth century basis of *positivism* in IS, "At the base of logical positivism is the famous *Verification Principle:* only assertions that are in principle verifiable by observation or experience can convey factual information. Experience was thought to be the only source of meaning and the only source of knowledge" (Gregor, 2006, p. 615). Yet in the next paragraph Gregor depicts the *interpretivist* tradition, quoting Schwandt, as one dedicated to "understanding the complex world of lived experience from the point of view of those who live it" (1994, p. 118). Clearly, *experience* is important; yet here it is claimed as the basis for two opposing traditions. On the one hand, experience gives us access to clear factual information—the testimony of the senses as represented in the metrics of the measurement of external phenomena. On the other hand, it gives us access to the internal subjective and emotional world of sensation—how we are affected by what our senses reveal to us. In both cases, it is our experience that is the foundation upon which knowledge is built, yet the two sets of knowledge seem to be opposed to one another: the one focused exclusively outward upon an 'objective' external world; the other focused exclusively inward upon a 'subjective' internal world. The common factor—the *experiencer*—is deemed an epiphenomenal irrelevance in the 'external' world (Thompson, 2007, p. 5), and yet inescapably sovereign in the 'internal' one (Karakayali, 2015). Both descriptions—the 'realist' and 'idealist'—of the *experiencer* cannot be correct: they are mutually exclusive.

Much western philosophy—certainly since the seventeenth century—relies upon the distinction between the objective and the subjective presupposed by the realist, and upon a hierarchy between them, with the politics of our academic institutions (science over the humanities) foregrounding the one over the other, and prescribing what is 'True' and what is 'False' (evidence-based fact over imaginary opinion). Hence the eventual development and victory of verificationism. Yet the voices of those saying that all this goes too far have simply grown louder over the course of the twentieth century, giving rise, in IS, to interpretivism, by the end of it. It is, in short, the Enlightenment Project so heavily critiqued by Foucault (1970, 1972) that brought us this disjunction between the objective and the subjective in the nature of experience. Now, undoubtedly, much of the academic pursuit of knowledge rests upon the foregrounding of 'evidence-based fact' over 'imaginary opinion', and the achievements of the scientific method must be acknowledged as without parallel in human history. Nonetheless, reasoned argument without recourse to evidence-based fact remains a vibrant part of society, let alone the aesthetic, romantic and delightfully frivolous aspects of life without which it would surely not be worth living!? A good deal of experience, after all, is about enjoyment, entertainment, love, and friendship, all deemed merely meaningless "pseudo-statements" by the Vienna School of logical positivists (Carnap, 1932, pp. 60–61). As Mingers tells us, in IS, the problem with Business Schools attempting, since the 1960s, to attain academic rigor is that such positivist work is "rigorous in the sense of being highly quantitative and mathematical, but … far from the practical messy problems faced by real managers" (Mingers, 2015, p. 316).

Now, there is a growing school of thought in the information systems field—including John Mingers—leaning toward 'critical realism' as a philosophical approach through which we might begin to address some of the questions I have just raised. However, this is not the route that I have taken. Crucially, to my eyes, at the juncture of the personal and the material, where the experiencer is the receiver both of the objective evidence of the senses and the subjective evidence of sensation, there is also an experience of *time* that is different from the time that we measure. Critical realism does not seem, to me, to address this key issue. The time that we measure—as with all measurement—can, in theory, and in calculation, go in both directions. All measurement is reversible, by its nature. Yet of course, except in the novels of H. G. Wells, time is not reversible.

Experienced time, on the other hand, is something that is *lived*, and experienced—by the *experiencer*—as duration. It is, crucially, during the process of duration when *choices* are made, *by the experiencer*. The 'agency of the individual', as this is most often described—for all the (Foucauldian) social constraints of conditioning and interplay, and the physical constraints of material conditions—makes decisions during the unfolding of lived time. The individual agent takes part in key aspects of determination, through such choices. This fact renders existence profoundly *indeterminate* and our role as choosers in its unfolding existential. This lived time, where free will is exercised, in the words of the great early twentieth-century process philosopher Henri Bergson, is called the *durée réelle*, an understanding of our experience of time that places it at the heart of a unified appreciation of existence, both objective and subjective, an existence in relation to which we are neither epiphenomenal nor sovereign, an existence of which we are a key part, whilst still subject to its core material constraints (Bergson, 2005[1889]).

I ground my search for new philosophical foundations on which indigenous theory in IS can be built, therefore, upon the oeuvre of process philosopher Henri Bergson. Dubravka Cecez-Kecmanovic gave a useful introduction to process philosophy in her paper, 'From Substantialist to Process Metaphysics—Exploring Shifts in IS Research' (2016). Process methods are also increasingly popular in organization studies (Hernes, 2014; Hernes & Maitlis, 2010; Langley & Tsoukas, 2010; Poole, Van de Ven, Dooley, & Holmes, 2000). In adopting this approach, I am also consciously foregrounding the key importance of the field of IS within the wider transdisciplinary context, as one of the few areas of academic study devoted specifically to an understanding of the interface between 'us' and 'it', between the personal and the material. In IS, moreover, we study this interface in the special circumstances where the material is *manufactured*, and *technological*. In this foregrounding, crucially, a key element is an appreciation that there is personal time in the form of *durée*, for 'us', as well as clock time in the sequencing of bits and bytes, for 'it', and our understanding of the multifarious aspects of the relationship must reflect this. The importance of this shift in understanding, I believe, is, moreover, nothing less than existential, in our present time, as I will describe further in the next section.

Infomateriality

From this philosophical work, a new notion has been gradually arising that I am calling *infomateriality* (Kreps, 2018a, 2018b, 2019; Kreps et al., 2020). It is distinguished from Orlikowski's sociomateriality in that:

(a) it is not trying to merge two 'opposites'—the social and material—founded, as it is, in an ontology that does not distinguish between them so fundamentally;
(b) although sharing this process ontology with Giddens' worldview (Giddens mentions "*durée*" as foundational in his introductory remarks (1984, p. 3)), it focuses more upon the physical and the durational and less upon ephemeral social 'structures';
(c) it acknowledges (a non-suffocating context of) Foucauldian networks of power relations and the mutual definition of the self in social contexts, against which individual agency must struggle, but whilst at the same time merging such relations with the physical constraints of our embeddedness within natural and built environments, such that

the two are not entirely distinguishable for social science on the one hand, and physical science on the other; and

(d) it acknowledges the key element of human meaning already inherent within 'information' (Checkland 1988, p. 239; Kreps, 2017) as causal (Markus & Rowe, 2018), whilst laying stress upon the anthropogenic shift of digital transformation: that we live in a world we have made and are rapidly remaking (Kreps, 2018b).

Infomateriality may be understood, therefore, as a condition of contemporary human societies in which our exchange of information, and the digital tools with which we now undertake that exchange—all the more so in the age of 'lockdown'—have become constitutive of the physical context in which we live. More than simply the instantiation of urban or virtual environments, the means by which we 'wayfind' through infomaterial environments are conditioned by the tools that we use, as much as by the environments themselves. The distinctions between mental and physical have become blurred, the realms interpenetrating to the extent that considering them separately becomes a distraction from clear understanding. The boundaries between Popper's 'Worlds,' as described by Gregor, are blurred in a process-philosophy supported view. She tells us, Popper's "World 1 is the objective world of material things; World 2 is the subjective world of mental states; and World 3 is an objectively existing but abstract world of man-made entities: language, mathematics, knowledge, science, art, ethics, and institutions" (Gregor, 2006, p. 615). World 3, in the Anthropocene (Crutzen, 2002), has become constitutive of much of the World 1 that we experience, where more than half the human population (World Bank, 2019) live in urban environments ("After water, concrete is the most widely used substance on the planet" (Watts, 2019)), and a sixth great mass extinction (Davis, Faurby, & Svenning, 2018) and complete reconfiguration of global climate (WMO, 2018) is underway. Popper's distinctions between the objective—World 1—and the mental—World 2—are blurring by the day, and only even possible when one's philosophical world view persists in ignoring the realities of duration and free will, in the doublethink where mental states are not material whilst at the same time being impacted by the material (in our perceptions) and having impact upon the material (in our choices).

The philosophical implications of the Anthropocene represent, in short, a foundational challenge to those such as the logical positivists, for whom free will is a part of theological debate.

How this notion of infomateriality may play out in our understanding of digital experience has been the subject of a funded research project which I describe in the next section.

Understanding Digital Events

As a British Academy Mid-Career Fellow, I ran a 12-month project across 2018 entitled *Understanding Digital Events: A philosophical and sociological study of virtual experience in the everyday*. This research—the UDE project—explored analysis techniques by which a philosophical approach based in a study of process philosophy might be incorporated into sociological studies of engagement with digital interfaces, and the techniques of designers and User Experience practitioners. The UDE project built upon my previous work exploring the philosophy of Henri Bergson and how it might be understood today in light of the advent of complexity theory: *Bergson, Complexity, and Creative Emergence* (Kreps, 2015), and a growing fear that the products of the information systems field are built upon philosophical foundations that are profoundly anti-environmental and detrimental to human survival, a view expressed in the short polemic, *Against Nature: The Metaphysics of Information Systems* (Kreps 2018).

The project fieldwork constituted three phases: recruiting participants, managing diary studies and undertaking interviews, and analysis of results. Recruitment was undertaken locally, through word-of-mouth and personal contacts. A former colleague, now retired, was invited to become a participant, and in turn invited members of a community group to which they now belong. Recent and current students at the University were invited—through LinkedIn, or in class—and the manager of a youth group focused on sports fitness, undertaking a separate project with another colleague, publicized the project in the group and helped recruit a number of others. In this way members of the community surrounding the University were drawn into the research, bringing in both those directly connected with it, and those connected only by association. Sufficient for a small qualitative pilot study, this cohort of participants nonetheless reflected quite a broad cross-section of the local population. Diary studies were undertaken by each of these participants, detailing their mental and physical experiences using digital technologies, and were kept for 4 weeks, 3 entries a week, by most participants. Some of the participants were also invited to a semi-structured interview. To ensure ease of participation, an online web application accessible from any Internet-connected device was created to collect the entries. This combined a number of qualitative and quantitative questions to obtain insight into each participant's daily experiences, identifying their patterns of use and perceptions of digital technologies. Finally, a sub-set of the participants were interviewed, in part to follow-up on points raised in their diaries, and in general to explore further their attitudes to privacy, surveillance, and choice. All this data was then anonymized and imported into NVivo, where I have been able to query it in order to reveal many fascinating insights.

Grover's Four Aspects in the UDE Data

Varun Grover, at the SIGPHIL workshop at ICIS in San Francisco in December 2018, presented a fascinating story of how there has been a distinct shift from a 'world of IT' up until the end of the 1990s, to a 'world of the digital' since the turn of the millennium. He outlined how in fact these are two very different worlds, and the theories developed and used in the IS field for understanding the world of IT are no longer adequate. The digital, as he outlined, is infrastructural, and subject to a unique economics. There are, for Grover, four 'fundamental aspects' of the digital that render it different from IT: (i) Embeddedness, (ii) Decoupling, (iii) Representation, and (iv) Generativity (Grover, 2018). The research undertaken for this chapter involved taking Grover's four 'fundamental aspects,' generating a range of synonyms for each of them and then querying the data from the UDE project with NVivo's own synonym function, to find comments from the diaries and the interviews that related to each of the four aspects. The rest of this chapter focusses upon this, and what implications it has for the notion of infomateriality.

Each aspect is readily understood with a few phrases (from Grover, 2018) and some keywords, as shown in Table 6.1, and the synonyms brought many candidate comments in the data for each aspect to the fore.

A key approach to understanding the four aspects, revealed in Grover's presentation in 2018, is to use them to tell stories that link all four. Grover gave two examples:

1. "a device in car (embeddedness) delivered to insurance company (decoupling) can better tap customer behavior (representation) and combine with information—like potholes—to provide new products for customers (generativity)."
2. "Weight Watchers captures information from bar-codes, human health (embedded) delivered through its app/website (decoupling) to represent eating habits (behaviors) and generate new products (through social connections, point system, food delivery)." (Grover, 2018)

Table 6.1 Grover's 'Four Fundamental Aspects'

Embeddedness	The digital "enhances affordances of physical objects; function is no longer constrained by form."	Synonyms such as Enclose, Contain, Connect
Decoupling	The digital "separates content from (packaging) delivery system; all forms of content can be syndicated."	Synonyms such as Dissociate, Differentiate, Separate, Syndicate
Representation	The digital represents "behaviors and states of people"; there is representation of "experience (time, space, interactions)." The digital enables representation to be enhanced at "various levels of granularity."	Synonyms such as Profile, Symbolize, Express, Embody
Generativity	The digital generates novelty through "networks and people; there is combinatorial complexity and unanticipated innovation."	Synonyms such as Develop, Release, Upgrade, Initiate, Expand, Engender, Create

When queried in the manner described above, the Understanding Digital Events project data revealed many such stories. A sample of just four of them are quoted below (with the anonymized participant in brackets):

- "Having a device attached to the internet in your pocket is very useful and reassuring in a difficult situation. You can keep up to date with the current situation and find ways round problems." (B2-24)
- "Apart from the use of my mobile I also interacted with my smart-watch mainly to check my activity throughout the day. It helps since I work in an office and I sit at my desk most part of the day I get reminders once an hour to stand up to take a rest and a deep breath but I also try and hit my daily target for calories consumed. I find this activity quite fun." (A3-16)
- "Also today I used my Garmin fitness watch to track my time on a Parkrun. This then links to Strava the website for tracking recording and sharing information about runs [representation] (and rides too if you cycle). Strava is really helpful and easy to use and for a large event like the Macc Parkrun there were 60 or so runners with Strava accounts and you can see how you performed against them. If you run a route several times you can see how your performance has improved (or not). Part of the enjoyment here of course is the flow of endorphins after the run." (B2-20)
- "I'll be travelling this weekend so I used my mobile to check the places where I'm planning to go. I used Google maps and pinpointed several locations. Depending on the traffic and how much time is left I might be able to see them all. The app is easy to use and it will provide live traffic warnings which will help me decide whether or not I will try a different route (sometimes the app actually provides alternatives when there is heavy traffic)." (A3-16)

It is clear that Grover's four aspects can very easily be used to analyze this data. The four quotations from the data are repeated, below, with the four aspects in square brackets picking out the embeddedness, decoupling, representation and generativity in each quote:

- "Having a device attached to the internet in your pocket [embeddedness] is very useful and reassuring in a difficult situation. You can keep up to date [decoupling] with the current situation [representation] and find ways round problems [generativity]." (B2-24)
- "Apart from the use of my mobile I also interacted with my smartwatch [embeddedness] mainly to check my activity [decoupling] throughout the day [representation]. It helps since I work in an office and I sit at my desk most part of the day I get reminders once an hour to stand up to take a rest and a deep breath but I also try and hit my daily target for calories consumed [generativity]. I find this activity quite fun." (A3-16)
- "Also today I used my Garmin fitness watch [embeddedness] to track my time on a Parkrun. This then links to Strava the website [decoupling] for tracking recording and sharing information about runs [representation] (and rides too if you cycle). Strava is really helpful and easy to use and for a large event like the Macc Parkrun there were 60 or so runners with Strava accounts and you can see how you performed against them. If you run a route several times you can see how your performance has improved (or not) [generativity]. Part of the enjoyment here of course is the flow of endorphins after the run." (B2-20)
- "I'll be travelling this weekend so I used my mobile to check the places where I'm planning to go [embeddedness]. I used Google maps [decoupling] and pinpointed several locations [representation]. Depending on the traffic and how much time is left I might be able to see them all. The app is easy to use and it will provide live traffic warnings which will help me decide whether or not I will try a different route (sometimes the app actually provides alternatives when there is heavy traffic) [generativity]." (A3-16)

Discussion

One purpose of this chapter, therefore—to show in primary research data that Grover's four fundamental aspects can be accurately used to analyze narratives of digital experience—I believe is adequately shown in just the four examples above. The main purpose of this chapter, however, is to show that the notion of the 'infomaterial' may offer a philosophical grounding upon which Grover's four 'fundamental aspects' can rest. Together, I believe, they may represent a powerful philosophical and theoretical tool for understanding the digital world.

How then, does this play out? Table 6.2 shows how the four aspects sit within the context of the notion of infomateriality.

Table 6.2 Grover's 'Four Fundamental Aspects' and Infomateriality

Embeddedness	the enhancing of the affordances of physical objects by the digital	• Physical objects (digital devices) external to the experiencer in perception become internally coherent with the play of agency when the experiencer makes choices through interactions with these digital devices. • In the context of the digital devices embedded both in our material environment and in those very networks of social and power relations, the notion of embeddedness sits right at the core of the process-philosophy underpinning of the infomaterial: digital devices become limbs.
Decoupling	whereby all content can be repurposed and repackaged	• Emphasizes—and instantiates—the relationality of process philosophy: that the universe we inhabit is not one of 'fixed things' external to us, but one of continuous flow in which the objects that we perceive are but temporary eddies, brief clusters of energy in a quantum continuum. Molecules and atomic particles are constantly on the move and being exchanged. Almost every cell in our bodies is replaced over cycles as short as a few days or as long as a few years. • Celebrated quantum physicist Louis de Broglie wrote that Bergson was "Bohr before Bohr, Heisenberg before Heisenberg" in his depiction of a universe of flow forty years before quantum mechanics (de Broglie, 1969, p. 47; Kreps, 2015, 2018b).

(*continued*)

Table 6.2 (continued)

Representation	whereby the digital represents experience	• Brings the internal into the external, displays the durational states of the experiencer in a range of facets both at macro and at micro levels, personal and agglomerate. • The 'profile' and the 'digital footprint' of the individual. • The calculations of the artificial intelligence algorithms by which our 'behavioral futures' are marketed, meanwhile, rely upon the agglomeration—the macro-view of our many wishes, relations, preferences, and whims—in sum (Zuboff, 2019). • Our own self-worth, shockingly, seems increasingly dependent upon the number of 'likes' each imprint accrues in the digital 'front room' representing our selfhoods, unless our own sense of purpose can be strengthened (Rowe, 2018). The more automated we become, the more the correlation between 'likes' and self-worth is strengthened (Burrow & Rainone, 2017).
Generativity	whereby innovations emerge from embedded, decoupled representations	• Probably the key differentiator between the 'world of IT' and the 'world of the digital.' • New ways of doing things, unforeseen outcomes, and whole new processes are emerging, and there is a pressing need to understand how best we, as aware human beings, should navigate this new realm. • The accent upon the *experiencer* in infomateriality renders the digital political.

Conclusion

Infomateriality, therefore, I argue, offers a Type 1 IS theory, in Gregor's terms—"theory for analyzing" (Gregor, 2006, p. 620)—and an attempt, in Grover's terms, "To engage in two-way interaction between digital phenomena and abstract theory" (Grover, 2018). Infomateriality makes *better* use of the notion of *durée* regarded as foundational by Giddens, and therefore also has deeper roots and clearer philosophical positions than Barad or the apolitical approach of actor-network theory. The reality of free will at the nexus of the personal and the material in fact underpins—and could greatly strengthen—these other theories.

Key, therefore, to an understanding of the impact of the condition of infomateriality on human societies are:

1. If free will is as key to the nature of consciousness and the unfolding of reality as process philosophy suggests, then our digital experience can be regarded as positive inasmuch as it enhances or makes or allows space for us to be sufficiently mindful and reflective to reach decision points that instantiate and enact free will. Conversely, our digital experience can be regarded as negative inasmuch as we are rushed or pressured into accepting or acquiescing to situations that, given sufficient time, we would not have chosen, or are in effect tricked by hidden processes, into activities and situations we would not have chosen had we been aware of all the facts. The age of surveillance capitalism, in other words, in which the digital has arguably been hijacked by forces of capitalist oligarchy, is inescapably negative: it seeks to *automate* us (Zuboff, 2019).
2. The relationality of a world understood from a process philosophy approach, in which the multiple interrelationships of a shifting universe become the focus, in contrast to the distinctiveness of 'fixed things,' is redolent of a world understood in terms of actor-network theory. Process philosophy understandings can indeed be read as underpinning the notion that 'objects' can be understood as '*actants*.'
3. Technological artefacts in our lives, therefore, need to be seen—to continue the theatrical metaphor inherent in the term, '*actant*'—as *co-directors* of our lives (Coeckelbergh, 2019). The choices, in other

words, that we make, when exercising our free will, are constrained and circumscribed not merely by the fundamentals of gravity, body shape, atmosphere, etc., but by the (digital) tools we ourselves have made.

Infomateriality, then, takes a position in favor of Slow Tech (Patrignani & Whitehouse, 2014, 2018) and an ethical approach to our interactions with technology—promoting our ability to choose, and therefore solidly against the automation of our behavior through the hooks and nudges (Eyal, 2014) that feed into the "behavioral futures market" that trades in our data in pursuit of behavioral modification for commercial gain (Zuboff, 2019, p. 8). Its appreciation of the Anthropocene places it firmly in favor of Green IT/Green IS/Sustainable ICT. It is profoundly aligned with Tech for Good. It is also, I would argue, a far better means of understanding the nature of the digital, than any of the theories devoted to understanding the world of IT.

The data from the Understanding Digital Events project, moreover, clearly both supports Grover's ideas and underlines the difference of digital to IT. As Grover urges us, we should "engage with theories that do NOT strictly fit with current conceptualization of the digital phenomena" (Grover, 2018), and I have argued in this chapter that my notion of 'infomateriality' is an attempt to conceptualize this new milieu that bears serious consideration in the IS field. The interface of the personal and the technological is a philosophically rich nexus, and Allen Lee's "phenomena that emerge when the two interact," (Lee, 2001, p. iii), and Ray Paul's IS that "emerges from the usage and adaptation" (Paul, 2007, p. 193), point us towards greater understanding of the nexus as a key aspect of our contemporary existence.

References

Barad, K. (2007). *Meeting the universe halfway*. London: Duke University Press.

Bergson, H. (2005 [1889]). *Time and free will.* Adamant Media Elibron Classics reproduction of 1913 edition, Trans. F.L. Pogson, New York: George Allen and Unwin.

Burrow, A. L., & Rainone, N. (2017). How many likes did I get?: Purpose moderates links between positive social media feedback and self-esteem. *Journal of Experimental Social Psychology, 69*, 232–236.

Carnap, R. (1932). The elimination of metaphysics through logical analysis of language. *Erkenntnis, 1*, 60–81.

Cecez-Kecmanovic, D. (2016). From substantialist to process metaphysics—Exploring shifts in IS research. In L. Introna, D. Kavanagh, S. Kelly, W. Orlikowski, & S. Scott (Eds.), *Beyond interpretivism? New encounters with technology and organization IFIP WG 8.2 Working Conference on Information Systems and Organizations, IS&O 2016 Dublin, Ireland, December 9-10, 2016 Proceedings*. Cham, Switzerland: Springer.

Coeckelbergh, M. (2019). Technology, narrative, and performance in the social theatre. In D. Kreps (Ed.), *Understanding digital events: Bergson, Whitehead, and the experience of the digital*. London: Routledge.

Crutzen, P. J. (2002). Geology of mankind. *Nature, 415*, 23.

Davis, M., Faurby, S., & Svenning, J. (2018). Mammal diversity will take millions of years to recover from the current biodiversity crisis. *PNAS, 115*(44), 11262–11267.

de Broglie, L. (1969). The concept of contemporary physics and Bergson's Ideas on Time and Motion. In: *Bergson and the evolution of physics*, Pete A.Y. Gunter (Ed. and Trans.) Knoxville: University of Tennessee Press.

Eyal, N. (2014). *Hooked: How to build habit-forming products*. London: Penguin.

Foucault, M. (1970). *The order of things*. London: Routledge.

Foucault, M. (1972). *The archaeology of knowledge*. London: Routledge.

Furley, D. (1967). *Two studies in the Greek atomists*. Princeton: Princeton University Press.

Giddens, A. (1984). *The constitution of society: Outline of the theory of structure*. Berkeley CA: University of California Press.

Gregor. (2006). The nature of theory in information systems. *MIS Quarterly, 30*(3), 611–642.

Grover, V. (2018). Indigenous theory: Caveats of "Moving to the Edges" *SIGPHIL Presentation at International Conference on Information Systems (ICIS)* San Francisco, USA, 13th-16th December 2018

Grover, V., & Lyytinen, K. (2015). New state of play in information systems research: The push to the edges. *MIS Quarterly, 39*(2), 271–296.

Hernes, T. (2014). *A process theory of organization*. Oxford: OUP.

Hernes, T., & Maitlis, S. (2010). *Process, sensemaking, and organizing*. Oxford: OUP.

Israel, J. (2002). *Radical Enlightenment: Philosophy and the Making of Modernity 1650-1750*. London: OUP Oxford.

Karakayali, N. (2015). Two ontological orientations in sociology: Building social ontologies and blurring the boundaries of the 'social'. *Sociology, 49*(4), 732–747.

Kreps, D. (2015). *Bergson, complexity and creative emergence*. London: Palgrave.

Kreps, D. (2017). Matter and memory and deep learning. In Y. Hirai, H. Fujita, & S. Abiko (Eds.), *Diagnoses of Bergson's matter and memory: Developments towards the philosophy of temporal experience, sciences of consciousness, aesthetics, and ethics* (pp. 196–225). Tokyo: Shoshi Shinsui.

Kreps, D. (2018a). *Against nature: The metaphysics of information systems*. London: Routledge.

Kreps, D. (2018b). Infomateriality. *International Conference on Information Systems (ICIS)* San Francisco, USA, 13th-16th December 2018

Kreps, D. (Ed.). (2019). *Understanding digital events: Bergson, whitehead, and the experience of the digital*. London: Routledge.

Kreps, D., & Kimppa, K. (2015). Theorising Web 3.0. *Information Technology and People, 28*(4), 726–741.

Kreps, D., Rowe, F., & Muirhead, J. (2020). Understanding digital events: Process philosophy and causal autonomy. *Proceedings of the 53rd Hawaii International Conference on System Sciences* 7-11/1/2020 https://scholarspace.manoa.hawaii.edu/handle/10125/64492

Langley, A., & Tsoukas, H. (2010). Introducing 'Perspectives on process organization studies'. In T. Hernes & S. Maitlis (Eds.), *Process, sensemaking, and organizing* (pp. 1–26). Oxford: Oxford University Press.

Lee, A. S. (2001). Editorial. *MIS Quarterly, 25*(1), iii–vii.

Markus, L., & Rowe, F. (2018). Is IT changing the world? Conceptions of causality for information systems theorizing. *MIS Quarterly, 42*(4), 1255–1280.

Mingers, J. (2015). Helping business schools engage with real problems: The contribution of critical realism and systems thinking. *European Journal of Operational Research, 242*, 316–331.

Oates, W. (1940). *The stoic and epicurean philosophers: The complete extant writings of Epicurus, Epictetus, Lucretius and Marcus Aurelius*. New York: Random House Modern Library.

Orlikowski, W. (2002). Knowing in practice: Enacting a collective capability in distributed organising. *Organisation Science, 13*(3), 249–273.

Orlikowski, W. (2005). Material works: Exploring the situated entanglement of technological performativity and human agency. *Scandinavian Journal of Information Systems, 17*(1), 183–186.

Orlikowski, W. (2006). Material knowing: The scaffolding of human knowledgeability. *European Journal of Information Systems, 15*(1), 460–466.

Patrignani, N., & Whitehouse, D. (2014). Slow Tech: A quest for good, clean and fair ICT. *J. Inf. Commun. Ethics Soc., 12*(2), 78–92.

Patrignani, N., & Whitehouse, D. (2018). *Slow tech and ICT: A responsible, sustainable and ethical approach*. London: Palgrave.

Paul, R. (2007). Challenges to information systems: Time to change. *European Journal of Information Systems, 16*, 193–195.

Pickering, A. (1995). *The mangle of practice: Time, agency and science*. Chicago, IL: University of Chicago Press.

Poole, M. S., Van de Ven, A., Dooley, K., & Holmes, M. (2000). *Organizational change and innovation processes: Theory and methods for research*. Oxford: Oxford University Press.

Rowe, F. (2018). Being critical is good, but better with philosophy! From digital transformation and values to the future of IS research. *European Journal of Information Systems, 27*, 380–393.

Spinoza, B. (2003). *Complete works*, Trans. Samuel Shirley, London: Hackett Publishing Co.

Thompson, E. (2007). *Mind in life: Biology, phenomenology, and the sciences of mind*. London: Belknap Harvard.

Watts, J. (2019). Concrete: The most destructive material on Earth. *The Guardian*. Retrieved February 26, 2019, from https://www.theguardian.com/cities/2019/feb/25/concrete-the-most-destructive-material-on-earth

WMO. (2018). WMO Greenhouse Gas Bulletin (GHG Bulletin)—No. 14: The State of Greenhouse Gases in the Atmosphere Based on Global Observations through 2017 https://library.wmo.int/index.php?lvl=notice_display&id=20697#.W_bLW0X7T5U

World Bank. (2019). Retrieved January 31, 2019, from https://data.worldbank.org/indicator/sp.pop.grow

Zuboff, S. (2019). *The age of surveillance capitalism*. London: Profile Books.

7

Design Science Theorizing: The Contribution of Practical Theory

Göran Goldkuhl and Jonas Sjöström

Introduction

The IS community has extensively debated the idea of theoretical contributions in design science (DS). Hevner, March, Park, and Ram (2004) focus on artifacts as the primary outcome of such research. The essence of design science is considered to be building and evaluating artifacts. Design science is contrasted to "behavioral science", with an emphasis on development and justification of theories. The authors thus exclude theory development from design science studies. This view on design science

G. Goldkuhl (✉)
Department of Informatics and Media, Uppsala University, Uppsala, Sweden

Department of Management and Engineering, Linköping University, Linköping, Sweden
e-mail: goran.goldkuhl@liu.se

J. Sjöström
Department of Informatics and Media, Uppsala University, Uppsala, Sweden
e-mail: jonas.sjostrom@im.uu.se

N. R. Hassan, L. P. Willcocks (eds.), *Advancing Information Systems Theories*, Technology, Work and Globalization, https://doi.org/10.1007/978-3-030-64884-8_7

essence and outcome and its contrasts to behavioral science are also favored in the "Memorandum on design-oriented information systems research" (Österle et al., 2011). In this opinion piece from the German-speaking IS community, the results from such research are considered to be artifacts and "instructions for action … that allow the design and operation of IS" (ibid., p. 8). The word "theory" does not appear in their text at all.

However, there are many scholars in the IS/DS community who have argued for theorizing as an integral part of design science; e.g., Venable (2006), Gregor and Jones (2007), Baskerville and Pries-Heje (2010), Winter (2014), Niehaves and Ortbach (2016). The focus is mainly on design theory as a result of design science studies. In later works by Alan Hevner, the view on DS theorizing has shifted (Gregor & Hevner, 2011, 2013). The new position is characterized as "agnostic to the need for design theory" (Gregor & Hevner, 2011, p. 4).

Editors from AIS basket journals (Baskerville, Lyytinen, Sambamurthy, & Straub, 2011) later disputed the Österle et al. (2011) Design Science Memorandum and articulated general criteria for acceptance of DS articles in those journals: "The strongest design science papers will likely to gain publication in top journals because they make *significant contributions to theory* as well as practice (ibid., p. 14; our emphasis). The editors further elaborated on DSR contributions as "*discovery of novel theory* related to IS phenomena through design science procedures" or "*significant extension of existing theory* through its exercise in a design-and-build research cycle" (ibid.; our emphasis). The opponents to the Memorandum conclude "the 'Memorandum' underemphasizes *the importance of theory* and scholarly contribution *as key criteria for acceptance*" (ibid.; our emphasis). We conceive of the editors' arguments as a clear demand for theory as a contribution from DS studies, at least if it should be published in a top journal.

In a JAIS editorial, Baskerville, Baiyere, Gregor, Hevner, and Rossi (2018) adopt a longitudinal perspective on knowledge development and further reflect about DS contributions by contrasting artifacts v. theory as outcomes of design science, stating that "a contribution to design (prescriptive, technical) knowledge can be recognized as a sufficient contribution when the newness and usefulness of an artifact can be demonstrated, although there may be limited conceptualization and theorizing." They do, however, state that design theory (prescriptive, scientific knowledge) is a desirable goal.

While we do agree with the state-of-the-art as expressed by Baskerville et al. (2018), we still see a white spot on the DS theorizing map. In practice-oriented DS (e.g., Goldkuhl & Sjöström, 2018; Iivari, 2015; Sein, Henfridsson, Purao, Rossi, & Lindgren, 2011), researchers engage with practitioners. In doing so, there are multiple ways to theorize. One focal area is prevalent in design science: Exploring and conceptualizing a class of problems and articulating abstract solutions to the problem class. Problems and solutions belong to some demarcated domain. Another focal area, less explored in the DS discourse, is the development of instrumental knowledge to facilitate the inquiry into the practice to be changed. We refer to these two focal areas as domain knowledge vs. instrumental knowledge (Davison, Martinsons, & Ou, 2012).

Domain knowledge refers to design knowledge aimed at solving a particular class of design problems. A domain is a type (class) of practices, for example, social work practice. A domain (as a type of practice) consists of typical stakeholders, activities and expected outcomes. Class of problems is typical challenges in such a domain and corresponding solutions (artifacts) are designed responses to such challenges/problems. Instrumentalities aid researchers and practitioners in the process of conducting an inquiry into a problematic situation and its reconfiguration through the design of artifacts. The idea of instrumentalities has gained much attention in the information systems field, albeit often using different terminology. Instrumentalities are constructive knowledge such as practical theories and other cognitive resources that govern our focus and actions in the inquiry process. For example, there is a large stream of research on information systems development (ISD) methods. If we, for instance, look at the history of IS research in Scandinavia (Iivari & Lyytinen, 1998), a lot of researchers have engaged in ISD projects, and in doing so, they developed methods and other instrumentalities for organizational change and systems development. On a broader note, ISD methods research has been the focus of a large part of the IS community (Henderson-Sellers, Ralyté, Ågerfalk, & Rossi, 2014). In the DS discourse, with few exceptions, the discussion about knowledge outcomes focuses on domain-specific knowledge to solve a particular class of problems.

In this chapter, we problematize the demands for theory contribution in papers from design science studies. An exaggerated demand for full-blown theory contributions may hamper the role of design science studies in published IS research. Focusing on the notion of instrumentality, we present a nuanced and constructive view of how design science studies may benefit from the use of, and contribute to the development of, instrumental/practical theory. We draw from the pragmatist notion of inquiry to ground our arguments. We also articulate ideas on the accumulation of ideas over time in design science studies.

The chapter proceeds as follows. First, we discuss the needs for theory contributions from DS/IS with reference to similar discussions in management science. Second, we propose the notion of design inquiry as an approach to knowledge development and theorizing in design science studies. Third, we introduce the notion of practical theory, a pragmatist way of understanding and representing "instrumentalities". Fourth, we present a case example to demonstrate the concept of practical theory and how such a theory can emerge in a design inquiry. Fifth, we describe in more detail how practical theories can be used in and developed through the different phases of design science studies. The chapter is concluded with a discussion including implications for research and practice.

Design Science Cases and Theorizing

In this section, building on the account of DS theorizing in the introduction section, we explore ideas on theorizing and publishing. This discussion will benefit from similar debates on the needs of theory in published papers in management science (Corley & Gioia, 2011; Eisenhardt, 1989; Hambrick, 2007; Sutton & Staw, 1995; Weick, 1995). At the end of the section, drawing from the discussion, we propose four implications for design science theorizing.

In a discussion on acceptance criteria for journal papers "What theory is not", Sutton and Staw (1995) identify five elements that may occur in papers that should not be confused with theory: references to extant literature, data, list of variables/constructs, diagrams, and hypotheses. As a contrast, they argue for "strong theory", which answers why-questions by explaining and predicting causal relations between phenomena.

In a reply, "What theory is not—theorizing is", Weick (1995) nuances the claims by Sutton and Staw (1995). One main argument by Weick (1995) is to acknowledge the *progression in developing theory*. Theories exist in different states of progress and therefore "theory work can take a variety of forms, because theory itself is a continuum" (ibid., p. 387). As we interpret Weick's argument, the presentation of a theory should include elements of its emergence and all the five elements mentioned by Sutton and Staw (1995) are building blocks in the theory development. We would also add that three of them can be seen as constituents of a theory (constructs, diagrammatic descriptions of relationships between constructs, statements that still are hypothetical), although they may not be seen as exhaustive constituents. This means that a description (in a scholarly paper) made with theoretical ambitions should contain such elements of "interim struggles" (Weick, 1995 p 389) with theory. A presentation of a theoretical contribution should not only contain the theory as such. It should be warranted through a description of how it has emerged. Weick (1995, p. 389) states: "The process of theorizing consists of activities like abstracting, generalizing, relating, selecting, explaining, synthesizing, and idealizing. These ongoing activities intermittently spin out reference lists, data, lists of variables, diagrams, and lists of hypotheses." Researchers need to justify theory empirically and theoretically and demonstrate its conceptual rigor.

Hambrick (2007) takes a critical position to claims that every journal article should contribute to theory. He calls it an "obsession with theory" (ibid., p. 1346) and states "the requirement that every paper must contribute to theory is not very sensible; it is probably a sign of our academic insecurity" (ibid.). One important argument is that the "idolization of theory" (ibid., p. 1347) will prevent other important results from research to be described in publications such as rich empirical descriptions. He asks for "papers that identify compelling empirical patterns that cry out for future research and theorizing. They might be rich qualitative descriptions of important but unexplored phenomena that, once described, could stimulate the development of theory and other insights" (ibid.,

p. 1350). Scholars may theorize based on earlier published rich case descriptions and tentative abstractions. Hambrick also puts forth the risk that a too strong focus on novel theories will de-emphasize empirical grounding of proposed theories: "We in management, however, are so riveted on new and revised theories, and so dismissive of simple generation of facts and evidence, that our revealed ethos is that we care much more about what's fresh and novel than about what's right" (ibid., p. 1350).

In our view, Hambrick's objections rely on a strong cumulative view of scientific knowledge and research work. Scientific knowledge is developed in small steps and not every publication can embrace the full range from hunches and ideas through empirics to an exhaustive theoretical account.

Eisenhardt (1989) has described the development of theory based on case studies in management. She is inspired by case study methodology, grounded theory and other qualitative approaches in social sciences. Eisenhardt elaborated a way of working with theory development based on case studies. She does not advocate a strict inductive development of constructs and theoretical elements. She acknowledges the possibility of an *a priori* formulation of constructs as it may improve measurement in the case study. However, she emphasizes, that such initially formulated constructs should not be guaranteed a place in resultant theory. Every construct in such a theory must earn its place there based on a thorough data collection and data analysis.

Theorizing can sometimes be set in opposition to practical usefulness. Corley and Gioia (2011) argue that these two aspects do not need to be antagonistic. On the contrary, theorizing in management should be conducted with a focus on practical relevance and usefulness. "There is nothing about the nature of either theory or practice that prevents them from being served simultaneously" (ibid., p. 28). With a clear foundation in pragmatist epistemology, they argue that "our theories should be problem driven—that is, in some fashion addressing a problem of direct, indirect, or long-linked relevance to practice, rather than narrowly addressing the (theoretical) 'problem'" (ibid., p. 22). This entails also naturally to engage in dialogues with practitioners. Corley and Gioia (2011) also argue for a broad view of theoretical usefulness. A theoretical contribution should be potentially useful to both these target groups. Ågerfalk and Karlsson (2020) also confer.

In conclusion to this section, we propose four implications for theorizing in design science. *First*, with inspiration from Eisenhardt (1989), we can state that it should be *possible to develop a theory* based on DS case studies. The development of a resultant theory can be performed along abductive lines as we interpret her suggestions. We make these conclusions since we conceive of a design science study, in most situations, as a special kind of a case study. We call this a *design case study*. It is so since it is an in-depth study in a problematic situation with a focus on designing artifacts as a response to identified problems. This is similar to stating that certain action research studies can be considered and labeled action cases (Braa & Vidgen, 1999).

Second, with inspiration from Hambrick (2007), we think that it is *challenging to generate and formulate an elaborate theory* as an outcome from a DS study, and, thus that it should *not* be considered *necessary to create any exhaustive theory*. We will make nuances to this conclusion in our third conclusion below. Most design case studies will give great opportunities to develop rich empirical descriptions of identified problems and designed artifact features. Such descriptions, if they are expressed with sufficient details, can give other scholars possibilities to advance this work further.

Before we formulate our third stated implication, we need to introduce some conceptual distinctions. There are several roles of theories in design science studies. We distinguish between three different theory kinds in relation to their functions in design science: Introductory theory, emergent theory and contributed theory. An *introductory theory* is a theory selected from a scholarly knowledge base to be used in a design science study. This means that it is an externally provided theory. An *emergent theory* is a theory that is developed during a DS study and is used as instrument in that process. An emergent theory can be influenced by introductory theory or it can be a novel instrument generated on the basis of addressed design issues. An emergent theory can be transformed into an explicit theoretical contribution as a result of the design science study. Such a theory outcome is labeled *contributed theory*. We will elaborate further on these three theory kinds in the sections on practical theory below.

Third, with inspiration from Weick (1995), we think that it is *desirable to formulate elements of a theory in progress* as one result of a DS study. A design science study should in our view be conducted as alternating between designing artifacts and theorizing (see next section). In theorizing, we should work with emergent theory as abstractions from the design process and with one aim to furnish theoretical tools to that design process. Such an emergent theory can also be made to a contributed theory from the design science study. This, taken together, implies that what has emerged as useful abstractions during a DS study should be described and reported as important results from such a study. We should, of course, give a rich description of the design case as mentioned in the second conclusion above. However, it should be considered as a valuable DS contribution to *enunciate the abstract reasoning* that has taken place as an integral part of the DS study. This can be (1) new, useful conceptual abstractions and/or (2) refined and/or (3) corroborated conceptual elements from introductory theory. We do not claim that every DS study should produce a coherent and comprehensive theory as a result, but our firm conviction is that *a report from the theorizing process with emergent theory may contribute to the scholarly community's cumulative theory evolution*. This follows the pragmatist position of theory as in a state of continual becoming (Dewey, 1938; Friedrichs & Kratochvil, 2009).

Fourth, with inspiration from Corley and Gioia (2011), we think that *theorizing should take its vantage point in problems in practice* and an *aim for knowledge contributions as useful for both scholars and practitioners*. In design science studies, it is a natural position to start the inquiry from problems and needs in practice and let the design of an artifact be a sensible response to such problems and needs. It is also a natural position to not only produce useful artifacts but also abstracted knowledge that may be useful to both scholars and practitioners.

Design Inquiry and Theorizing

A design science study creates one or more artifacts and knowledge related to those artifacts. We mean that design here is best understood as an inquiry process. In the philosophy of pragmatism, knowledge development is considered as inquiry processes (Cronen, 2001; Dalsgaard, 2017; Dewey, 1910, 1938). A pragmatist inquiry can be contrasted to a traditional view of knowledge development as making a map of reality. *Inquiry means knowledge development interwoven with improvement of some practice.* Knowledge is created for the sake of improvement. The classical view of inquiry explicates it as a movement from a problematic situation to the resolution of that situation (ibid.), similar to Simon's (1996, p. 111) view that "Everyone designs who devise courses of action aimed at changing existing situations into preferred ones. The intellectual activity that produces material artifacts is no different fundamentally from the one that prescribes remedies for a sick patient or the one that devises a new sales plan for a company or a social welfare policy for a state." Inquiry involves the creation of knowledge, and this is not done as a purpose of its own, but as instrumental to the wider purposes of improvement of some human conditions. Such an improvement is created through knowledgeable actions, and if adequate knowledge and insightful actions were not there, improvement could not happen.

Knowledge development as part of inquiry processes goes beyond an epistemology of descriptive and explanatory knowledge as is often considered as archetypical for scientific development. Such types of knowledge can perfectly well be part of pragmatist inquiries, but there is also knowledge of other epistemic types, such as diagnostic, normative and constructive knowledge. Descriptive and explanatory knowledge may earn their places in the knowledge development of an inquiry if they are useful for the settlement of a problematic situation to a resolved situation.

Change and improvement mean development of some human practice and such development is not possible without knowledge. This is why knowledge development is a constituent and interlaced part of this broader kind of practical development. Knowledge development in an inquiry cannot be separated from the improvement aims and actions of that inquiry. It is purposeful knowledge development. Van Strien (1997, p. 684) describes it as "understanding reality is made subservient to changing reality".

An inquiry process may take different shapes. There exist classical descriptions of generic stages in Dewey (1910, 1938). However, Dewey (ibid.) himself acknowledges that inquiry practice in itself is subject to improvement and change. In the classical inquiry description (ibid.), it is conceptualized as a transformation of a problematic situation into a settled and determinate one. This transformation passes through different inquiry stages: problem formulation, proposal formulation, abstract reasoning and testing of proposals. This description should be considered as a proper vantage point for inquiry as design but some specific adaptions should be expected. We will later in this chapter elaborate further on the design process as inquiry.

One crucial question is what makes a design process part of a design science study? How is design part of research and what makes it scientific? There exist arguments about innovation and novelty concerning designed artifacts as preconditions for making design scientific (Hevner et al., 2004; Iivari, 2007), but these arguments will not be scrutinized here. We take another route forward, the route of abstraction and theorizing. There are several scholars that describe design-oriented research in IS as an alternation between a concrete design layer and an abstract reflection layer although labels are shifting (e.g. Baskerville, Kaul, & Storey, 2015; Goldkuhl & Lind, 2010; Lee, Pries-Heje, & Baskerville, 2011; Sjöström & Ågerfalk, 2009; Winter, 2014). We have elaborated this in Fig. 7.1. We have given the abstract layer the label "theorizing" and the concrete design layer is called "situational design inquiry".

What we call theorizing is part of a broader design inquiry process (the design science study), but it can itself be considered as an inquiry. It is an inquiry as it is a transformation from a research problem/idea to developed abstract and useful knowledge as a resolution of that original and sometimes transformed research challenge. We describe a design science study as a research inquiry consisting of two interrelated and mutually dependent and supportive sub-inquiries: (1) A situational design inquiry

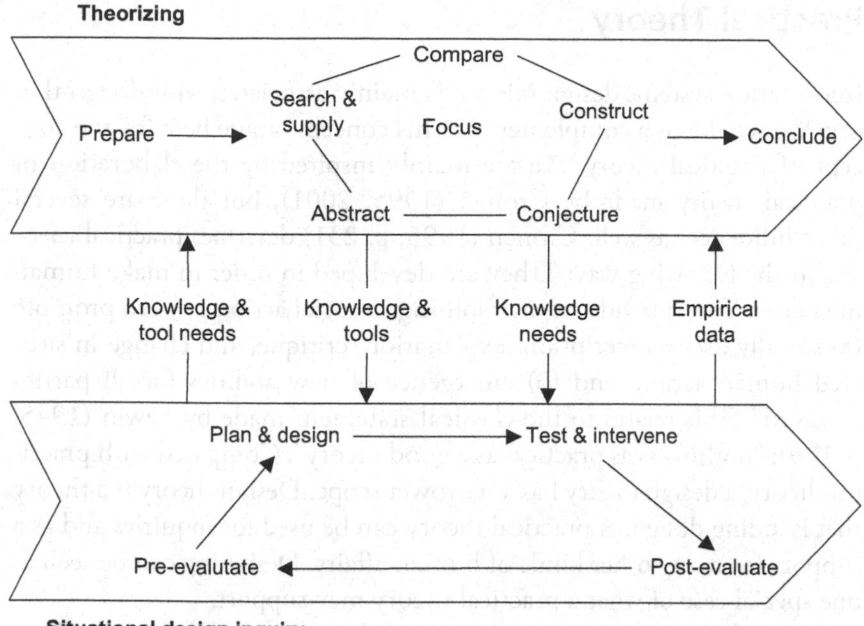

Fig. 7.1 A design science study as an alternation between situational design inquiry and theorizing (from Goldkuhl & Sjöström, 2018)

where a problematic practice situation is transformed, through the design of one or more artifacts, to an improved practice situation and, (2) a theoretical sub-inquiry where a research problem is transformed into abstract and useful knowledge contributions.

Our view relates to Baskerville et al. (2015), who provide a view of four modes of inquiry. They base their view on dichotomies: idiographic vs. nomothetic and design vs. science. They explain DSR as an intellectual activity shifting between focus on design and focus on science and alternating between the situational problem and the abstract problem. The two polarities lead to four modes of inquiry. Our view, albeit with a slightly different angle, acknowledges that we need to understand DS as a continual shift between these modes of inquiry and that they are mutually supportive.

Practical Theory

Information systems design science is mainly associated with design theory. We would, as a complement to this concept, argue here for the concept of practical theory. We are mainly inspired by the elaboration of practical theory made by Cronen (1995, 2001), but there are several other influences as well. Cronen (1995, p. 231) describes practical theories in the following way: "They are developed in order to make human life better. They provide ways of joining in social action so as to promote (a) socially useful description, explanation, critique, and change in situated human action; and (b) emergence of new abilities for all parties involved". This relates to the classical statement made by Lewin (1945, p. 129): "nothing is as practical as a good theory". Compared with practical theory, a design theory has a narrower scope. Design theory is a theory that is aiding design. A practical theory can be used for inquiries and as a support for many other kinds of human affairs. Designing can be seen as one special case of what a practical theory may support.

Practical theories are aimed for both scholars and practitioners. They can be used in inquiries, according to the broad scope described above, and in other kinds of management of social affairs. Cronen (2001, p. 30) motivates the use of a practical theory in the following way: "Its use should, to offer a few examples, make one a more sensitive observer of details of action, better at asking useful questions, more capable of seeing the ways action are patterned, and more adept at forming systemic hypotheses and entertaining alternatives".

There are other terminologies for this type of concept or similar types. Jung et al. (2010) talk about "perception-action theories" to be used in design practices by both design researchers and design practitioners. Davison et al. (2012) describe the use of "instrumental theories" in action research. Van Strien (1997) does not explicitly use the phrase "practical theory", but he describes well the instrumental use of practically adapted theories related to psychological practices. Craig and Tracy (1995) use the terminology "practical theory" but they use also "normative theory" as almost a synonym. Friedrichs and Kratochvil (2009) describe the development of theory as a process of pragmatic abduction with the main purpose of producing theory with useful knowledge. The use and application of what is called practical theory have been described in different areas such as public administration (Miller & King, 1998), education (Feldman, 2000), international relations (Bohman, 2002), social work (Moore, 2004), entrepreneurship (Rae, 2004), communication (Barge, 2004), nursing (Stevenson, 2005) and information systems (Goldkuhl, 2007, 2011; Goldkuhl & Sjöström, 2018; Hultgren & Goldkuhl, 2013; Sjöström, 2010). This referenced body of knowledge on practical theory (and similar notions) is used here as a basis for our articulation of practical theory in relation to design science studies.

A practical theory, as a "knowledge artifact", is produced and used. We describe it here in relation to inquiry processes in general. In the next section, we will be more specific and relate it more clearly to design science studies and its possible roles in situational design inquiry and theorizing. We will here use the three theory roles described above (introductory theory, emergent theory, contributed theory).

A practical theory is thus considered as an instrument in an inquiry process. However, the use of practical theory in an inquiry may change that specific theory. Thus, we distinguish between an *introductory* practical theory and an *emergent* practical theory. An introductory theory is what exists before an inquiry and is selected by the inquirer for use. Such a theory can (1) be selected before an inquiry starts based on a comprehension of what the inquiry will pertain or (2) it can later be selected due to "discoveries" and insights during the inquiry process. There may be several introductory theories used as complements during an inquiry.

A practical theory can emerge during an inquiry process. Such an emergence process can be *inductively* driven, where empirical knowledge is abstracted to theoretical constructs, typically following the principles of coding in grounded theory (Strauss & Corbin, 1998). An emergent practical theory can also be the result of a reflective application of an introductory practical theory where certain re-adjustments, reformulations, and additions are made to the introductory theory. Empirical insights from the inquiry can lead to such a transformative development of a practical theory. Such a theory development process is rather an abductive process as being influenced by both extant theory and generated empirical data (Friedrichs & Kratochvil, 2009; Goldkuhl &

Cronholm, 2010; Kelle, 2005; Thornberg, 2012; Van Maanen, Sørensen, & Mitchell, 2007).

An emergent theory can also be considered as a theory outcome from a theorizing work related to an inquiry process. Such a product is called a *contributed* practical theory. A practical theory that has emerged during an inquiry process can be made to a contributed theory through some final adjustments; for example, further abstraction, conceptual clarification and making it a proper transferrable knowledge product. Such a contributed practical theory needs, of course, to be presented in a scholarly publication where (1) its inquiry origin and context is well described as a proper empirical grounding and (2) together with a conceptually positioning of it in relation to relevant extant theories (theoretical grounding); cf. Goldkuhl and Cronholm (2010).

As mentioned above, an emergent practical theory can have its origin from an existing introductory theory. This means also that such an introductory theory can be further developed through active use in and continual refinement through an inquiry process. The inquiry process can give more evidence to the applicability and usefulness of different parts of the theory, while other parts become refined or even refuted. This illustrates the continual progression of theoretical knowledge as one key principle in pragmatist thinking (Dewey, 1938).

The use of a practical theory in an inquiry process should lead to an insightful understanding of the specific domain as the study object. As Cronen (1995) states, the use of a practical theory should make one a more "sensitive observer of details of action" and "more capable of seeing the ways action are patterned". This means that an in-depth situational understanding should emerge through the application of a practical theory. One needs to distinguish between a practical theory and its resulting situational knowledge. A practical theory is formulated in general, abstract terms, and is aimed for use in studying different practices. The emergent situational knowledge is aimed to clarify a specific study object. The formulations of such a situational understanding must use terms clearly related to elements in that studied practice even if there may be some influence of abstract terms from practical theory in order to enhance apprehension. It is also important to note that a situational understanding is not developed in a strict deductive fashion from a practical theory that is formulated in more abstract terms. It is rather the case of mutual fertilization between practical theory and situational knowledge (Van Strien, 1997). As mentioned above, empirical insights can influence the emergence of a changed/new practical theory.

The development of structured, situational knowledge is sometimes characterized as "theory" formulation. This is, for example, made through the use of "local theory" related to action research (Elden, 1983), "program theory" related to evaluation (Rogers, Petrosino, Huebner, & Hacsi, 2000), and "mini-theory" in psychological treatment (Van Strien, 1997). Even if we can acknowledge the point in characterizing this kind of structured, low-level situational knowledge as a theory, we will avoid this terminology here in order to prevent any confusion with practical theory.

A Case Example

We illustrate the use and development of practical theories with examples from an e-government design science study. The two authors participated as action/design researchers in a case concerning personal assistants' support and service for clients with certain disabilities. Different aspects of this case have been reported previously (Goldkuhl, 2011; Sjöström, 2010; Sjöström & Goldkuhl, 2009). Representatives from 14 Swedish municipalities and the Swedish Social Insurance Agency (SIA) participated actively in the project. The administration of assignments and allowances to the personal assistants was experienced to be very cumbersome and complicated due to its regulatory character and complex stakeholder relationships. The DS study consisted of the design of integrated work processes and IT support in order to achieve more effectiveness, transparency, and simplicity in the studied practice.

Selection of Introductory Theory Tools

In order to design such improved work processes and IT artifacts, there was a strong need to first inquire the current work situation (i.e. to conduct "pre-evaluate" following our DS process model in Fig. 7.1). We made a thorough analysis of problems in the current situation. In this inquiry task, we applied a problem analysis method (Goldkuhl & Röstlinger, 1993, 2003). Through this analysis, we generated several situational problem models revealing difficulties and challenges in the work situation. This problem analysis method contains notation for problem diagramming, procedural rules for problem investigation and problem formulation, and a conceptualization of problems into causes, effects, and problem chains and circles. These instrumentalities for problem inquiry comprise thus not only modeling tools but also kinds of practical theories for performing problem analysis. Resulting problem diagrams from this inquiry can be found in Sjöström (2010).

Through this problem investigation, we identified complex relationships and interactions between the different stakeholders such as SIA, municipalities, assistance providers and clients. It gave rise to a need to describe and evaluate these relationships and interactions in a thorough way. We selected a practical theory to be used as an instrument in this description and evaluation. The Generic Exchange Model (Goldkuhl & Röstlinger, 2007)—GEM—was selected as such an introductory theory. GEM has its theoretical roots in speech act theory (Habermas, 1984; Searle, 1969) and the language action perspective in IS (e.g. Winograd & Flores, 1986). GEM conceptualizes and describes the interaction between two parties (supplier, client) in four generic exchange phases (initiation, agreement, fulfillment, completion). These four exchange phases constitute the accomplishment of a transaction between a supplier and a client. The exchange actions in the generic model are considered as of genuinely social character with accompanying commitments and expectations (Goldkuhl & Lind, 2007; Habermas, 1984; Sjöström & Goldkuhl, 2009; Weber, 1978).

The main idea was to use this generic model as a template and inspirational source in pre-evaluation and design. With the generic model as a template, situational models should be generated that describes the social interaction between different stakeholders in the egov case. GEM had been used earlier with good results in several other research projects, both in governmental and commercial settings. For a more detailed description of motives for the selection of GEM as an introductory theory in this DS case, see Goldkuhl (2011).

The Emergence of an Adapted Practical Theory

When applied, the Generic Exchange Model could not contribute to clear descriptions of stakeholder interactions and relationships. Social exchange was described, but it could not capture the complex regulatory character of this case in a sufficient way. These problems and needs in description led to theoretical reflection and reasoning. Could we find other ways to conceptualize and describe this kind of regulation-based interaction? Different action concepts from the Generic Exchange Model and the language action perspective were still considered useful, but we needed another way to comprehend this type of practice. A new way of looking at this kind of practice emerged through alternating between theoretical reflection and empirical analyses. We tried out different situational models and continually a new understanding emerged. Instead of seeing just one transactional practice, we divided the interaction into two different, but related, types of transactional practices; one *regulating* practice and one *regulated* practice. In parallel, a new generic model/practical theory and situational models of the assistance practice emerged with these two practice layers. The emergent practical theory was labeled "Generic Regulation Model" (GRM); see Goldkuhl (2011). Informed by this emergent theory, situational models were developed in process modeling workshops together with the participating practitioners. See Fig. 7.2 for a situational model from this case describing the transactional practices of assistance to disabled persons.

This model was used in the design inquiry in several ways; for more details see Sjöström and Goldkuhl (2009), Sjöström (2010), and Goldkuhl (2011). The situational model (including this way of thinking) was used in the pre-evaluate phase in order to better understand the problematic situation and to demarcate the further design inquiry. The model was also used in the design phase as guiding a socially informed design of IT artifacts.

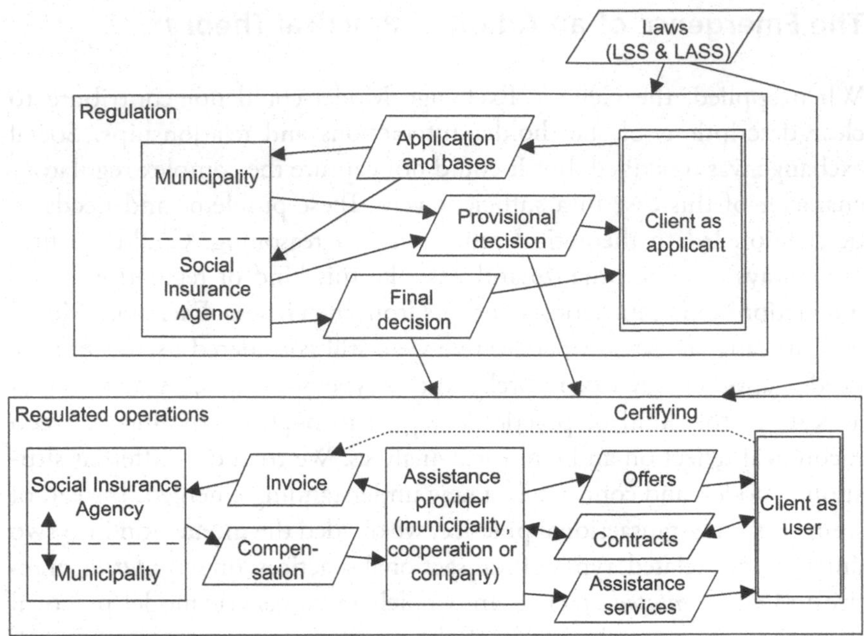

Fig. 7.2 A situational model of social/regulatory interaction in the assistance case (from Sjöström & Goldkuhl, 2009; Goldkuhl, 2011)

Formalization of a Contributed Theory

The new theoretical understanding (in terms of GRM) that had emerged during this DS study was closely tied to the actual case. We saw a potential of making this theoretical model an explicit and separate outcome from this DS study. This involved further theoretical and empirical grounding and conceptual refinement (Goldkuhl, 2011). We conducted some conceptual analysis of the applied regulation concept in relation to scholarly literature. Resonating with Lee and Baskerville (2003), we also revisited some previous case studies in egov and used this empirical material for the application of new constructs and ways of modeling. Through such theoretical and empirical confrontation, we could advance this emergent practical theory on regulation in e-government. We made some further abstraction and refinements to the GRM model. This model consists now of three explicit layers/practices: (1) Regulating general practice (legislation as issuing general regulations), (2) regulating specific practice (issuing decisions/individual rules through application of legislation), (3) regulating practice (applying general and specific rules). The GRM model (Fig. 7.3) was made available in a scholarly publication (Goldkuhl, 2011) as both a contribution to the scholarly knowledge base and a general practice contribution. In this chapter, the emergence of this model through the design case as well as its role as a practical theory was described. These and other parts of the design case study (including design experiences and considerations) have been described in other publications (Sjöström, 2010; Sjöström & Goldkuhl, 2009).

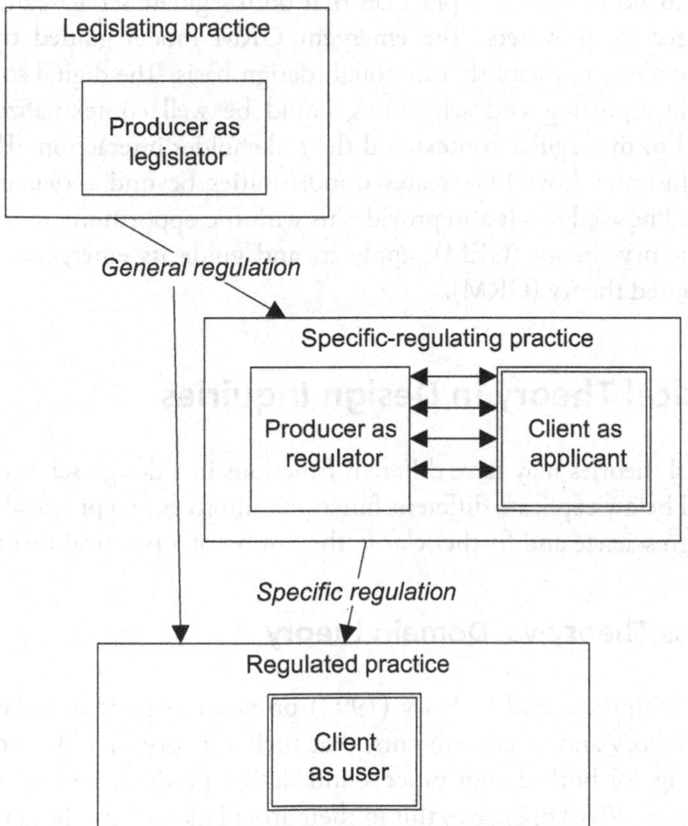

Fig. 7.3 The Generic Regulation Model (from Goldkuhl, 2011)

Importantly, relating to the example case above, the project goal was to develop a solution for time reporting for personal assistance in a Swedish municipal context. The typical DSR approach would seek to develop domain knowledge for that design context, i.e., time reporting and scheduling, potentially narrowing down the problem to the specific context in some way. An ADR approach, for instance, would probably seek to articulate design principles to support design given the time reporting/scheduling problem class. The development of GRM, however, is an equally important theoretical result of the DS case, and thus constitutes an example of how DS studies can enable different types of theorizing: First, a domain theory for time reporting/scheduling design. Second, a practical theory to aid inquiry into practices that both regulate services and operate as service providers. The emergent GRM model guided the pre-evaluate phase to establish a thorough design basis. The digital solutions for time reporting and scheduling could be well contextualized and adapted to the regular context and the stakeholder interaction. The case thus illustrates how DS creates opportunities beyond articulation of domain knowledge—it also provides us with the opportunity to select an introductory theory (GEM), apply it, and guide its emergence into a contributed theory (GRM).

Practical Theory in Design Inquiries

Practical theories may have different functions in a design science study. We will below explicate different functions and roles of a practical theory in design science and further clarify the notion of a practical theory.

Process Theory vs. Domain theory

Walls, Widmeyer, and El Sawy (1992) propose, in their articulation of design theory and its constituents, that such a theory should encompass knowledge of both design process and design product. Baskerville and Pries-Heje (2010) oppose to this in their articulation of the design theory concept. They claim the importance of separating a design practice theory and a design theory on design artifacts. We can agree that it is possible to theorize these different aspects of design separately, that is, design practice/process vs. results and effects of design. The latter is here labeled *domain theory* as it may include knowledge of artifacts (as designed responses to practice problems), possible uses by users and other actors (as effects from artifact uses), and other elements and interactions in a practice (as contextual domain knowledge). Domain theory also includes theories about vantage points for design, that is, practices interpreted with problems and needs for improvements, specifically concerning enduement of potential artifacts. One example of a domain theory is the Generic Regulation Model as described in our case example. The former theory type can be called *design process theory*. Following the division of the design inquiry in Fig. 7.1, a design process theory can pertain to this whole process or how to perform different parts of it (i.e., pre-evaluate, plan & design, test & intervene, post-evaluate). One example of a design process theory is a theory for problem investigation as mentioned in our case example.

However, in an inquiry-oriented practical theory (as we put forth here), there is no need to make any sharp differentiation between process theory and domain theory. A practical theory can have a process emphasis or a domain emphasis. A theory with a focus on the design process may naturally give references to what is to be designed through such a process. The design process will be in the foreground, while objects and domain to be designed will be in the background. A practical domain theory may also contain knowledge about how to conduct an inquiry in this domain, i.e., knowledge about inquiry processes. The Generic Regulation Model provides a way to think about regulation in governmental practices.

This position to theories is a natural consequence of our pragmatist interpretation of it as an instrument (Cronen, 2001; Van Strien, 1997). A theory is seen as an instrumentality as other tools in the design inquiry process, such as models and methods. If we look at the instrumentality sibling of an IS design method, such knowledge artifacts usually contain both process elements (how to design) and product elements (what to design); see, for example, Henderson-Sellers et al. (2014) for descriptions of contents of design methods.

As described above, we adhere to the terminology of "practical theory", and thus that the applied theory should be practical and useful in the design inquiry course of action. We include any theory that can be made practical in the design inquiry, even if this was not the intentions of the originators of that theory. Many theories have the potential of being practical and useful, as noted by Craig and Tracy (1995). Fishman (1999, p. 289) accentuates that "the pragmatist views all theoretical concepts as potential tools to be employed in practical problem solving".

General Roles of Practical Theory in Design Inquiry

A practical theory should serve the progression of the design inquiry. A design science study is a *cooperative effort* involving different stakeholders (researchers, practitioners). As a research endeavor, it concerns interested members of the scientific community. This stakeholder dependence implies that transparency and communication should permeate the design inquiry. The use of a practical theory is one supportive element in a DS study as a *communicative practice*. An extant practical theory, as represented in a scholarly knowledge base, is an open knowledge artifact accessible to those interested. An emergent practical theory, created during a DS study, should be reported as a contributed theory to the scientific community to have it reviewed and valued, and made accessible to others for use and further development. The introductory theory in our case example (GEM) was previously reported (Goldkuhl & Röstlinger, 2007). In the design case, a new emergent theory (GRM) was created when the introductory theory was found insufficient in this design situation. The presentation of GRM, in a scholarly publication (Goldkuhl, 2011), positioned this new practical theory clearly in relation to previous scholarly knowledge as, for example, GEM.

A practical theory is useful in a design inquiry when it is generative for creating different kinds of descriptions that are considered valuable in the progression of this collective inquiry process. Situational descriptions, such as problem descriptions, goal models, design proposals, and effect evaluations, are essential means for communication and collaboration (Cronen, 2001; Davison et al., 2012; Stevenson, 2005). Open communication is a driving force for making design inquiry a *reflective and collaborative practice* (Craig & Tracy, 1995).

It should be added that such a design inquiry is fundamentally influenced by its aims for improvement and that this "colours the [participants'] course of thinking right from the beginning. Because of the improvement-directed character of practice, every step of its methodological cycle is guided by norms and values" (Van Strien, 1997, p. 686). Design inquiries are suffused with values and practical theories should guide the development of *normative rationality* during that process (Craig & Tracy, 1995).

Practical Theories Used in Different Phases of Situational Design Inquiry

Practical theories can be used to support different phases of the situational design inquiry in a DS study. Such theories can be supportive in the pre-evaluation of problematic situations in practices, in the designing of artifacts, and in the post-evaluation of tested/used artifacts in changed practices. Davison et al. (2012) describe that (what they call) instrumental theory can be useful in corresponding phases of diagnosis, intervention planning and change evaluation in Canonical Action Research. Sein et al. (2011, p. 41) suggest, in Action Design Research, the use of theories "to structure the problem … to identify solution possibilities … and to guide design".

Theories used in the initial phase (i.e. pre-evaluate) of a situational design inquiry can thus be called "diagnostic theories" (Davison et al., 2012; Van Strien, 1997). Their use "enables orientation in the social world" and they can be helpful "to understand complex social phenomena and/or to explain observed social regularities" (Friedrichs & Kratochvil, 2009, p. 706). The new practical theory, GRM, was created, as a response to apprehended complexities in the design case, and as such it is an adapted theoretical tool to be helpful in clarifying those complexities. A practical theory combined with resonating modeling tools can support the inquirers to conduct modeling of work processes in order to clarify how work is done and what works well and what does not work well (Davison et al., 2012; Goldkuhl & Sjöström, 2018). A practical theory can be supportive in the assessment and critique of a current practice (Craig & Tracy, 1995; Cronen, 2001; Stevenson, 2005).

In the phase of Pre-evaluate, the search for identification and clarification of problematic situations is important. The claims in Davison et al. (2012, p. 768) for the use of "instrumental theory" in CAR can be transferred to Pre-evaluate in situational design inquiry: "In early-stage diagnosis, an instrumental theory tends to play an important role in helping the researchers to develop a thorough understanding of the organizational problem(s)". Craig and Tracy (1995) talk about a dilemmatic perspective, that is, identifying dilemmas with contrary impulses and ideas about suitable actions in a practice. A practical theory may have guiding roles in searching for problems, but first of all, it is vital for the inquirers to have a curious and open mind to the studied practice. One should start the inquiry with an open mind for what is in the situation. Theories of any kind should not be imposed on the situation and prevent discoveries. Cronen (2001, p. 28) claims that "the particulars of a specific situation provide the best initial orientation."

In the orientation and search for the understanding of a problematic situation, different elements of practical theories can be used. First and foremost, the concepts of the theory can be a searchlight for developing an understanding of the situation. It is important to acknowledge that the building blocks of a theory and in all knowledge development are concepts. If we use extant theories, we should first be aware of and use its constituent concepts and second to use statements of causal or other relational types (Friedrichs & Kratochvil, 2009). Concepts that are fetched from extant theories should be used in a sensitizing way (Blumer, 1954; Bowen, 2006).

Sensitizing concepts are not only important in the Pre-evaluate phase. They are also important in Plan & design and Post-evaluate. As emphasized by Jung et al. (2010), there is a close connection between perception and intervention in designing. They claim the importance of a perceptual field in the design process: "We define a Perceptual Field as sensing organized around a purposeful activity. With the notion of a perceptual field, we want to refer to what one notices when one is engaged in the activity of designing. This noticing can refer to things in the environment or to internal states and feelings" (ibid., p. 237).

Designing involves (1) interpretive understanding, (2) normative reflection and (3) instrumental rationality when choosing design options that may fulfill enounced values (Craig & Tracy, 1995). A practical theory can serve these different aspects of designing through (1) interpretive schemes and sensitizing concepts, (2) articulated fundamental values and procedures for a situational normative analysis and (3) providing a repertoire of designerly action procedures for linking enounced values and features of design. The designed artifact may have clear traces from theoretical inspiration during design. Influence from theory will make the designed artifact a "theory-ingrained artifact" (Sein et al., 2011).

Practical theories may also have functions in the concluding phase of Post-evaluate in situational design inquiry. A designed artifact put to test can be seen as a design hypothesis, and such a hypothesis may be part of an emergent theory developed during the DS study. Certain use effects are predicted to occur according to the theorized understanding of the artifact and its place in a practice situation. An evaluation of the test and use of an artifact should be informed by such an emergent theory in order to judge if the outcomes are in accordance with the design hypotheses within theory. This means also that post-evaluation should generate data for assessing theories and thus validating/modifying/rebutting such theories (Davison et al., 2012).

Selection and Development of Practical Theory in Design Science Studies

Encounters with problematic situations and design challenges may stimulate a search for theories that may support further inquiries. Theories are selected from a knowledge base due to their abilities to serve further design-oriented knowledge development. A selected theory may be further developed through its active use in design inquiry. Cronen (2001, p. 30) describes a practical theory to be a malleable and evolvable theory: "A practical theory should allow for further development of old methods and creation of new ones". "A practical theory should grow in the richness of its instrumentalities". The initial selection of GEM, in our design case, is an illustrative example of a need-based search for and selection of a theoretical tool that was considered potentially useful. This theoretical tool was, however, not found as fitting to the investigated circumstances. This misfit gave rise to the development of a better adapted practical theory (GRM). However, certain conceptual elements from the previous theory (GEM) has been transferred into the new theory (GRM). There was a clear knowledge progression of practical theories; one developed based on a former one.

Cumulative progression, interlaced with practical use, should be seen as the natural way of theory building. Inquiry and theorizing are described as co-dependent processes. Stevenson (2005, p. 197) writes about the close linkage between inquiry and theory-building: "Practical theory is not the end product of practical inquiry. The two create one another and the process involved consists of 'loops' in which inquiry informs theory and theory informs inquiry". Concepts are central in this dialectic development of theory and practical understanding/intervention. Concepts are instrumental and evolvable throughout the design science process. Friedrichs and Kratochvil (2009, p. 717) describe these interdependent processes as "the mutual adaptation of conceptual framework, field of research, and empirical findings". These interdependent processes are centered around concepts in the following way: "Our concepts constitute our field of research. What we observe in that field will, in turn, elucidate or modify our understanding of the concepts" (ibid.). This implies a hermeneutic-sensitizing nature of practical theories and inquiry work. The movement from empirics to abstracted theory is well described by Van Strien (1997, p. 693) as "based on a cognizance of the 'essential' in a certain category of situations and problems: the 'general in the specific', we could say. The description and analysis of typical cases that can serve as a heuristic aid in interpreting a new case is the method par excellence of this kind of hermeneutic understanding." This corresponds well to a view of the progression of practical theories through an accumulation of (design) cases with various applications of those theories.

When a practical theory is used to inform one or several phases of the situational design inquiry process in a design science study, this will naturally act back on the applied theory. The theory may be appropriated in use and the learnings from this use will inform further development of this theory. New conceptual distinctions may be added to the theory and other conceptual modifications and refinements will be done. This means that a theoretical development is a natural outcome from a DS study; that is, what we have called a contributed theory from a DS study. This does not necessarily mean the development of a full-blown novel theory. A valuable theoretical outcome can thus be a proposed modification of an applied extant/introductory theory. Another possible theoretical outcome can be an embryonic theory that has emerged as a novel conceptualization through a design inquiry. In the design case example, a new practical theory (GRM) emerged as a response to apprehended work complexities. In a later theory stage, this new theory was further abstracted and generalized in order to further improve it as a transferrable and useful knowledge product.

The Multi-functional and Cumulative Character of Practical Theory in a Community of Practice

We have emphasized a practical theory to be instrumental. It should be so in several respects, that is, it should be a multi-functional instrument. Many demands can be raised concerning a practical theory. We have above stated many functions for a practical theory in design inquires. We summarize this by stating that a practical theory should be supportive for orientation in a problematic situation, interpretation, and clarification of understanding of such problematic situations, description and assessment of work processes, articulation of goals and values to govern design and intervention, artifact designing, post-evaluation of test and use and further conceptual and theoretical development.

Jung et al. (2010, p. 234) envision an instrumental and multi-functional theory: "How could theory be structured to grow a culture of designerly inquiry … that supports us in what we do, helps us in communicating our understanding and lets us expand and improve our knowing and doing?" Besides the list of supportive functions mentioned above, we can add such a communicative role in community growth. This communicative role is important in that such a theory should be one instrument and stimulus for the evolution of a community of reflective designers (design-oriented researchers, design practitioners). Such a common instrument may function as a means for developing reflective designer skills and sharing experiences that can be added as exemplars of published cases and further development of this common designers' tool. A practical theory as being a tool that is used by different actors makes it naturally an evolvable object. A tool that is used by different actors leads to different experiences, which can be shared and lead to possible modifications of the tool. Instrumental knowledge for and about the world should evolve in parallel with the world as in a continual state of becoming. As emphasized by Friedrichs & Kratochvil (2009, p. 713): "Pragmatism reinstates the provisional character and historical contingency of scientific knowledge".

The cumulative progression of instrumental theorizing includes also dialectical, integrative and prospective streaks. It starts from actions with experiences and learnings, and after reflection and conceptual redesign, it turns back to application and use. The learnings from use are integrated with other experiences and insights into a new holistic understanding. This renewed instrumental knowledge includes forethoughts of its possible applications. Similar views on the development of practical theories are expressed by Stewart & Zediker (2000, p. 12): "people … *connect* new concepts and insights to learnings they've already appropriated, *integrate* the parts of a new understanding into a coherent whole, and search for how to *apply* new ideas". The development of the GRM model, in the design case, was initiated by experiences of (1) an overwhelmingly complex workpractice and (2) an unsuccessful attempt to apply an extant theoretical tool (GEM) in order to develop a foundational understanding for design. After reflection and theoretical emergence of a new embryonic theory (GRM), the focus went back to the inquiry object (the assistance practice) and an application of these new ideas and concepts to check how useful they might be.

We can also add to this multi-functional character of a practical theory its potential function of holding knowledge of a reconstructed and idealized practice (Craig & Tracy, 1995; Stewart & Zediker, 2000). A practical theory, if process-oriented, can hold knowledge of design inquiry practices and a domain-oriented theory can hold knowledge of such a domain practice. Craig & Tracy (1995, p. 248) state, from a practical discipline perspective, that "theory is conceived as a rational reconstruction of practices for the purpose of informing further practice and reflection". This includes describing concepts, language, values, roles, actions, and objects in such practices. Craig & Tracy (1995, p. 265) describe further this reconstructive role of a practical theory: "Rational reconstruction involves not just generalization but idealization and rationalization of practices. Theory construction, therefore, requires critique, revision, and elaboration of the reasoned basis of techniques, problems, and situated ideals involved in a practice". To this, we must add and emphasize that a rational reconstruction of practices should not be made as an uncritical mapping of current work. In a pragmatist and design science spirit, it should involve a prospective articulation of possibilities and ideals.

Conclusion

Our main contribution to the DS theorizing discourse is an elaborated view on a previously poorly explored topic: Developing instrumentalities in the form of practical theory in DS. We have discussed the concept of practical theory in a DS context and elaborated on the guided emergence of such theory in the mutually dependent processes theorizing and situational design inquiry. Through rich engagement with practice, DS researchers are typically in a good position to use and develop practical theories. Our perspective does not imply a lack of design focus. Rather, we acknowledge the inescapable situational aspect of design work. Design does not start with prescriptions. Practical theories guide our attention in understanding a situation. In doing so, they *facilitate* design, without necessarily being explicitly prescriptive about the solution. Through the use of practical theories, we increase our chances to make well-informed design decisions and to document and communicate the rationale of those decisions to others. Thus, practical theories may improve rigor in DS research—similar to the call for increased rigor in action research through a stronger focus on instrumentalities used in the research process (Davison et al., 2012).

The provided perspective on DS theorizing has implications for DS research. First, it provides an opportunity for additional theorizing in DS, on top of the domain-centric theorizing ambitions that are often articulated at the inception of a DS project. Use and refinement of instrumentalities can be integrated into any DS initiative. Second, practical theory (by definition) should be understandable and useful both for practitioners and researchers. By shifting our attention towards practical theory, we thus address the prevalent 'practice–academia divide' in the IS field. Third, our perspective on DS theorizing also has implications for cumulative research in the IS field. As illustrated by the GRM case example, practical theory typically has a broad scope. GRM is potentially useful to inquire into a vast range of organizations—at least in the public sector context. The broad scope of GRM and other practical theories increase their applicability in different contexts. Thus, there is an opportunity for the DS community to test and refine practical theories over time, even when DS is carried out in different practical domains. This means a continual and use-based refinement of practical theories, where each (design) case study contributes to such a cumulative knowledge growth. Looking at DS case studies in this way makes the reporting from such design science studies as naturally consisting of both theoretical contributions and empirical contributions comprising design illustrations and rich descriptions of experiences and other case data (Ågerfalk & Karlsson, 2020). The theoretical contributions do not need to be a full-blown and exhaustive description of a novel theory. As problematized above in this chapter, this would, for most DS papers be exaggerated and counter-productive claims. Instead, publications from DS cases should contribute theoretically through demonstration of the application of practical theories and their further advancement. The emphasis resides naturally in how theoretical constructs are instrumental in different parts of a design inquiry process. This means concepts tied to designerly actions.

The thoughts presented here do not imply that we find other forms of DS contributions less valuable. On the contrary, we appreciate the call for epistemological pluralism (Purao et al., 2008) in DS. We do find, however, that the emergence of practice-oriented DS genres (Iivari, 2015; Peffers, Tuunanen, & Niehaves, 2018) needs to be accompanied by practice-oriented epistemological reflections. This chapter, drawing from a pragmatist philosophy and other practice-oriented research streams, provides such reflection.

References

Ågerfalk, P. J., & Karlsson, F. (2020). Theoretical, empirical and artefactual contributions in information systems research: Implications implied. This volume.

Barge, J. K. (2004). Articulating CMM as a practical theory. *Human Systems, 15*, 187–198.

Baskerville, R., Baiyere, A., Gregor, S., Hevner, A., & Rossi, M. (2018). Design science research contributions: Finding a balance between artifact and theory. *Journal of AIS, 19*(5), 358–376.

Baskerville, R., Kaul, M., & Storey, V. (2015). Genres of inquiry in design-science research: Justification and evaluation of knowledge production. *MIS Quarterly, 39*(3), 541–564.

Baskerville, R., Lyytinen, K., Sambamurthy, V., & Straub, D. (2011). A response to the design-oriented information systems research memorandum. *European Journal of Information Systems, 20*, 11–15.

Baskerville, R., & Pries-Heje, J. (2010). Explanatory design theory. *Business & Information Systems Engineering, 5*, 271–282.

Blumer, H. (1954). What is wrong with social theory? *American Sociological Review, 19*(1), 3–10.

Bohman, J. (2002). How to make a social science practical: Pragmatism, critical social science and multiperspectival theory. *Millennium: Journal of International Studies, 31*(3), 499–524.

Bowen, G. (2006). Grounded theory and sensitizing concepts. *International Journal of Qualitative Methods, 5*(3), 12–23.

Braa, K., & Vidgen, R. (1999). Interpretation, intervention, and reduction in the organizational laboratory: A framework for in-context information system research. *Accounting, Management & Information Technology, 9*, 25–47.

Corley, K., & Gioia, D. (2011). Building theory about theory building: What constitutes a theoretical contribution? *Academy of Management Review, 36*(1), 12–32.

Craig, R. T., & Tracy, K. (1995). Grounded practical theory: The case of intellectual discussion. *Communication Theory, 5*(3), 248–272.

Cronen, V. (1995). Practical theory and the tasks ahead for social approaches to communication. In W. Leeds-Hurwitz (Ed.), *Social approaches to communication*. New York: Guildford Press.

Cronen, V. (2001). Practical theory, practical art, and the pragmatic-systemic account of inquiry. *Communication Theory, 11*(1), 14–35.

Dalsgaard, P. (2017). Instruments of inquiry: Understanding the nature and role of tools in design. *International Journal of Design, 11*(1), 21–33.

Davison, R. M., Martinsons, M. G., & Ou, C. (2012). The roles of theory in canonical action research. *MIS Quarterly, 36*(3), 763–786.

Dewey, J. (1910). *How we think*. Boston: D C Heath & Co..

Dewey, J. (1938). *Logic: The theory of inquiry*. New York: Henry Holt.

Eisenhardt, K. M. (1989). Building theories from case study research. *Academy of Management Review, 14*(4), 532–550.

Elden, M. (1983). Democratization and participative research in developing local theory. *Journal of Occupational Behaviour, 4*(1), 21–33.

Feldman, A. (2000). Decision making in the practical domain: A model of practical conceptual change. *Science Education, 84*(5), 606–623.

Fishman, D. (1999). *The case for pragmatic psychology*. New York: New York University Press.

Friedrichs, J., & Kratochvil, F. (2009). On acting and knowing: How pragmatism can advance international relations research and methodology. *International Organization, 63*, 701–731.

Goldkuhl, G. (2007). What does it mean to serve the citizen in e-services?—Towards a practical theory founded in socio-instrumental pragmatism. *International Journal of Public Information Systems, 2007*(3), 135–159.

Goldkuhl, G. (2011). Generic regulation model—The evolution of a practical theory for e-government. *Transforming Government: People, Process and Policy, 5*(3), 249–267.

Goldkuhl, G., & Cronholm, S. (2010). Adding theoretical grounding to grounded theory—Towards multi-grounded theory. *International Journal of Qualitative Methods, 9*(2), 187–205.

Goldkuhl, G., & Lind, M. (2007). Grounding business interaction models: Socio-instrumental pragmatism as a theoretical foundation. In P. Rittgen (Ed.), *Handbook of ontologies for business interaction* (pp. 69–86). Hershey, PA: Idea Group.

Goldkuhl, G., & Lind, M. (2010). A multi-grounded design research process. *DESRIST-2010 Proceedings*, LNCS 6105, Berlin: Springer.

Goldkuhl, G., & Röstlinger, A. (1993). Joint elicitation of problems: An important aspect of change analysis. In D. Avison et al. (Eds.), *Human, organizational and social dimensions of Information systems development*. North-Holland: IFIP. wg. 8.2

Goldkuhl, G., & Röstlinger, A. (2003). The significance of workpractice diagnosis: Socio-pragmatic ontology and epistemology of change analysis. *Proceedings of the International workshop on Action in Language, Organisations and Information Systems* (ALOIS-2003), Linköping University.

Goldkuhl, G., & Röstlinger, A. (2007). Clarifying government—Citizen interaction: From business action to generic exchange. *Proceedings of the 4th Scandinavian Workshop on e-Government*, Örebro.

Goldkuhl, G., & Sjöström, J. (2018). Design science in the field: Practice design research. *Proceedings Desrist-2018*, LNCS 10844, pp. 67–81, Springer, Berlin.

Gregor, S., & Hevner, A. (2011). Introduction to the special issue on design science. *Journal of Information Systems and e-Business Management, 9*, 1–9.

Gregor, S., & Hevner, A. (2013). Positioning and presenting design science research for maximum impact. *MIS Quarterly, 37*(2), 337–355.

Gregor, S., & Jones, D. (2007). The anatomy of a design theory. *Journal of AIS, 8*(5), 312–335.

Habermas, J. (1984). *The theory of communicative action 1. Reason and the rationalization of society*. Cambridge: Polity Press.

Hambrick, D. (2007). The field of management's devotion to theory: Too much of a good thing? *Academy of Management Journal, 50*(6), 1346–1352.

Henderson-Sellers, B., Ralyté, J., Ågerfalk, P., & Rossi, M. (2014). *Situational method engineering*. Berlin: Springer.

Hevner, A., March, S., Park, J., & Ram, S. (2004). Design science in information systems research. *MIS Quarterly, 28*(1), 75–115.

Hultgren, G., & Goldkuhl, G. (2013). How to research e-services as social interaction: Multi-grounding practice research aiming for practical theory. *Systems, Signs & Actions, 7*(2), 104–120.

Iivari, J. (2007). A paradigmatic analysis of information systems as a design science. *Scandinavian Journal of Information Systems, 19*(2), 39–64.

Iivari, J. (2015). Distinguishing and contrasting two strategies for design science research. *European Journal of Information Systems, 24*, 107–115.

Iivari, J., & Lyytinen, K. (1998). Research on information systems development in Scandinavia—Unity in plurality. *Scandinavian Journal of Information Systems, 10*(1 & 2), 135–186.

Jung, M., Sonalkar, N., Magobunje, A., Bannerjee, B., Lande, M., Han, C., et al. (2010). Designing perception-action theories: Theory-building for design practice. In *The eighth Design Thinking Research Symposium* (DTRS8), Sydney.

Kelle, U. (2005). 'Emergence' vs. 'Forcing' of empirical data? A crucial problem of 'Grounded Theory' reconsidered. *Forum: Qualitative Social Research, 6*(2), Article 27.

Lee, A., & Baskerville, R. (2003). Generalizing generalizability in information systems research. *Information Systems Research, 14*(3), 221–243.

Lee, J. S., Pries-Heje, J., & Baskerville, R. (2011). Theorizing in design science research. *Proceedings DESRIST-2011*, LNCS 6629, pp. 1–16, Springer, Berlin.

Lewin, K. (1945). The Research Center for Group Dynamics at Massachusetts Institute of Technology. *Sociometry, 8*(2), 126–136.

Miller, H., & King, C. (1998). Practical theory. *American Review of Public Administration, 8*(1), 43–60.

Moore, D. B. (2004). Managing social conflict—The evolution of a practical theory. *Journal of Sociology & Social Welfare, 31*(1), Article 6.

Niehaves, B., & Ortbach, K. (2016). The inner and the outer model in explanatory design theory: The case of designing electronic feedback systems. *European Journal of Information Systems, 25*, 303–316.

Österle, H., Becker, J., Frank, U., Hess, T., Karagiannis, D., Krcmar, H., et al. (2011). Memorandum on design-oriented information systems research. *European Journal of Information Systems, 20*(1), 7–10.

Peffers, K., Tuunanen, T., & Niehaves, B. (2018). Design science research genres: Introduction to the special issue on exemplars and criteria for applicable design science research. *European Journal of Information Systems, 20*(2), 129–139.

Purao, S., Baldwin, C. I., Hevner, A. R., Storey, V. C., Pries-Heje, J., Smith, B., et al. (2008). *The sciences of design: Observations on an emerging field.* In Working paper 09-056. Harvard Business School.

Rae, D. (2004). Practical theories from entrepreneurs' stories: Discursive approaches to entrepreneurial learning. *Journal of Small Business and Enterprise Development, 11*(2), 195–202.

Rogers, P., Petrosino, A., Huebner, T., & Hacsi, T. (2000). Program theory evaluation: Practice, promise, and problems. *New Directions for Evaluation, 87*, 5–13.

Searle, J. R. (1969). *Speech acts. An essay in the philosophy of language*. London: Cambridge University Press.

Sein, M., Henfridsson, O., Purao, S., Rossi, M., & Lindgren, R. (2011). Action design research. *MIS Quarterly, 35*(1), 37–56.

Simon, H. A. (1996). *The sciences of the artificial*. Cambridge, MA: MIT Press.

Sjöström, J. (2010). *Designing information systems. A pragmatic account*. PhD diss., Uppsala University.

Sjöström, J., & Ågerfalk, P. J. (2009). An analytic framework for design-oriented research concepts. *Proceedings of AMCIS-2009*, San Francisco.

Sjöström, J., & Goldkuhl, G. (2009). Socio-instrumental pragmatism in action. In B. Whitworth & A. De Moor (Eds.), *Handbook of research on socio-technical design and social networking systems*. IGI, Hershey.

Stevenson, C. (2005). Practical inquiry/theory in nursing. *Journal of Advanced Nursing, 50*(2), 196–203.

Stewart, J., & Zediker, K. (2000). Practically theorizing theory and practice. *The Practical Theory, Public Participation and Community Conference*, Waco

Strauss, A., & Corbin, J. (1998). *Basics of qualitative research. Techniques and procedures for developing Grounded Theory* (2nd ed.). Newbury Park: Sage.

Sutton, R., & Staw, B. (1995). What theory is not. *Administrative Science Quarterly, 40*, 371–384.

Thornberg, R. (2012). Informed Grounded Theory. *Scandinavian Journal of Educational Research, 56*(3), 243–259.

Van Maanen, J., Sørensen, J., & Mitchell, T. (2007). The interplay between theory and method. *Academy of Management Review, 32*(4), 1145–1154.

Van Strien, P. (1997). Towards a methodology of psychological practice: The regulative cycle. *Theory & Psychology, 7*, 683–700.

Venable, J. (2006). The role of theory and theorising in design science research. *Proceedings DESRIST-2006*, Claremont.

Walls, J. G., Widmeyer, G. R., & El Sawy, O. A. (1992). Building an information systems design theory for vigilant EIS. *Information Systems Research, 3*(1), 36–59.

Weber, M. (1978). *Economy and society*. Berkeley: University of California Press.

Weick, K. (1995). What theory is not, theorizing is. *Administrative Science Quarterly, 40*(3), 385–390.

Winograd, T., & Flores, F. (1986). *Understanding computers and cognition: A new foundation for design*. Norwood: Ablex.

Winter, R. (2014). Towards a framework for evidence-based and inductive design in information systems research. In M. Helfert et al. (Eds.), *Proceedings EDSS-2013* (CCIS 447) (pp. 1–20). Berlin: Springer.

8

Pathways to IT-Rich Recontextualized Modifying of Borrowed Theories: Illustrations from IS Strategy*

Mohammad Moeini, Robert D. Galliers, Boyka Simeonova, and Alex Wilson

Introduction

The extent of theory borrowing in IS is criticized for leaving IT un(der) theorized (Grover and Lyytinen 2015), while the heavy reliance on theories not founded on IT-related constructs is deemed "distracting" (Benbasat and Zmud 2003, p. 192). Accordingly, a strong and active stream of research encourages developing both indigenous (Grover and

*An earlier version of this chapter appeared as Moeini, M., Simeonova, B., Galliers, R.D. & Wilson, A. (2020) Theory borrowing in IT-rich contexts: Lessons from IS strategy research. *Journal of Information Technology, 35*(3), pp. 270–282.

M. Moeini (✉)
Warwick Business School, Coventry, UK
e-mail: mo.moeini@wbs.ac.uk

R. D. Galliers
Bentley University, Waltham, MA, USA

Loughborough University, Loughborough, UK
e-mail: rgalliers@bentley.edu

N. R. Hassan, L. P. Willcocks (eds.), *Advancing Information Systems Theories*, Technology, Work and Globalization, https://doi.org/10.1007/978-3-030-64884-8_8

Lyytinen 2015) and middle-range (Hassan and Lowry 2015) theories that are unlikely to be products of theory borrowing from other fields. Such theories will be "above and beyond the theories we import from other fields" (Markus and Saunders 2007, p. iv). While the merits of such theorizing are apparent, theory borrowing in IS will not, and should not, cease completely, particularly due to its appeal (Grover and Lyytinen 2015) and potential usefulness (e.g., see Weber 2003, p. x). Therefore, in addition to the need to have more indigenous theories, attention to the quality of borrowing is also crucial.

We build on the premise that theory borrowing is legitimate if constructs are owned, and that borrowed theories are adapted appropriately to our unique subject matter (Hassan 2011; Markus and Saunders 2007). Drawing from the studies within (Hong et al. 2014) and outside IS (Oswick et al. 2011), such borrowing can be characterized as IS-recontextualization. To demarcate our unique subject matter, we note that whether there is an intellectual core to the field and what constitutes it have been debated (King and Lyytinen 2006; Whinston and Geng 2004). While acknowledging critical commentaries (e.g., Galliers 2006), in this chapter we consider what IS recontextualization would look like if the borrowed materials are not silent about the IT artifact.

Particularly, we seek to increase our understanding of IT-rich recontextualized modifying of borrowed theories. Three key modes of theory borrowing in IS have been articulated: instantiating, extending, and modifying (Grover and Lyytinen 2015). By definition, *instantiating* is unlikely to provide a basis for offering much IS-related theoretical insights as theory is foreign to IS and the recontextualization efforts are only minor. *Extending* encourages theory development to go beyond the borrowed theory by adding new constructs and relationships that are IS-related. It is in *modifying*, where "The model modifies constructs, configurations, and/or logic from the borrowed theory to the IS context" (Grover and Lyytinen 2015, p. 278), that IT-rich recontextualization becomes key.

Towards this aim, we build on the literature on theorizing the IT artifact (Benbasat and Zmud 2003; Grover and Lyytinen 2015; Orlikowski

---

B. Simeonova • A. Wilson
Loughborough University, Loughborough, UK
e-mail: b.simeonova@lboro.ac.uk

and Iacono 2001; Straub 2012), develop a framework of IT-rich recontextualized modifying of borrowed theories, and demonstrate this framework within the domain of IS strategy. The first dimension of the framework recognizes two approaches: specification and distinction. The specification approach concerns how the resulting theory can confer rich insights about the IT artifact, towards enabling non-nominal views of IT (Orlikowski and Iacono 2001). The distinction approach involves how to make the borrowed instance uniquely different from other possible IS instances (internal distinction) and non-IS instances (external distinction) of the same theory. The second dimension of the framework recognizes that modifying can occur across two key elements of the borrowed theory: constructs and relationships (Bacharach 1989). Construct reconceptualization is concerned with relabeling, redefining, and redimensionalizing constructs to offer rich insights about the IT artifact; this is usually as the object of a perception, ability, or action of a social actor, captured by the nomological network surrounding the IT artifact (Benbasat and Zmud 2003). Moreover, relationship rejustification is chiefly about explaining theoretical mechanisms for the influence of one construct on another by explicitly referring to particular aspects of the IT artifact.[1]

To illustrate the adoption of these IT-rich recontextualized modifying approaches, we use several examples from IS strategy (ISS) research. Our focus on ISS has several motivations: there is much discourse around IS as a conglomerate of subtopics, and more in-depth discussions of each subtopic can provide a rich understanding that can guide future research in that topic. ISS has been seen to be a top concern of CIOs over the decades (e.g., Luftman et al. 2006), and it is a particularly important topic in the current digital era in which many businesses are becoming increasingly digitally enabled and transformed (Barley et al. 2017; Davison and Ou 2017; Vial 2019; Baptista et al. 2020). Moreover, ISS is generally recognized as adopting a nominal view to the IT artifact (see Grover and Lyytinen 2015, p. 279; also, Orlikowski and Iacono 2001, p. 128), so we seek examples of capturing IT in the ISS research.

Our contribution is twofold. First, we contribute to studies on embedding IT in IS research (e.g., Akhlaghpour et al. 2013; Grover and Lyytinen

[1] For simplicity, in this chapter we focus only on borrowing variance theories that aim at establishing a causal path between constructs.

2015; Orlikowski and Iacono 2001; Straub 2012) by further clarifying and formalizing IT-rich IS-recontextualized modifying in terms of *approach* (specification and distinction) and *locus* (reconceptualization and rejustification). Second, we contribute to ISS research by exploring several specific examples of how the IT artifact is embedded in it. In doing so we add to our understanding of the state of ISS research by demonstrating that IT is not always nominal in this domain. Overall, we seek to reinvigorate the debate on theory borrowing within IS (Truex et al. 2006) and provide some further directions for moving away from nominal views of IT in ISS.

IT-Rich Recontextualized Modifying of Borrowed Theories: A Framework

When theories from outside a reference discipline are borrowed, they are subject to recontextualization. As Oswick et al. (2011, p. 323) put it, "Effectively, recontextualization involves a process of repackaging, refining, and repositioning a discourse (or text) that circulates in a particular community for consumption within another community". Researchers who have studied similar borrowing modes, including contextualization (Hong et al. 2014), domestication (Oswick et al. 2011), and modifying (Grover and Lyytinen 2015), consider that some core constructs and relationships might be added or removed from the borrowed theory; however, what we mean by IT-rich recontextualization is much narrower. In this chapter, we delve into how unique IS insights can be infused in the constructs and relationships of the borrowed theory.

To increase our understanding of IT-rich recontextualized modifying of borrowed theories, we develop a 2 × 2 framework by separating the recontextualization *approach* and *locus* (Fig. 8.1) by drawing from past research on developing IS theories (Benbasat and Zmud 2003; Grover and Lyytinen 2015; Orlikowski and Iacono 2001; Straub 2012). The recontextualization approach distinguishes specification and distinction. The recontextualization locus considers that specification and distinction can happen in two core elements of a theory: constructs and relationships

		Approach	
		Specification	Distinction
Lo cu s	Construct Reconceptualization	1. Modifying the borrowed constructs for specific IS phenomena centered around the IT artifact.	3. Explaining how some aspects of the modified conceptualizations are unique to IS phenomena concerning specific IT artifacts.
	Relationship Rejustification	2. Modifying the borrowed causal links by drawing upon IS phenomena centered around the IT artifact.	4. Explaining how some aspects of the modified causal links are unique to IS phenomena concerning specific IT artifacts.

Fig. 8.1 A framework of IT-rich recontextualized modifying of borrowed theories

(Bacharach 1989). Given our speculation that, when borrowing theories in IS, the IT artifact often remains nominal until the empirical part of a paper; this framework excludes recontextualized operationalizations to emphasize recontextualized theorizing. Below, we explain each dimension of the framework in detail.

Below, after a separate discussion of each dimension, we will discuss each quadrant of the framework.

Specification

Specification refers to the efforts made to describe the IT artifact embedded in the surrounding IS phenomena. Specification recognizes that if the IT artifact is absent in a theory, the study might be subject to the "error of exclusion" (Benbasat and Zmud 2003, p. 189). From this view, a research piece might be criticized for having "no reference to any specific technology" (Orlikowski and Iacono 2001, p. 128). Specification is therefore a necessary (but not sufficient) condition for adopting the non-nominal views of IT (e.g., the ensemble view). While such specificity is recognized as a key characteristic of native IS theories (Straub 2012), we explore specificity in theory borrowing.

Before we proceed to discuss some possible avenues for specification, we acknowledge the *specificity dilemma*. Specification involves a trade-off between too narrow and too broad IT artifact descriptions, where striking the right balance is a quandary (DeSanctis and Poole 1994; Markus and Silver 2008). Like any other system, which is composed of sub-systems that could again be decomposed into lower-level sub-systems, IT artifacts (and their attributes) could be decomposed into more detailed elements—an issue referred to as the repeating decomposition problem (DeSanctis and Poole 1994; Markus and Silver 2008). The specificity dilemma is also visible in the argument that IT is usually unspecified in IS studies (Orlikowski and Iacono 2001), but when it is specified, it often provides "overly narrow views of technology" (ibid., p. 122). The specificity dilemma escalates into challenges in specifying the nomological network surrounding the IT artifact. In the ISS context, for example, Wade and Hulland (2004, p. 128) explain the challenges with defining IT resources: "Broadly defined resources have the advantage of being readily generalized beyond a specific research situation but can lose their explanatory value when applied to overly narrow or specific situations". Yet "Narrow definitions help to fine-tune our understanding of specific resources and their effect on competitive position and performance in given settings" (ibid., p. 129). While acknowledging this dilemma, our view is that it should not stop all research from attempting to be specific. Specification will enable researchers to demonstrate which aspects of an IT artifact are salient, and without it "we have no language for making clear distinctions between types of constraints and affordances" (Leonardi and Barley 2008, p. 166).

We base our view of the IT artifact on Markus and Silver's (2008, see p. 625) explanation of IT artifacts as technical objects with material and immaterial real things that are independent for their existence on people's perceptions of them. We recognize six ways in which the IT artifact could be specified: the taxonomic nature of the IT artifact, the IT artifact being either an IT or an IS product (including specific system types), key subcomponents of the IT artifact, holistic attributes of the IT artifact, the IT artifact being either whole or an aggregate of subcomponents (the features and functionalities of the IT artifact), and finally, purposeful nonspecificity (e.g., when the findings are generalizable).

1. Specifying the taxonomic nature of the IT artifact: We adopt a pluralistic view of the IT artifact (recognized as the IS artifact—Lowry et al. 2017) and consider that an IT artifact could belong to one taxon or a combination of taxa, including, but not limited to, technical systems (hardware, software, and algorithms embedded in them), information and data, techniques and methodologies, policies, certifications, standards, and patents. When an IT artifact is labeled accordingly, attention is shaped to a particular (although often vaguely defined) subset of artifacts. In ISS, for example, Melville et al. (2004, p. 294), as part of explaining technical resources, distinguish between hardware and software: "(1) IT infrastructure, i.e., shared technology and technology services across the organization, and (2) specific business applications that utilize the infrastructure, i.e., purchasing systems, sales analysis tools, etc.". Other ISS examples are: an IT infrastructure specified as a bundle of business applications (Tallon and Pinsonneault 2011), decision-making algorithms (Newell and Marabelli 2015), data warehouses (March and Hevner 2007), data visualization dashboards (Zammuto et al. 2007), enterprise systems (Tan et al. 2004), blockchain (Beck et al. 2018), business intelligence (Shollo and Galliers 2016), data analytics techniques and methodologies (Chen et al. 2012; Günther et al. 2017), information security policies (Cram et al. 2017), security certifications (Hsu 2009), electronic data interchange (EDI) standards (Zhu et al. 2006), and software patents (Mykytyn et al. 2002).

2. Specifying the IT artifact as an IT vs. IS product: Two related but differing forms of technical objects are IT and IS (in terms of products not the research disciplines—Hassan 2006). While IT can be any digital technology, including personal computers, the Internet, and digital gadgets, IS is a particular category of IT used by organizations to represent their business processes (Weber 2003). Delving into the latter, IS products are often recognized by *specific system types*. Some particular instances of system types in ISS are interorganizational information systems (Johnston and Vitale 1988; Saraf et al. 2007) including EDI (Mukhopadhyay et al. 1995), e-marketplaces (Bakos 1991) and multi-sided platforms (Pagani 2013), customer relationship management (CRM) systems (Setia et al. 2013), knowledge management (KMS) systems (Joshi et al. 2010; Pavlou and El Sawy 2006, 2010), enterprise software (Ceccagnoli et al. 2012; Tan et al. 2004), strategic decision support systems (March and Hevner 2007; Sabherwal and Chan 2001), and enterprise social media platforms (Rode 2016; von Krogh 2012).

3. Specifying key subcomponents of the IT artifact: The IT artifact can be regarded as a holistic package or an assemblage of key subcomponents. Orlikowski and Iacono (2001, p. 131) explain that: "IT artifacts are usually made up of a multiplicity of often fragile and fragmentary components, whose interconnections are often partial and provisional and which require bridging, integration and articulation in order for them to work together". An example of this view is decomposing an ERP system into different functional modules (Benders et al. 2006).

4. Specifying holistic attributes of the IT artifact: Some holistic attributes (characteristics) of the IT artifact or its key subcomponents can be used to describe them. For example, in defining commodity-like IT assets, Nevo and Wade (2010) discuss *customizability* as a core feature of some off-the-shelf IT, although they argue that this customizability is still imitable and does not make the assets rare or non-substitutable. Likewise, the *flexibility* of software, hardware, and networks are used in conceptualizing the IT infrastructure flexibility construct (Tallon and Pinsonneault 2011). *Information quality* and *information completeness* are used in studying customer service systems (examined by Setia et al. 2013). *Complexity* (Meyer and Curley 1991) which "contributes to making it difficult for organizations to assimilate and effectively use technology" (Piccoli and Ives 2005, p. 761). In an interorganizational systems context, *integration* is referred to "the extent to which the IS applications of a focal firm work as a functional whole in conjunction with the IS applications of its business partners" (Saraf et al. 2007, p. 324). Some other examples are *ubiquity* (Carr 2003), *fragility* and *unreliability* (Butler and Gray 2006), *scalability* (O'Leary 2013; Tallon and Pinsonneault 2011), *maturity* (Karimi et al. 1996; Venkatesh et al. 2007), *newness* (Barki et al. 1993; Swanson and Ramiller 2004), *uniqueness* (Piccoli and Ives 2005), and *representation*, which is deemed as "the essence of an information system" (Weber 2003, p. viii).

5. Specifying features/functionalities of the IT artifact: The IT artifact can also be viewed as a collection of various interrelated features/functionalities[2] that, individually or taken together, could enable or

[2] Features and functionalities can be distinguished. Features are the "building blocks or component parts of a technology" (Griffith 1999, p. 475) and "The functionality of a system refers to the range

support users to perform certain tasks. Decomposing an IT artifact into features/functionalities allows researchers to avoid having a black box view of technology (Benbasat and Zmud 2003; Jasperson et al. 2005; Orlikowski and Iacono 2001). In ISS, for example, an IS application is defined as "a piece of software functionality that is developed and installed on specific IT platform(s) to perform a set of one or more business tasks independently of other surrounding IS components" (Saraf et al. 2007, p. 321). Below, we list five approaches to discussing features/functionalities when specifying the IT artifact.

a. Core vs. optional features/functionalities of the IT artifact: Features/functionalities could be core or optional in an IT artifact (Griffith 1999). Features can be "a bundle of inherent or typically-occurring" or alternatively "a constellation of selected possible features" (Griffith and Northcraft 1994, p. 274). In this vein, DeSanctis et al. (1994, p. 322) distinguish the "basic set of functions or core features that a given technology should support" from optional features that might be included in "Some, but not all, implementations of the technology". Likewise, Bui (2017) distinguishes the essential and supportive elements of enterprise architecture programs. In ISS, for example, by referring to the *customizability* of the off-the-shelf IT artifacts, researchers consider the possibility of having both core and optional features (Nevo and Wade 2010; Piccoli and Ives 2005).

b. Autonomous vs. user-controlled features/functionalities of the IT artifact: A system can be in the design mode or in the use mode (Orlikowski 1992). In the design mode, some automated functionalities can be programmed into the systems. Running the system by end users means allowing it to perform these functionalities, and changing the functionalities requires the involvement of the designers or system administrators. Other functionalities require the volitional (or accidental) use of certain features by the end users in the use mode. For example, the automatic donated data anonymization by a mobile app, as discussed by O'Leary (2013), is an autonomous functionality beyond users' control, but the peer production and collaboration features of social software in

of operational tasks it supports" (DeSanctis et al. 1994, p. 322). In this chapter, we use these terms interchangeably.

organizations, discussed by von Krogh (2012), require an active use of the available functionalities by the end users. Accordingly, we recognize that features/functionalities range from "what the system does" (Ein-Dor and Segev 1993, p. 168) to affordances or "what technology enables users to do with it" (Markus and Silver 2008, p. 617).

c. Enabling vs. inhibiting features/functionalities of the IT artifact: Features/functionalities embedded in the IT artifact can enable or constrain human action (Leonardi and Barley 2008). For example, while a system might be vulnerable to unintended access by unauthorized users, the user interface design of an enterprise system might highlight access identifiability to prevent users from violating access policies (Vance et al. 2013).

d. Functionalities vs. dysfunctionalities: Functionalities refer to positive applications of IT artifacts. For example, Pavlou and El Sawy (2010) discuss three functionalities of collaborative work systems for new product development (i.e., conveyance, presentation, and convergence). Likewise, in discussing the notion of IT-enabled capabilities, Sambamurthy et al. (2003) discuss information technologies with an ability to support key functionalities: agility, digital option generation, and entrepreneurial alertness. However, using the IT artifact might lead to dysfunctionalities. Jasperson et al. (2005, p. 529) state that "Individuals can apply features in nonproductive ways or they may be overwhelmed by the presence of too many features, resulting in an inability to choose among feature sets or to apply the features effectively". For example, Stein et al. (2019) suggest that using datafication to track worker activities could hinder the pursuit of meaningfulness in work.

e. Micro vs. macro features/functionalities of the IT artifact: Indeed, specifying features/functionalities is vulnerable to the specificity dilemma explained above, because features could be decomposed into other features (see Griffith and Northcraft 1994, p. 273). For example, the *integration* feature of an IT artifact (Barki and Pinsonneault 2005) is composed of some computational and communication (e.g., connectivity) features. To help overcome this dilemma, we distinguish micro and macro features/functionalities. Micro features/functionalities are related to the tasks performed by different end-users using the IT artifact, and several such features/functionalities could exist for one technology. For example, one could perform several tasks (e.g., import, clean up, and visualize) on a specific dataset using a decision-making dashboard. Macro features/functionalities, however, are a few high-level ones for each IT artifact. In the ISS context, for example, such macro features/functionalities can be discussed at the level of the strategic purpose associated with the IT artifact, where a strategic purpose refers to the pathways to achieve (sustainable) competitive advantage and performance. We recognize several strategic purposes for the IT artifact as follows:

- **No strategic purpose:** In some studies, the strategic purpose of the IT artifact is unclear. IT is considered as a tool to make a local improvement, hence an isolated system (Galliers 1988).
- **Implementing strategy:** The IT artifact could be a means for implementing specific business strategies. This corresponds to the reactive category in Galliers (1988), the business strategy as the driver perspectives in Henderson and Venkatraman (1993), and, in general, the traditional view of strategic alignment (Renaud et al. 2016). In ISS, Clemons and Row (1991) argue that IT can implement strategy by reducing transactions costs and coordinating strategic resources.
- **Enabling strategic products:** The IT artifact might be embedded in digital products of strategic value. In ISS, the importance of product digitalization is discussed (e.g., Bharadwaj et al. 2013; Clemons and Row 1991).
- **Shaping strategy:** IT can be a driver of strategy, where it is IT that leads business strategy (McFarlan 1984). This corresponds to the IT strategy as the enabler view in Henderson and Venkatraman (1993). In ISS, Sambamurthy et al.'s (2003) view of IT as a digital option generator contributes to the view of IT as driving strategy.
- **Fused with strategy:** As captured by the fusion perspective underlying the notion of digital business strategy (Bharadwaj et al. 2013), IS and business strategy co-evolve or are the same thing. For example, the use of IS can be seen as strategy practice (Arvidsson et al. 2014), strategizing is seen as an integral aspect of business strategy (Galliers 2011), and aligning can be seen as a process in which IS and business functions interact dynamically (Karpovsky and Galliers 2015).

- **Strategizing tool:** The IT artifact can be used for making strategic decisions (e.g., March and Hevner 2007) or strategy ideation, e.g., in a bottom-up fashion (e.g., Baptista et al. 2017). For example, Baptista et al. (2017) demonstrate how various features of organizational social media can be used to influence the strategic activities of organizations in an open strategy context (Morton et al. 2019, 2020; Tavakoli et al. 2017).
- **Creating a new strategic landscape:** IT can create a competitive environment/industry sector that did not exist before. Some examples are platform ecosystems (Ceccagnoli et al. 2012) that extend e-marketplaces (Bakos 1991) and blockchains that have introduced a new economy landscape (Beck et al. 2018).

6. Purposeful nonspecificity of the IT artifact: When research claims are generalizable to all IT artifacts, researchers might opt to justify the lack of IT specificity. For example, a study on the strategic use of IT might argue that the findings would hold for any type of IT in organizations, hence IT nonspecificity. We did not find an example of such argument in ISS, perhaps implying that most researchers consider their findings to generalize only to some certain kinds of IT.

In sum, we have identified some avenues for specifying the IT artifact. Richer specifications can be provided by combining multiple approaches, for example, by referring to specific features/functionalities of a particular system type. Other ways of specifying IT are feasible but fall beyond the scope of this chapter. For example, IT can be specified based on its usage context (e.g., manufacturing, healthcare), organizational level of use (e.g., inter-organizational networks), or intended/realized user groups (e.g., CxOs, middle managers, IT personnel, laypeople).

Distinction

Distinction refers to efforts dedicated to delineating the various instances of borrowed constructs and relationships by highlighting what is unique to the particular IT artifact (or the surrounding IS phenomena) being studied. Distinction leverages specification, for example, in order to differentiate features and functionalities of two IT artifacts. We discuss two approaches of internal and external distinction.

1. Internal distinction: Internal distinction is within IS and builds on the idea that "Information technologies differ with respect to their intrinsic characteristics" (Piccoli and Ives 2005, p. 760). To make an internal distinction, specification can be used to capture the variations between different ITs rather than adopting a homogeneous view of IT. Some ISS examples follow. Oh and Pinsonneault (2007, see p. 249) capture how different IT applications in the manufacturing industry are associated with different strategic values of cost reduction, quality improvement, and revenue growth. Ray et al. (2005) distinguish between tacit and explicit IT capabilities: "Tacit, path dependent, and socially complex IT capability (shared knowledge) explains variation in process performance. Explicit IT resources (technical IT skills, generic information technologies, and IT spending) do not" (ibid., p. 643). Internal distinction can also be carried out about varying perceptions, abilities, and actions about similar IT artifacts. For example, Scott and Vessey (2002) distinguish the way two companies managed similar IT artifacts (variants of SAP R/3 ERP) and how this led to significantly different firm outcomes.

2. External distinction: External distinction can be made between the IS and non-IS versions of the borrowed theories. It recognizes that "we should look to how uniquely the theory base applies to IS" (Straub 2012, p. v). Giving the example of the replaceability of "IS" with "marketing" in some IS strategy theories, Orlikowski and Iacono (2001) suggest that the IS and non-IS distinction made in a theory must be evident. In simpler cases of external distinction, one can speak of certain characteristics of the IT artifact that make the unmodified instances of a borrowed theory inadequate for understanding IS phenomena (Grover and Lyytinen 2015), requiring further recontextualization efforts. We recognize four external distinction approaches.

a. Distinguishing IT from non-IT instances of the same theories: This involves explaining what is unique to an IS take on the borrowed theory. For example, in ISS, Wade and Hulland (2004) offer a theoretical "basis for comparison between IS and non-IS resources, and thus can facilitate cross-functional research" (p. 109).

b. Distinguishing IT from non-IT technologies: The way a theory is borrowed to examine an IT artifact can be distinguished from such borrowing for a non-information technology. Following this approach means specifying the ways in which IS modifications of the borrowed theory are different from non-IS ones. For example, an ISS study can discuss the specific strategic purposes only associated with IS, but no other technological assets, such as manufacturing and R&D facilities. In an extreme case, which falls outside our focus on theory borrowing, one can "identify a set of generic characteristics of information technology that cause the existing theories about technology fail" (Weber 2003, p. vii).

c. Recognizing the similarity (hence the concurrent value added) to non-IT technologies: When a theory that was not originally developed to understand technology-centric phenomena is borrowed and modified to explain an IS phenomenon, the generalizability of this modification to the undersanding of technologies other than IT (e.g., manufacturing technologies) can be explicitly discussed. For example, perceived usefulness and perceived ease of use (Davis 1989), developed by modifying the theory of reasoned action, are useful not only for examining information technology adoption but also for understanding the adoption of other technologies.

d. Distinguishing the IS angle from other angles for the same IT artifact: Various cognate disciplines, such as computer science, have a vested interest in the IT artifact (Hassan and Will 2006). By building on the business and social aspects of the specified IT artifact, an ISS study can specify how the adopted view of an IT artifact is linked to, but departs from, the examination of similar IT artifacts from a computer science viewpoint.

Above, we discussed the specification and distinction approaches. Next, we will take a closer look at how these approaches can be applied in recontextualizing constructs and relationships.

Specification and Distinction in Construct Recontextualization

Constructs refer to "conceptual abstractions of phenomena that cannot be directly observed" (Suddaby 2010, p. 346). For an IT-rich recontextualized modifying of borrowed constructs, one should "adapt them to reflect a unique IT or IS component" (Benbasat and Zmud 2003, p. 193) towards "clarifying the IS nuances involved" (ibid.). Below, we distinguish such opportunities into reconceptualization form and content.

1. Construct Recontextualization Form: The specification of an IT artifact, towards IT-rich recontextualized modifying of borrowed constructs, can appear in four complementary forms: construct labels, definitions, dimensions, and exemplifications.

a. Recontextualized construct labels: The labels of the borrowed constructs can be revised to hint at the specified IT artifact. For example, Grover and Lyytinen (2015, p. 278) observe that "a construct like 'asset specificity' (AS) in transaction cost economics (TCE) is used to study outsourcing but modified to 'system specificity'". More specific references to an IT artifact are feasible.

b. Recontextualized construct definitions: A good practice in theorizing is ensuring construct clarity (Barki 2008; Suddaby 2010), especially by offering formal definitions (Wacker 2004). Construct definitions can be modified by making explicit reference to the specified IT artifact. General rules for offering a formal conceptual definition (Wacker 2004) apply. For example, the rule suggesting that "Definitions should include only unambiguous and clear terms. Put another way, do not use vague or ambiguous terms" (Wacker 2004, p. 638) could be regarded as an invitation to avoid having a non-specified (nominal) IT artifact. For example, besides relabeling the "assets" construct to "IT assets", Piccoli and Ives (2005, p. 753) provide an IT-rich definition of the modified construct:

> IT assets available to the firm include hardware components and platforms (e.g., a private satellite network), software applications and environments (e.g., a proprietary revenue management system using custom developed models), and data repositories (e.g., a database of historical customer behavior).

Distinction in construct reconceptualization can be achieved by explaining how the conceptual territory covered by a construct has been modified because of the specified IT artifact.

c. Recontextualizing construct dimensions: A multidimensional construct can be modeled in different ways, including the latent (higher-order) and aggregate (multiplicative or additive) approaches (Law et al. 1998). The recontextualization of a borrowed multidimensional construct can involve adding or removing dimensions by referring to IT matters. For example, the adaptation of the notion of shared domain knowledge in the examination of the social dimension of alignment by Reich and Benbasat (2000) is reconceptualized to consider the two dimensions of "IT-knowledgeable business managers and business-knowledgeable IT managers" (p. 84).

d. Recontextualization by exemplification: Exemplification is an important part of conceptualization. Jaccard and Jacoby (2009, see p. 76) consider providing concrete examples vital for developing theoretical ideas. In particular, where construct labels and definitions are non-specified, rich exemplification provides a further opportunity for specification. In ISS, for example, Piccoli and Ives (2005) blend in several examples in defining IT concepts. For instance, in defining the visibility of IT, they refer to two specific IT artifacts: "The visibility dimension can be conceptualized as a continuum spanning from internal systems (e.g., Harrah's Entertainment's engine for data analysis) to public systems (e.g., Lands' End Live: Web-based chat with customer service agents)" (p. 760). Examples can also be used to make a distinction between various takes on the same construct.

2. Construct Recontextualization Content: The content of such recontextualization constructs can be linked to a specified IT artifact in four ways: by focusing on the IT artifact on its own, or by referring to the functional affordances, perceptions, and abilities/actions surrounding the IT artifact.

a. Conceptualizing the IT artifact (or its properties) on its own: A construct could be merely about the IT artifact as a technical object (i.e., a "thing"). Moreover, a construct could be about the properties of the IT artifact, such as the maturity of the IT artifact, the newness of the IT artifact (state-of-the-art, legacy, obsolete), and the flexibility of the IT artifact. For example, in ISS, digital design, defined as information quality for a customer service unit, is theorized to influence customer service capabilities (information quality and completeness) (Setia et al. 2013).

b. Conceptualizing the functional affordances of the IT artifact: Functional affordances are "the possibilities for goal-oriented action afforded to specified user groups by technical objects" (Markus and Silver 2008, p. 622). Affordances have causal potential but are not deterministic (Markus and Silver 2008). Affordances link technologies to their users as "one cannot talk about a complex technology without reference to the social setting, just as it makes limited sense to talk about a door handle without discussing the people opening the open doors" (Zammuto et al. 2007, pp. 752–753). Affordances might emerge out of users' actual usage from the technology. This is akin to the view that "Features of a technology are interpreted (and possibly adapted) by individual users so as to constitute a *technology-in-use*" (Jasperson et al. 2005, p. 529, original emphasis). The results of enacting affordances "may depart from those prescribed by developers or implementers, and across time and context" (Orlikowski 1995, p. 8). In ISS, Chatterjee et al. (2015) examine the role of the core IT affordances in firms (namely, collaborative affordance, organizational memory affordance, and process management affordance) on organizational improvisational capabilities via organizational virtues.

c. Conceptualizing the IT artifact as an object of a perception: Traditionally, perceptions of IT are used as a proxy for examining IT (Orlikowski and Iacono 2001). For example, while affordances are real things (Markus and Silver 2008), they might or might not be perceived by users (Volkoff and Strong 2013), hence affordance perception is key to affordance actualization (Anderson and Robey 2017).

d. Conceptualizing the IT artifact as an object of actions and abilities: Beyond perceptions, constructs might capture the actions and abilities of social actors (e.g., people or organizations). The nomological network of Benbasat and Zmud (2003) with the IT artifact at its core, broadly discusses the management, use, and impact of IT. Inspired by this network, we recognize several action/ability themes around the IT artifact in the ISS context:

- **Planning the specified IT artifact(s).** Planning might involve specifying mission, vision, and objectives, or evaluating the realized strategy. It can focus on aligning/fusing IT and business strategies. For example, in ISS, Earl (1993) discusses five different approaches to strategic information systems planning, and Lederer and Hannu (1996) offer a theory of strategic planning.
- **Acquiring the specified IT artifact(s).** Acquiring an IT artifact can be achieved in different ways. Acquisition might be achieved by investing in IT. Acquisition might focus on sourcing (explore, outsource, offshore, provisioning) IT. Sourcing from the market can involve identifying and evaluating various IT opportunities. In ISS, for example, Lacity et al. (2009) review various outsourcing studies. Acquisition might be achieved by developing IT (create, design, innovate, explore) in-house or as off-shore, outsourced, or in spin-offs. For example, Willcocks et al. (1999) discuss the outsourcing of logistics information systems.
- **Implementing the specified IT artifact(s).** Implementation focuses on putting the acquired solution in place and maintaining it until seamless operation is realized. In ISS, Grant (2003) examines the impact of global enterprise system implementation on strategic alignment, and Dulipovici and Robey (2013) discuss the misalignment resulting from the implementation and use of a KMS.
- **Using the specified IT artifact(s).** Utilizing IT can involve using, leveraging, and exploiting IT. IT use is a crucial concept because while "affordances are simply *potentials* for action, purposeful use of the system entails *actualizing* affordances" (Burton-Jones and Volkoff 2017, p. 470, original emphases). In ISS, Arvidsson et al. (2014) examine system use as strategy practice and Roberts et al. (2016) consider IT use for sensing new organizational innovation opportunities.
- **Structuring the/by the specified IT artifact(s).** The social actor might be shaped by IT (e.g., embedded structures and surrounding regulations) and/or the social actor might shape IT. For example, Pagani (2013) discusses how the structure of value networks (especially in terms of control points) are reshaped by digital innovations, to the point that vertically integrated networks could be reshaped into two-sided markets by disruptive innovations.

- **Governing the specified IT artifact(s).** Action and capabilities of social actors can be about determining the rules and regulations surrounding IT. For example, in ISS, Wu et al. (2015) demonstrate how IT governance is linked to organizational performance via strategic alignment.
- **Protecting the specified IT artifact(s) (or protection from the specified IT).** Protecting the IT artifact considers that social actors might attempt to keep the IT artifact away from the competition or other opponents. In ISS, for example, researchers argue that maintaining a competitive advantage via having proprietary IT requires patenting or secrecy, both of which are difficult to achieve (Mata et al. 1995). Protection from IT might mean involvement with cyber-security measures. Other privacy and security issues, such as hacking, are also relevant here; for example, security (particularly, network security) is considered to be one of the key IS competencies (Cragg et al. 2011).

Specification and Distinction in Relationship Rejustification

IT-rich constructs can appear in two structural positions in a model: (1) as independent, dependent, and path-modifying variables and (2) as causal links.

1. Recontextualization of Independent, Dependent, and Path-modifying Variables: The specification of the IT artifact (or its properties) on its own as an independent variable can lead to having IT-pure constructs that stem from a technology imperative view of "technology as the primary independent variable" (Orlikowski and Iacono 2001, p. 123). For example, in ISS, Ray et al. (2005) hypothesize the flexibility of IT infrastructure to influence customer service performance. However, the IT-specificity of the construct does not mean that they should be purely about the IT artifact, and it is important to consider that some materialism is possible without intending determinism (Leonardi and Barley 2008). As Leonardi and Barley (2008, p. 159) state, "scholars [...] often conflate the distinction between the material and social with the distinction between determinism and voluntarism." Further, IT-rich recontextualizing of path modifiers (such as moderators) would mean that the nature, direction, and/or strength of a path between an independent and a dependent variable is changed because of an IT artifact or its surrounding phenomena.

2. Recontextualization of Causal Links: When building a theory, the relationship between two constructs is explained using one or more causal mechanisms, where "Causality involves real physical, psychological, and/or social processes that connect inputs and outputs under certain conditions" (Markus and Rowe 2018, p. 1263).

While the views on causality in IS research are several, to illustrate IT-rich modifications of borrowed theories, for brevity, we focus on a common view, which is considering the causal links underlying borrowed relationships as intervening variables (Mahoney 2001). In simple terms, if a variance proposition stipulates that construct A influences construct B, construct C might be used as the causal link. This usually means that two conceptual assumptions (that A impacts C, and C influences B) are explained but are taken for granted as true. Put differently, C is a hidden mediator. More than one causal link is usually specified. The borrowed causal link(s) can be recontextualized using the specified IT artifact/IS phenomena.

Back to our view on specification, we suggest that the causal links could be explicit about the role of the specified IT artifact, or the perceptions, affordances, actions/abilities surrounding it in linking an independent variable to a dependent variable. For example, effective use, defined as "the effective actualization of affordances arising from the relation between the system and its users" (Burton-Jones and Volkoff 2017, p. 470), is considered to link IT use to desired outcomes (ibid., see p. 482). Further examples from ISS follow. Peppard and Ward (2004, p. 187) discuss some specific mechanism through which IS competencies underlying IS capability can lead to firm performance:

> The underlying IS competencies will determine the extent to which IT opportunities are incorporated in business strategy, the effectiveness of business operations through systems and technology support, how well the IT infrastructure is designed and resourced, the level of performance achieved by IT operations and the quality of its services, and the ability of

an organization to deliver specific, measurable business benefits from IS/IT investment and deployment.

In explaining the path between possessing certain resources and superior firm performance through resource scarcity, Bharadwaj (2000, p. 173) characterizes the IT infrastructure resources that are causally ambiguous, hence scarce, as:

> IT infrastructures that enable firms to (1) identify and develop key applications rapidly, (2) share information across products, services, and locations, (3) implement common transaction processing and supply chain management across the business, and (4) exploit opportunities for synergy across business units.

Clemons and Row (1991, p. 283) state that "IT can reduce the basic transactions costs involved in the vertical flow of goods and services along a value chain. This includes direct savings, such as the costs of searching", also that "IT can be used to coordinate strategic resources in similar or complementary activities in different value chains. These horizontal interactions may be intra-organizational or inter-organizational" (ibid., p. 284).

To achieve distinction, first, given the nature of the specified IT artifact and its surrounding nomological network, new causal links for connecting two constructs could be introduced. In ISS, Nevo and Wade (2010) update the direct causal link assumed between a resource and firm performance by opening up the notion of resource bundles (see Barney and Arikan 2001, p. 144) and theorizing on the power of resources to enable other organizational resources to add value: " IT assets, despite their wide availability and commodity-like nature, can play a strategic role when combined with organizational resources for the purpose of creating IT-enabled resources" (ibid., p. 177).

Second, one can achieve distinction by challenging the borrowed relationship using the specification of the IT artifact: Conceptual assumptions are made to explain why there is a relationship between two constructs (Whetten 2002). When theories are borrowed without being challenged, their underlying assumptions are inherited (Alvesson and

Sandberg 2011). However, the borrowed conceptual assumptions can be relaxed, denied, and revised (Rivard 2014). We note that such modifications to the underlying assumptions can build on the nature of the specified IT artifact. In ISS, Mata et al. (1995), and later Wade and Hulland (2004), argue that the inimitability mechanism discussed in RBV is not a strong one for the IT artifacts; this is because the IT artifacts usually have high visibility (Piccoli and Ives 2005), which defeats secrecy and cannot be easily patented. Particularly, Mata et al. (1995, p. 500) make an internal distinction between the technical and managerial IT skills and argue that, among various IT attributes:

> only IT managerial skills are likely to be a source of sustained competitive advantage. IT management skills are often heterogeneously distributed across firms. Moreover, these skills reflect the unique histories of individual firms, are often part of the "taken for granted" routines in an organization, and can be based on socially complex relations within the IT function, between the IT function and other business functions in a firm, and between the IT function and a firm's suppliers or customers.

Moreover, a clear example of external distinction is performed by Mata et al. (1995) by referring to the susceptibility of IT artifacts to reverse engineering and difficulties in patenting them. From their view, the causal link of resource inimitability by patenting competitive advantage that works for some technologies does not work for IT applications as they are difficult to patent, and patents do not protect enough against imitation.

Conclusion

In this chapter, we build on the literature that theorizes the IT artifact and develop a 2 × 2 framework for the IT-rich recontextualized modification of borrowed theories in IS (see Table 8.1). The framework considers two loci of construct reconceptualization and relationship rejustification for implementing such specification and distinction. It provides some

Table 8.1 Summary of the discussed IT-rich recontextualized modifying approaches

Framework dimension	Discussed elements
Specification	1. Specifying the taxonomic nature of the IT artifact 2. Specifying the IT artifact as an IT vs. IS product 3. Specifying key subcomponents of the IT artifact 4. Specifying holistic attributes of the IT artifact 5. Specifying features/functionalities of the IT artifact 6. Purposeful nonspecificity of the IT artifact
Distinction	1. Internal distinction: Distinguishing from the specification for other IT artifacts 2. External distinction a. Distinguishing IT from non-IT instances of the same theories b. Distinguishing IT from non-IT technologies c. Recognizing the similarity (hence the concurrent value added) to non-IT technologies d. Distinguishing the IS angle from other angles for the same IT artifact
Specification and distinction in construct reconceptualization	1. Form of construct reconceptualization a. Recontextualized construct labels b. Recontextualized construct definitions c. Recontextualized dimensions d. Contextual exemplification 2. Content of construct reconceptualization a. Specifying the IT artifact on its own b. Specifying functional affordances of the IT artifact c. Specifying the IT artifact as an object of a perception d. Specifying the IT artifact as an object of actions and abilities
Specification and distinction in relationship rejustification	1. Recontextualization of independent, dependent, and path modifying variables 2. Recontextualization of causal links

pathways of embedding more IT in IS research, particularly, the IS strategy literature.

We make two contributions. First, through formalizing specification and distinction and discussing it for theorizing ISS, we provide a deeper appreciation of embedding the IT artifact in borrowed theories in IS. This has important implications for how IS scholars adapt theories from other disciplines. It renders visible a wide range of options (see Table 8.1) available to IS researchers. The addition of elements within specification and distinction provides a language to describe and how borrowed theory is constructed. Indeed, enabling researchers to "own" borrowed theory through specification or distinction unlocks—and stratifies—different ways to develop theory. Moreover, this builds on existing work to evaluate the stasis and future potential of the IS field.

The framework enables us to appreciate how some recontextualized theories are "IT-light" while others are "IT-rich". IT-rich studies are specific about unique aspects of IT/IS phenomena and demonstrate a deep engagement with the IT artifact (Orlikowski and Iacono 2001). For example, a study on IT investment, if the focus is only on a general dollar amount (the "proxy-capital" view in Orlikowski and Iacono 2001) is IT-light. Such a study can be more IT-rich if, rather than just reporting an aggregated dollar amount, it can use specification and examine the value invested in various system types, such as security and cloud systems. It could also perhaps use the distinction strategy to theorize the effective balance of such investment portfolios.

Further, we suggest that whether a theory is borrowed or native and whether a theory provides a light or rich understanding of the IT artifact are separate (although interrelated) matters. On the one hand, even an extension of a borrowed theory might not be IT-rich as it might extend with constructs remote from the IT artifact and its surrounding nomological network. The extension is an IS contribution if it enriches our understanding of IT/IS phenomena; otherwise, the contribution is likely to have significant overlaps with other disciplines. On the other hand, even instantiation can be IT-rich. The true difference between being IT-rich or IT-light is in implementation, that is, to what extent construct conceptualizations and theoretical assumptions are enriched with IT-related phenomena. The more a study attempts to domesticate a theory (Hong et al. 2014; Oswick et al. 2011), the less the resulting product will look like a simple instance of the borrowed theory.

Our second contribution is to ISS research by exploring several specific examples of how the IT artifact has been captured from a non-nominal view. In doing so, while Orlikowski and Iacono (2001) (similarly Akhlaghpour et al. 2013 and Grover and Lyytinen 2015) were looking for how researchers have viewed the IT artifact, we have looked for "spelled-out" specifications of the IT artifact that clearly demonstrate how researchers have captured the material nature of the IT artifact when linking it to social phenomena in the context of ISS. Recognizing IT-light vs IT-rich borrowing enables us to note that even when the same theories are borrowed, the level of IT-rich recontextualization can vary significantly. For example, the same theory of RBV has been borrowed in several different ways. In terms of constructs, assets capabilities and competencies have been specified on a range of nominal IT attributes to specific systems. Differentiation strategy has been used for assets by mentioning six characteristics. In terms of causal links, some have referred to unique characteristics of IT that would confer competitive advantage or sustained performance, such as increased knowledge management. Future studies could explore the variance in the IS recontextualization by various studies that borrow the same theories.

Through owning borrowed theories, the IS field can reinforce its identity by addressing the (false) dichotomy of focusing on theorizing the IT artifact to "belong" within IS vs viewing IT as a bolt-on to other allied disciplines.

References

Akhlaghpour, S., Wu, J., Lapointe, L., & Pinsonneault, A. (2013). The ongoing quest for the IT artifact: Looking back, moving forward. *Journal of Information Technology, 28*(2), 150–166.

Alvesson, M., & Sandberg, J. (2011). Generating research questions through problematization. *Academy of Management Review, 36*(2), 247–271.

Anderson, C., & Robey, D. (2017). Affordance potency: Explaining the actualization of technology affordances. *Information and Organization, 27*(2), 100–115.

Arvidsson, V., Holmström, J., & Lyytinen, K. (2014). Information systems use as strategy practice: A multi-dimensional view of strategic information system implementation and use. *The Journal of Strategic Information Systems, 23*(1), 45–61.

Bacharach, S. B. (1989). Organizational theories: Some criteria for evaluation. *The Academy of Management Review, 14*(4), 496–515.

Bakos, J. Y. (1991). A strategic analysis of electronic marketplaces. *MIS Quarterly, 15*(3), 295–310.

Baptista, J., Wilson, A. D., Galliers, R. D., & Bynghall, S. (2017). Social media and the emergence of reflexiveness as a new capability for open strategy. *Long Range Planning, Open Strategy: Transparency and Inclusion in Strategy Processes, 50*(3), 322–336.

Baptista, J., Stein, M. K., Klein, S., Watson-Manheim, M. B., & Lee, J. (2020). Digital work and organisational transformation: Emergent digital/human work configurations in modern organisations. *The Journal of Strategic Information Systems, 29*(2). https://doi.org/10.1016/j.jsis.2020.101618

Barki, H. (2008). Thar's gold in them Thar constructs. *ACM SIGMIS Database, 39*(3), 9–20.

Barki, H., & Pinsonneault, A. (2005). A model of organizational integration, implementation effort, and performance. *Organization Science, 16*(2), 165–179.

Barki, H., Rivard, S., & Talbot, J. (1993). Toward an assessment of software development risk. *Journal of Management Information Systems, 10*(2), 203–225.

Barley, S. R., Bechky, B. A., & Milliken, F. J. (2017). The changing nature of work: Careers, identities, and work lives in the 21st century. *Academy of Management Discoveries, 3*(2), 111–115.

Barney, J. B., & Arikan, A. M. (2001). The resource-based view: Origins and implications. In M. A. Hitt, R. E. Freeman, & J. S. Harrison (Eds.), *Handbook of strategic management* (pp. 124–188). Oxford: Wiley-Blackwell.

Beck, R., Müller-Bloch, C., & King, J. L. (2018). Governance in the blockchain economy: A framework and research agenda. *Journal of the Association for Information Systems, 19*(10), 1020–1034.

Benbasat, I., & Zmud, R. W. (2003). The identity crisis within the IS discipline: Defining and communicating the discipline's core properties. *MIS Quarterly, 27*(2), 183–194.

Benders, J., Batenburg, R., & Van der Blonk, H. (2006). Sticking to standards; technical and other isomorphic pressures in deploying ERP-systems. *Information & Management, 43*(2), 194–203.

Bharadwaj, A. S. (2000). A resource-based perspective on information technology capability and firm performance: An empirical investigation. *MIS Quarterly, 24*(1), 169–196.

Bharadwaj, A., El Sawy, O. A., Pavlou, P. A., & Venkatraman, N. (2013). Digital business strategy: Toward a next generation of insights. *MIS Quarterly, 37*, 471–482.

Bui, Q. N. (2017). Evaluating enterprise architecture frameworks using essential elements. *Communications of the Association for Information Systems, 41*, 6. https://doi.org/10.17705/1CAIS.04106

Burton-Jones, A., & Volkoff, O. (2017). How can we develop contextualized theories of effective use? A demonstration in the context of community-care electronic health records. *Information Systems Research, 28*(3), 468–489.

Butler, B. S., & Gray, P. H. (2006). Reliability, mindfulness, and information systems. *MIS Quarterly, 30*(2), 211–224.

Carr, N. G. (2003, May). IT doesn't matter. *Harvard Business Review*, 5–12.

Ceccagnoli, M., Forman, C., Huang, P., & Wu, D. J. (2012). Cocreation of value in a platform ecosystem: The case of enterprise software. *MIS Quarterly, 36*(1), 263–290.

Chatterjee, S., Moody, G., Lowry, P. B., Chakraborty, S., & Hardin, A. (2015). Strategic relevance of organizational virtues enabled by information technology in organizational innovation. *Journal of Management Information Systems, 32*(3), 158–196.

Chen, H., Chiang, R. H., & Storey, V. C. (2012). Business intelligence and analytics: From big data to big impact. *MIS Quarterly, 36*(4), 1165–1188.

Clemons, E. K., & Row, M. C. (1991). Sustaining IT advantage: The role of structural differences. *MIS Quarterly, 15*(2), 275–292.

Cragg, P., Caldeira, M., & Ward, J. (2011). Organizational information systems competences in small and medium-sized enterprises. *Information & Management, 48*(8), 353–363.

Cram, W. A., Proudfoot, J. G., & D'Arcy, J. (2017). Organizational information security policies: A review and research framework. *European Journal of Information Systems, 26*(6), 605–641.

Davis, F. D. (1989). Perceived usefulness, perceived ease of use, and user acceptance of information technology. *MIS Quarterly, 13*(3), 319–340.

Davison, R. M., & Ou, C. X. J. (2017). Digital work in a digitally challenged organization. *Information and Management, 54*(1), 129–137.

DeSanctis, G., & Poole, M. S. (1994). Capturing the complexity in advanced technology use: Adaptive structuration theory. *Organization Science, 5*(2), 121–147.

DeSanctis, G., Snyder, J. R., & Poole, M. S. (1994). Meaning of the interface: A functional and holistic evaluation of a meeting software system. *Decision Support Systems, 11*(4), 319–335.

Dulipovici, A., & Robey, D. (2013). Strategic alignment and misalignment of knowledge management systems: A social representation perspective. *Journal of Management Information Systems, 29*(4), 103–126.

Earl, M. J. (1993). Experiences in strategic information systems planning. *MIS Quarterly, 17*(1), 1–24.

Ein-Dor, P., & Segev, E. (1993). A classification of information systems: Analysis and interpretation. *Information Systems Research, 4*(2), 166–204.

Galliers, R. D. (1988). Information systems planning in the United Kingdom and Australia: A comparison of current practice. In *Oxford surveys in information technology* (pp. 223–255). New York: Oxford University Press, Inc.

Galliers, R. D. (2006). 'Don't worry, be happy ...' a post-modernist perspective on the information systems domain. In J. L. King & K. Lyytinen (Eds.), *Information systems: The state of the field* (John Wiley Series in Information Systems). John Wiley & Sons Inc.

Galliers, R. D. (2011). Further developments in information systems strategizing: Unpacking the concept. In R. D. Galliers & W. L. Currie (Eds.), *The Oxford handbook of management information systems: Critical perspectives and new directions* (pp. 329–345). Oxford: Oxford University Press.

Grant, G. G. (2003). Strategic alignment and enterprise systems implementation: The case of Metalco. *Journal of Information Technology, 18*(3), 159–175.

Griffith, T. L. (1999). Technology features as triggers for sensemaking. *The Academy of Management Review, 24*(3), 472–488.

Griffith, T. L., & Northcraft, G. B. (1994). Distinguishing between the forest and the trees: Media, features, and methodology in electronic communication research. *Organization Science, 5*(2), 272–285.

Grover, V., & Lyytinen, K. (2015). New state of play in information systems research: The push to the edges. *MIS Quarterly, 39*(2), 271–296.

Günther, W. A., Mehrizi, M. H. R., Huysman, M., & Feldberg, F. (2017). Debating big data: A literature review on realizing value from big data. *The Journal of Strategic Information Systems, 26*(3), 191–209.

Hassan, N. (2006). The relationship of IT to IS: An inquiry into the technoscientific nature of the IS field. *AMCIS 2006 Proceedings*, 434.

Hassan, N. R. (2011). Is information systems a discipline? Foucauldian and Toulminian insights. *European Journal of Information Systems, 20*(4), 456–476.

Hassan, N. R., & Lowry, P. B. (2015). Seeking middle-range theories in information systems research. *International Conference on Information systems (ICIS 2015)*, Fort Worth, TX, December, 13–18.

Hassan, N. R., & Will, H. J. (2006). Synthesizing diversity and pluralism in information systems: Forging a unique disciplinary subject matter for the information systems field. *Communications of the AIS, 17*(7), 152–180.

Henderson, J. C., & Venkatraman, N. (1993). Strategic alignment: Leveraging information technology for transforming organizations. *IBM Systems Journal, 32*, 4–16.

Hong, W., Chan, F. K., Thong, J. Y., Chasalow, L. C., & Dhillon, G. (2014). A framework and guidelines for context-specific theorizing in information systems research. *Information Systems Research, 25*(1), 111–136.

Hsu, C. W. (2009). Frame misalignment: Interpreting the implementation of information systems security certification in an organization. *European Journal of Information Systems, 18*(2), 140–150.

Jaccard, J., & Jacoby, J. (2009). *Theory construction and model-building skills: A practical guide for social scientists*. The Guilford Press.

Jasperson, J. S., Carter, P. E., & Zmud, R. W. (2005). A comprehensive conceptualization of post-adoptive behaviors associated with information technology enabled work systems. *MIS Quarterly, 29*(3), 525–557.

Johnston, H. R., & Vitale, M. R. (1988). Creating competitive advantage with interorganizational information systems. *MIS Quarterly, 12*(2), 153–165.

Joshi, K. D., Chi, L., Datta, A., & Han, S. (2010). Changing the competitive landscape: Continuous innovation through IT-enabled knowledge capabilities. *Information Systems Research, 21*(3), 472–495.

Karimi, J., Gupta, Y. P., & Somers, T. M. (1996). Impact of competitive strategy and information technology maturity on firms' strategic response to globalization. *Journal of Management Information Systems, 12*(4), 55–88.

Karpovsky, A., & Galliers, R. D. (2015). Aligning in practice: From current cases to a new agenda. *Journal of Information Technology, 30*(2), 136–160.

King, J. L., & Lyytinen, K. (2006). *Information systems: The state of the field*. Chichester and Hoboken, NJ: J. Wiley & Sons.

Lacity, M. C., Khan, S. A., & Willcocks, L. P. (2009). A review of the IT outsourcing literature: Insights for practice. *The Journal of Strategic Information Systems, 18*(3), 130–146.

Law, K. S., Wong, C.-S., & Mobley, W. H. (1998). Toward a taxonomy of multidimensional constructs. *The Academy of Management Review, 23*(4), 741–755.

Lederer, A. L., & Hannu, S. (1996). Toward a theory of strategic information systems planning. *The Journal of Strategic Information Systems, 5*(3), 237–253.

Leonardi, P. M., & Barley, S. R. (2008). Materiality and change: Challenges to building better theory about technology and organizing. *Information and Organization, 18*(3), 159–176.

Lowry, P. B., Dinev, T., & Willison, R. (2017). Why security and privacy research lies at the centre of the information systems (IS) artefact: Proposing a bold research agenda. *European Journal of Information Systems, 26*(6), 546–563.

Luftman, J., Kempaiah, R., & Nash, E. (2006). Key issues for IT executives 2005. *MIS Quarterly Executive, 5*(2), 81–99.

Mahoney, J. (2001). Beyond correlational analysis: Recent innovations in theory and method. *Sociological Forum, 16*(3), 575–593.

March, S. T., & Hevner, A. R. (2007). Integrated decision support systems: A data warehousing perspective. *Decision Support Systems, 43*(3), 1031–1043.

Markus, M. L., & Rowe, F. (2018). Is IT changing the world? Conceptions of causality for information systems theorizing. *MIS Quarterly, 42*(4), 1255–1280.

Markus, M. L., & Saunders, C. (2007). Looking for a few good concepts… and theories… for the information systems field, editor's comments. *MIS Quarterly, 31*(1), iii–vi.

Markus, L., & Silver, M. (2008). A foundation for the study of IT effects: A new look at DeSanctis and Poole's concepts of structural features and spirit. *Journal of the Association for Information Systems, 9*(10), 609–632.

Mata, F. J., Fuerst, W. L., & Barney, J. B. (1995). Information technology and sustained competitive advantage: A resource-based analysis. *MIS Quarterly, 19*(4), 487–505.

McFarlan, F. W. (1984). Information technology changes the way you compete. *Harvard Business Review*, reprint service, 98–103.

Melville, N., Kraemer, K., & Gurbaxani, V. (2004). Information technology and organizational performance: An integrative model of IT business value. *MIS Quarterly, 28*(2), 283–322.

Meyer, M. H., & Curley, K. F. (1991). An applied framework for classifying the complexity of knowledge-based systems. *MIS Quarterly, 15*(4), 455–472.

Morton, J., Wilson, A., Galliers, R., & Marabelli, M. (2019). Open strategy and information technology. In D. Seidl, R. Whittington, & G. Von Krogh (Eds.), *Cambridge handbook of open strategy* (pp. 171–189). Cambridge: Cambridge University Press.

Morton, J., Wilson, A. D., & Cooke, L. (2020). The digital work of strategists: Using open strategy for organizational transformation. *The Journal of Strategic Information Systems*. https://doi.org/10.1016/j.jsis.2020.101613

Mukhopadhyay, T., Kekre, S., & Kalathur, S. (1995). Business value of information technology: A study of electronic data interchange. *MIS Quarterly, 19*(2), 137–156.

Mykytyn, K., Mykytyn Jr., P. P., Bordoloi, B., McKinney, V., & Bandyopadhyay, K. (2002). The role of software patents in sustaining IT-enabled competitive advantage: A call for research. *The Journal of Strategic Information Systems, 11*(1), 59–82.

Nevo, S., & Wade, M. R. (2010). The formation and value of IT-enabled resources: Antecedents and consequences of synergistic relationships. *MIS Quarterly, 34*(1), 163–183.

Newell, S., & Marabelli, M. (2015). Strategic opportunities (and challenges) of algorithmic decision-making: A call for action on the long-term societal effects of 'datification'. *The Journal of Strategic Information Systems, 24*(1), 3–14.

O'Leary, D. E. (2013). Exploiting big data from mobile device sensor-based apps: Challenges and benefits. *MIS Quarterly Executive, 12*(4), 179–187.

Oh, W., & Pinsonneault, A. (2007). On the assessment of the strategic value of information technologies: Conceptual and analytical approaches. *MIS Quarterly, 31*(2), 239–265.

Orlikowski, W. J. (1992). The duality of technology: Rethinking the concept of technology in organizations. *Organization Science, 3*, 398–427.

Orlikowski, W. J. (1995). *Action and artifact: The structuring of technologies-in-use*. Sloan School of Management, Working Papers 3867-95, Massachusetts Institute of Technology.

Orlikowski, W. J., & Iacono, C. S. (2001). Research commentary: Desperately seeking the "IT" in IT research – A call to theorizing the IT artifact. *Information Systems Research, 12*(2), 121–134.

Oswick, C., Fleming, P., & Hanlon, G. (2011). From borrowing to blending: Rethinking the processes of organizational theory building. *Academy of Management Review, 36*(2), 318–337.

Pagani, M. (2013). Digital business strategy and value creation: Framing the dynamic cycle of control points. *MIS Quarterly, 37*(2), 617–632.

Pavlou, P. A., & El Sawy, O. A. (2006). From IT leveraging competence to competitive advantage in turbulent environments: The case of new product development. *Information Systems Research, 17*(3), 198–227.

Pavlou, P. A., & El Sawy, O. A. (2010). The "third hand": IT-enabled competitive advantage in turbulence through improvisational capabilities. *Information Systems Research, 21*(3), 443–471.

Peppard, J., & Ward, J. (2004). Beyond strategic information systems: Towards an IS capability. *The Journal of Strategic Information Systems, 13*(2), 167–194.

Piccoli, G., & Ives, B. (2005). IT-dependent strategic initiatives and sustained competitive advantage: A review and synthesis of the literature. *MIS Quarterly, 29*(4), 747–776.

Ray, G., Muhanna, W. A., & Barney, J. B. (2005). Information technology and the performance of the customer service process: A resource-based analysis. *MIS Quarterly, 29*(4), 625–652.

Reich, B. H., & Benbasat, I. (2000). Factors that influence the social dimension of alignment between business and information technology objectives. *MIS Quarterly, 24*(1), 81–113.

Renaud, A., Walsh, I., & Kalika, M. (2016). Is Sam still alive? A bibliometric and interpretive mapping of the strategic alignment research field. *The Journal of Strategic Information Systems, 25*(2), 75–103.

Rivard, S. (2014). Editor's Comments: The Ions of Theory Construction. *MIS Quarterly, 38*(2), iii–xiii.

Roberts, N., Campbell, D. E., & Vijayasarathy, L. R. (2016). Using information systems to sense opportunities for innovation: Integrating postadoptive use behaviors with the dynamic managerial capability perspective. *Journal of Management Information Systems, 33*(1), 45–69.

Rode, H. (2016). To share or not to share: The effects of extrinsic and intrinsic motivations on knowledge-sharing in enterprise social media platforms. *Journal of Information Technology, 31*(2), 152–165.

Sabherwal, R., & Chan, Y. E. (2001). Alignment between business and IS strategies: A study of prospectors, analyzers, and defenders. *Information Systems Research, 12*(1), 11–33.

Sambamurthy, V., Bharadwaj, A., & Grover, V. (2003). Shaping agility through digital options: Reconceptualizing the role of information technology in contemporary firms. *MIS Quarterly, 27*(2), 237–263.

Saraf, N., Langdon, C. S., & Gosain, S. (2007). IS application capabilities and relational value in interfirm partnerships. *Information Systems Research, 18*(3), 320–339.

Scott, J. E., & Vessey, I. (2002). Managing risks in enterprise systems implementations. *Communications of the ACM, 45*(4), 74–81.

Setia, P., Venkatesh, V., & Joglekar, S. (2013). Leveraging digital technologies: How information quality leads to localized capabilities and customer service performance. *MIS Quarterly, 37*(2), 565–590.

Shollo, A., & Galliers, R. D. (2016). Towards an understanding of the role of business intelligence systems in organisational knowing. *Information Systems Journal, 26*(4), 339–367.

Stein, M.-K., Wagner, E. L., Tierney, P., Newell, S., & Galliers, R. D. (2019). Datification and the pursuit of meaningfulness in work. *Journal of Management Studies, 56*(3), 685–717.

Straub, D. (2012). Editor's comments: Does MIS have native theories? *MIS Quarterly, 36*(2), III–XII.

Suddaby, R. (2010). Editor's comments: Construct clarity in theories of management and organization. *Academy of Management Review, 35*(3), 346–357.

Swanson, E. B., & Ramiller, N. C. (2004). Innovating mindfully with information technology. *MIS Quarterly, 28*(4), 553–583.

Tallon, P. P., & Pinsonneault, A. (2011). Competing perspectives on the link between strategic information technology alignment and organizational agility: Insights from a mediation model. *MIS Quarterly, 35*(2), 463–486.

Tan, C. W., Lim, E. T. K., Pan, S. L., & Chan, C. M. L. (2004). Enterprise system as an orchestrator of dynamic capability development: A case study of the IRAS and TechCo. In B. Kaplan, D. P. Truex, D. Wastell, A. T. Wood-Harper, & J. I. DeGross (Eds.), *Information systems research* (Vol. 143). IFIP International Federation for Information Processing, Boston, MA: Springer.

Tavakoli, A., Schlagwein, D., & Schoder, D. (2017). Open strategy: Literature review, re-analysis of cases and conceptualisation as a practice. *The Journal of Strategic Information Systems, 26*(3), 163–184.

Truex, D. P., Holmström, J., & Keil, M. (2006). Theorizing in information systems research: A reflexive analysis of the adaptation of theory in information systems research. *Journal of the AIS, 7*(12), 797–821.

Vance, A., Lowry, P. B., & Eggett, D. (2013). Using accountability to reduce access policy violations in information systems. *Journal of Management Information Systems, 29*(4), 263–290.

Venkatesh, V., Bala, H., Venkatraman, S., & Bates, J. (2007). Enterprise architecture maturity: The story of the veterans health administration. *MIS Quarterly Executive, 6*(2), 79–90.

Vial, G. (2019). Understanding digital transformation: A Review and a research agenda. *The Journal of Strategic Information Systems, 28*, 118–144.

Volkoff, O., & Strong, D. M. (2013). Critical realism and affordances: Theorizing IT-associated organizational change processes. *MIS Quarterly, 37*(3), 819–834.

von Krogh, G. (2012). How does social software change knowledge management? Toward a strategic research agenda. *The Journal of Strategic Information Systems, 21*(2), 154–164.

Wacker, J. G. (2004). A theory of formal conceptual definitions: Developing theory-building measurement instruments. *Journal of Operations Management, 22*(6), 629–650.

Wade, M., & Hulland, J. (2004). The resource-based view and information systems research: Review, extension, and suggestions for future research. *MIS Quarterly, 28*(1), 107–142.

Weber, R. (2003). Still desperately seeking the IT artifact. *MIS Quarterly, 27*(2), 183.

Whetten, D. A. (2002). Modelling-as-theorizing: A systematic methodology for theory development. In D. Partington (Ed.), *Essential skills for management research* (pp. 45–71). Thousand Oaks, CA: Sage.

Whinston, A. B., & Geng, X. (2004). Operationalizing the essential role of the information technology artifact in information systems research: Gray area, pitfalls, and the importance of strategic ambiguity. *MIS Quarterly, 28*(2), 149–159.

Willcocks, L. P., Lacity, M. C., & Kern, T. (1999). Risk mitigation in IT outsourcing strategy revisited: Longitudinal case research at LISA. *The Journal of Strategic Information Systems, 8*(3), 285–314.

Wu, S. P.-J., Straub, D. W., & Liang, T.-P. (2015). How information technology governance mechanisms and strategic alignment influence organizational performance: Insights from a matched survey of business and IT managers. *MIS Quarterly, 39*(2), 497–518.

Zammuto, R. F., Griffith, T. L., Majchrzak, A., Dougherty, D. J., & Faraj, S. (2007). Information technology and the changing fabric of organization. *Organization Science, 18*(5), 749–762.

Zhu, K., Kraemer, K. L., Gurbaxani, V., & Xu, S. X. (2006). Migration to open-standard interorganizational systems: Network effects, switching costs, and path dependency. *MIS Quarterly, 30*(Special Issue), 515–539.

9

Pluralist Theory Building: A Methodology for Generalizing from Data to Theory*

Sune Dueholm Müller, Lars Mathiassen, and Carol Saunders

Introduction

Information Systems (IS) researchers' interests vary from traditional topics such as systems development and technology management to the economics of IS, virtual teams, data analytics, social media, Internet of Things, and mobile technologies. To explore these wide-ranging topics, researchers draw on a broad array of reference disciplines, including

*Reprinted by permission Müller, S.D., Mathiassen, L. & Saunders, C. (2020) Pluralist theory building: A methodology for generalizing from data to theory. *Journal of the Association for Information Systems, 21*(1), pp. 23–49.

S. D. Müller
Aarhus University, Aarhus, Denmark
e-mail: sdm@processinnovation.dk

L. Mathiassen (✉)
Georgia State University, Atlanta, GA, USA
e-mail: lmathiassen@ceprin.org

N. R. Hassan, L. P. Willcocks (eds.), *Advancing Information Systems Theories*,
Technology, Work and Globalization, https://doi.org/10.1007/978-3-030-64884-8_9

psychology, anthropology, sociology, linguistics, mathematics, computer science, and management science. Thus, IS research is grounded in a wide range of research traditions and paradigms, each with its own theoretical assumptions and perspectives.

As the IS discipline matures, calls for more ambitious theorizing and native IS theories are mounting (Chiasson & Davidson, 2005; Grover & Lyytinen, 2015; Lee, 2001; Orlikowski & Iacono, 2001; Weber, 2003) in response to criticism of overreliance on reference discipline theories (Baskerville & Myers, 2002; Benbasat & Zmud, 2003). However, despite this growing interest in generalization and theorizing, there are only a few methodologies that provide guidance on how to build and present new IS theory (e.g., Remenyi & Williams, 1996; Carroll & Swatman, 2000; Weber, 2003, 2012; Rivard, 2014) and virtually none that address how to leverage the diversity of perspectives that is characteristic of the IS discipline. Although several papers combine perspectives in investigating and theorizing IS phenomena (e.g., Henfridsson, Mathiassen, & Svahn, 2014; Jasperson et al., 2002; Singh, Mathiassen, & Mishra, 2015), they do not elaborate and explain how the underlying research and theory building process may be applied by other researchers to advance knowledge. At the same time, the general literature on theory building is abstract and lacks practical guidance (e.g., Sutton & Staw, 1995; Weick, 1995). Lewis & Grimes (1999) is one notable exception, though our experiences, as elaborated below, indicate that it is difficult to translate their strategy for empirically driven theorizing into research practice. Thus, there is a need for comprehensive and practicable methodologies that IS researchers can use to build theory from data by leveraging the diversity of perspectives that characterize the discipline.

In this chapter, we propose and showcase Pluralist Theory Building as a research methodology that allows researchers to use multiple theoretical perspectives to build theory based on data. The methodology develops Lee and Baskerville's (2003) generalization framework with its four types of generalization into a practical process that allows researchers to move between empirical description and theory building. In addition, it draws on Mingers' (1997, 1999) pluralism grounded in critical realism to help

C. Saunders
University of Central Florida, Orlando, FL, USA
e-mail: csaunder@ucf.edu

researchers address complex real-world problems that are contingent upon a plurality of factors. Mingers' pragmatic approach to using multiple theoretical perspectives (2001) focuses on making sense of data rather than on philosophical concerns related to conflicting paradigms (Lewis & Grimes, 1999) and in doing so it helps researchers develop and articulate a deep empirical understanding as a strong foundation for building new theory. Such an approach requires, however, access to rich qualitative—and possibly quantitative—data about the studied real-world phenomena and their contexts.

We showcase the advantages of adding Pluralist Theory Building to the IS researcher's toolbox by drawing on a recent study (Müller, Mathiassen, Saunders, & Kræmmergaard, 2017) in which we theorize about politics during process innovation based on rich, multi-dimensional data. The study uses different perspectives on organizational politics (Bradshaw-Camball & Murray, 1991) to synthesize a comprehensive empirical account and build new theory. It also serves as an experiential basis for developing Pluralist Theory Building as an IS theory building methodology. As such, this chapter addresses Mingers' call for research into "alternative theoretical frameworks to provide practical guidance for multimethod design" (Mingers, 2001, p. 257) by combining Lee and Baskerville's generalization framework with multiperspective inquiry into a practical process for empirically based theory building using multiple theoretical perspectives.

The chapter proceeds as follows. We start by summarizing our experiences that led to developing the methodology in the Experiential Background section. Next, in the Theoretical Foundation section, we review the role of generalization and current use of pluralism in IS research. We then present Pluralist Theory Building in the Proposed Methodology section by describing its overall architecture, iterative steps, and key deliverables. In the Illustration and Guidelines section, we present experiences from our study of politics during process innovation (Müller et al., 2017) and draw on these to offer guidelines that detail the activities for each step of the methodology. Finally, we articulate the contribution of this research in the Discussion section by relating to other research methodologies and theory-building strategies within the IS discipline.

Experiential Background

Our study of politics during process innovation was an action research project (Chiasson, Germonprez, & Mathiassen, 2009; Mathiassen, 2002) in which we collected data over several years from multiple sources using mixed methods (including stakeholder interviews, participant observations, process maturity assessments, and archival documents). The resulting paper has been published recently in one of the leading IS journals. To draw on our experiences with building theory from both qualitative and quantitative data, we rely on documentation of the theorizing and related review process (including minutes of meetings between the researchers, feedback from journal reviewers and editors, and the evolving versions of the manuscript). Throughout the theory-building process, we leveraged the power of multiperspective inquiry.

Initially, we relied on metatriangulation (Lewis & Grimes, 1999) to apply multiple theoretical perspectives to develop a comprehensive empirical account and new theory. Although many researchers have successfully used metatriangulation to theorize based on literature, no studies have—since its publication—used the methodology to develop new theory based on data as suggested by the authors. Hence, we wanted to explore the practicality of metatriangulation for empirically based theory building. However, as we applied metatriangulation to our data, we found it difficult to translate its ideas about analyzing transition zones between perspectives and converting a metaparadigm perspective into practical theory building. After much trial and error, we therefore moved away from metatriangulation's rather abstract recommendations and instead approached multiperspective inquiry pragmatically to make sense of the data. As a result, we eventually drew on the IS literature on theorizing (Lee & Baskerville, 2003; Mingers, 1997, 1999) to develop our own approach to Pluralist Theory Building.

Our process innovation politics paper (Müller et al., 2017) presents an embedded case study of a company-wide process-innovation project across four business units in which we use contrasting theoretical perspectives on organizational politics (Bradshaw-Camball & Murray, 1991) to analyze and describe each unit's response to the project. Based on cross-case analyses and extant literature, we theorize how organizational actors engage in politics during process innovation efforts. As such, we leverage multiple perspectives in our theorizing efforts by generalizing from data to empirical accounts and theoretical statements, eventually arriving at new theory and—as shown in this chapter—a pluralist methodology for empirically based theory building. In the following, we draw on this background and key lessons learned to present and illustrate the proposed methodology and to discuss its contributions to IS research methodology. As our study provided the experiential background, we did not in a strict sense apply Pluralist Theory Building as presented here. However, for the sake of simplicity, we refer to our application of the methodology.

Theoretical Foundation

In the following, we provide an overview of the IS literature on generalization and pluralism as it relates to theory building. Rather than providing an exhaustive review of the literature on theory building, we describe the key sources we draw upon in Pluralist Theory Building.

Generalization: Between Description and Theory

The role of generalization in the context of theory building has been discussed by IS scholars for years. Whereas Weber defines theory as "a particular kind of model that is intended to account for some subset of phenomena in the real world" (Weber, 2012, p. 4), Seddon and Scheepers describe generalization as "the researcher's act of arguing, by induction, that there is a reasonable expectation that a knowledge claim already believed to be true in one or more settings is also true in other clearly defined settings" (Seddon & Scheepers, 2012, p. 7). Thus, theory building is a form of generalization.

IS researchers have addressed theory building, including Carroll and Swatman, who present a framework for interpretive theory building from qualitative data (Carroll & Swatman, 2000), and Remenyi and Williams, who explore the importance of qualitative data and narratives in developing theoretical conjectures and empirical generalizations (Remenyi & Williams, 1996). There are, however, few IS papers that provide comprehensive and practical guidance on how to build new IS theory (Weber, 2003, 2012). In key reference disciplines, there is more practical guidance as exemplified by (Eisenhardt, 1989) in the management literature, which describes the process of theory building in case study research from a positivist view, including performing within-case analysis and searching for cross-case patterns as a basis for generalizing knowledge claims to the level of hypotheses. The relative scarcity of comprehensive, practical guidance has led to calls for ambitious, IS-centered, and meta-level contributions to theory-building methodologies (Chiasson & Davidson, 2005; Grover, Lyytinen, Srinivasan, & Tan, 2008; Lee, 2001; Orlikowski & Iacono, 2001; Weber, 2003). In response, Kuechler and Vaishnavi developed a framework to support theory building in IS design science research (Kuechler & Vaishnavi, 2012).

Lee and Baskerville (2003) investigate the concept of generalization and present a classification of four types based on distinctions between, on the one hand, empirical versus theoretical statements and, on the other hand, what the researcher is generalizing from and to. Whereas empirical statements refer to data, measurements, observations, or descriptions about real-world phenomena, theoretical statements offer non-observable but theorized concepts and relationships (Lee & Baskerville, 2003). The four types of generalizations (Fig. 9.1) are: from empirical statements to other empirical statements (type EE generalization), from empirical statements to theoretical statements (type ET generalization), from theoretical statements to empirical statements (type TE generalization), and from theoretical statements to other theoretical statements (type TT generalization).

Type EE involves generalizing from one level of empirical statements to another in two different ways. First, there is generalizing data to a measurement, observation, or other description (simply referred to as description below) of the object of study. Second, there is generalizing the

	Generalizing to Empirical Statements	Generalizing to Theoretical Statements
Generalizing from Empirical Statements	EE Generalizing from Data to Description Generalizing data to a measurement, observation, or other description.	ET Generalizing from Description to Theory Generalizing a measurement, observation, or other description to a theory.
Generalizing from Theoretical Statements	TE Generalizing from Theory to Description Generalizing a theory, confirmed in one setting, to descriptions of other settings.	TT Generalizing from Concepts to Theory Generalizing a variable, construct, or other concept to a theory.

Fig. 9.1 Generalization framework. (Source: Adapted from Lee & Baskerville, 2003, p. 233)

resulting description beyond the domain or field setting from which the researcher collected data. In type ET, these descriptions (i.e., empirical statements) are generalized into theoretical statements. According to Yin, findings from case studies can for example be generalized to theoretical propositions (Yin, 2009). However, type ET is limited by the observed field settings in the sense that generalizing from empirical to theoretical statements is context bound (Lee & Baskerville, 2003, p. 236). This perspective is shared, among others, by Klein & Myers (1999) and Walsham (1995).[1] Drawing on Jaccard & Jacoby (2010), Rivard advocates alternating between abstractions and specific instances as a heuristic for developing propositions from data to explain studied phenomena (Rivard, 2014). Type TE involves generalizing from previously built and validated theory to an empirical statement that would be observable if the theory is used

[1] Walsham (1995) describes four types of generalization from IS case studies: development of concepts, generation of theory, drawing of specific implications, and contribution of rich insight.

in the specific context (Lee & Baskerville, 2003). Being able to claim that a theory is generalizable to a new setting ultimately requires validating it in the new context. This requires comparing what the theory describes or predicts to what is actually observed as happening in the new setting (Lee & Baskerville, 2003). Type TT involves generalizing from concepts to theory, for example, based on a synthesis of ideas from a literature review. Drawing on Bacharach's definition of theory, such theorizing efforts would ideally result in "a statement of relations among concepts within a set of boundary assumptions and constraints" (Bacharach, 1989, p. 496).

Although the Lee and Baskerville (2003) paper is recognized as a major contribution to the IS literature, it has been criticized by Tsang and Williams (2012), who propose an alternative classification of induction with five types of generalization. In addition to accusing Lee and Baskerville (2003) of self-contradiction in their conceptualization of generalization and criticizing the paper for a too narrow definition of induction, Tsang and Williams (2012) argue that Lee and Baskerville's definition does not correspond to the common conceptualization by either natural or social science researchers. In a rebuttal, Lee and Baskerville (2012) criticize Tsang and Williams (2012) for uncritically accepting the tenets of logical positivism and the notion of statistical inference, and they emphasize the value of diverse epistemological perspectives among researchers. In addition, they propose four judgment calls ("uniformity of nature", "sufficient similarity in relevant conditions", "successful identification of relevant variables", and "theory is true") that need to be made whenever generalizing a theory to a new setting and demonstrate how the process of such generalizing unfolds.

In our own efforts to build theory about politics during process innovation (Müller et al., 2017), we found Lee and Baskerville's (2003) four types of generalization very useful. However, their framework does not address the process of moving from one type to another, and universally agreed-upon conceptualizations and methodological guidance on the process of generalizing are still missing (Goeken & Börner, 2012). This has led some authors to characterize the treatment of generalization in IS as unsatisfactory (Seddon & Scheepers, 2012). Goeken and Börner (2012) call for methodological frameworks that provide researchers with practical guidance, and, although Seddon and Scheepers (2012) present a framework for justifying generalizations in IS research, including eight pathways for justifying knowledge claims, practical guidance of a less abstract kind is still needed. Hence, we have developed Lee and Baskerville's (2003) generalization framework into a practical process for theory building that leverages the power of multiperspective inquiry.

Pluralism: Between Single and Multiple Perspectives

Pluralist research involves the use of multiple perspectives in theory building, application, and validation. In this chapter, we focus on the former. Lewis and Grimes (1999), for example, provide an overview of multiparadigm inquiry and propose metatriangulation as a theory-building strategy with paradigms as heuristics. Their strategy enables researchers to juxtapose and link conflicting paradigm insights. By focusing on theoretical triangulation as a strategy of juxtaposing theoretical perspectives to analyze data and evaluate their explanatory power (Denzin, 1978), researchers are able to build theories that capture the complexity and paradoxical nature of organizational life (Lewis & Grimes, 1999). Tashakkori and Teddlie describe this kind of pluralism as an end to the so-called paradigm wars, applying perspectives from different philosophies to study particular research problems (Tashakkori & Teddlie, 1998).

Another proponent of pluralism is Tsang, who identifies different perspectives on generalization and describes and compares positivist, interpretivist, and critical realist views on generalizing from case study research (Tsang, 2014). While Scott and Briggs suggest pragmatism as the starting point for a pluralist methodology (Scott & Briggs, 2009), Tsang argues for critical realism as an appropriate basis for theory validation, empirical generalization, and theoretical generalization (Tsang, 2014). Consistent with this line of reasoning, Mingers (2001) advocates a pluralist methodology and rejects the incommensurability argument, that is, that perspectives bound to conflicting paradigms cannot be mixed in the same empirical account. Against this dogma, Mingers suggest that "paradigms are simply constructs of our thought. To hold that the world must actually conform to one of them is to commit the epistemic fallacy (limiting what may exist to our current knowledge) or, more generally, the anthropic fallacy (defining being or existence only in relation to human being)" (Mingers, 2001, p. 243). Hence, a pluralist methodology is required because "the real world is ontologically stratified and differentiated, consisting of a plurality of structures that generate the events that occur (and do not occur). Different paradigms each focus attention on different aspects of the situation, and so multimethod research is necessary to deal effectively with the full richness of the real world" (Mingers, 2001, p. 243). Based on this pragmatic and empirically focused approach, Mingers calls for the use of a plurality of perspectives in research and describes his philosophical position as critical pluralism (Mingers, 1997, 1999, 2001), which also forms the philosophical basis for our framework as advocated in current IS methodology discourse (Bygstad, 2010; Bygstad, Munkvold, & Volkoff, 2016; Dobson, 2001; Mingers, Mutch, & Willcocks, 2013).

Despite advantages of and support for pluralist research (Galliers, 1993; Landry & Banville, 1992; Lee, 1991; Robey, 1996), Mingers concludes in a literature review that only a small minority of papers in the main IS journals rely on pluralism (Mingers, 2003). One exception is Chiasson et al. (2009) in which the authors argue that pluralist research approaches generate both theoretical and practical knowledge. However, although Mingers and colleagues (Mingers, 1999; Mingers & Brocklesby, 1997) emphasize the desirability and feasibility of pluralist research and (Mingers, 1997) provides some guidelines, there is lack of knowledge about how to practice it effectively (Mingers, 1999). As a result, we combine our development of Lee and Baskerville's (2003) generalization framework into a practical process for theory building with Mingers' pluralist research strategy (Mingers, 1997, 1999, 2001; Mingers & Brocklesby, 1997) with a focus on multiple theoretical perspectives.

Proposed Methodology

Relying on theoretical triangulation as described by Denzin (1978), Pluralist Theory Building presupposes access to rich, multi-dimensional data and draws on the concepts of generalization and pluralism to build new theory (Lee & Baskerville, 2003; Mingers, 2001; Mingers et al.,

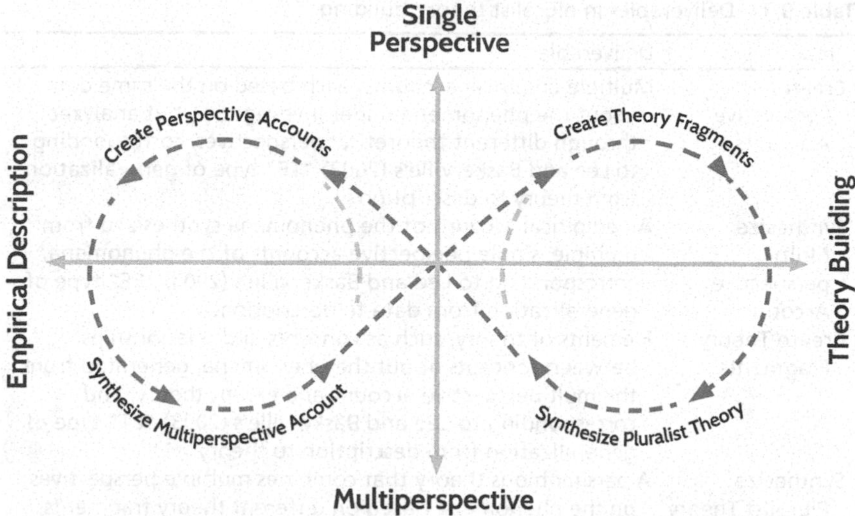

Fig. 9.2 Iterative steps in Pluralist Theory Building

2013). As such, the methodology leverages multiple theories within and across paradigms to move from data and empirical accounts to theory fragments and statements through four steps of creation and synthesis that iterate between empirical description and theory building, as illustrated in Fig. 9.2.

The process of theory building begins (step 1) with the analysis of data using contrasting theoretical perspectives to create multiple, single-perspective empirical accounts. The notion of "perspectives" should be interpreted rather broadly to include paradigms, for example, functionalism versus interpretivism (Bradshaw-Camball & Murray, 1991; Jasperson et al., 2002); theories, for example, path dependence versus path creation (Singh et al., 2015); actor perspectives, for example, Weltanschauung in Soft Systems Methodology (Checkland, 1986; Checkland & Scholes, 1990); and multi-level perspectives, for example, multi-level trust (i.e., individual, dyadic, team, and interorganizational trust) or multi-level resistance (Lapointe & Rivard, 2005). Also, the idea of creating contrasting empirical accounts necessitates the inclusion of at least two perspectives. The resulting single-perspective accounts are subsequently synthesized into one multiperspective empirical account (step 2) with a

Table 9.1 Deliverables in pluralist theory building

Step	Deliverable
Create Perspective Accounts	Multiple empirical accounts, each based on the same data about the phenomena under investigation but analyzed through different theoretical perspectives, corresponding to Lee and Baskerville's (2003) "TE" type of generalization from theory to description.
Synthesize Multi-perspective Account	An empirical account of the phenomena synthesized from multiple, single perspective accounts of the phenomena, corresponding to Lee and Baskerville's (2003) "EE" type of generalization from data to description.
Create Theory Fragments	Elements of theory, such as concepts and relationships between concepts about the phenomena, generated from the multiperspective account and extant theory, and corresponding to Lee and Baskerville's (2003) "ET" type of generalization from description to theory.
Synthesize Pluralist Theory	A parsimonious theory that combines multiple perspectives on the phenomena based on different theory fragments, corresponding to Lee and Baskerville's (2003) "TT" type of generalization from concepts to theory.

coherent storyline and compelling account. Next, researchers identify and analyze theoretical patterns underlying the multiperspective empirical account and draw on extant theory to create theory fragments (step 3). Finally, researchers synthesize these fragments into a new pluralist theory (step 4). Hence, Pluralist Theory Building involves iterating between four steps—Create Perspective Accounts, Synthesize Multiperspective Account, Create Theory Fragments, and Synthesize Pluralist Theory—across two dimensions. One dimension reflects the basic dualism in Lee & Baskerville (2003) between empirical statements and theoretical statements whereas the other draws on pluralist thinking to distinguish between single perspective and multiperspective views of the phenomena under investigation. As such, Fig. 9.2 illustrates Pluralist Theory Building as an iterative process with four steps that each involves the development of one or more key deliverables. These deliverables serve as output from one step and as input to the next, as summarized in Table 9.1.

Pluralist Theory Building begins with the Create Perspective Account step, which takes extant literature and two or more contrasting theoretical perspectives as input to an analysis of empirical data about the phenomena being studied and delivers multiple empirical accounts as output. These accounts feed into the subsequent Synthesize Multiperspective Account step, which focuses on establishing a coherent storyline across contrasting perspectives and yields one or more synthesized accounts (depending on the number of units of analysis). The Create Theory Fragments step serves to identify the basic building blocks (e.g., concepts and relationships) of the evolving theory by comparing identified patterns in the data and accounts to extant theory, and it yields theory fragments. The Synthesize Pluralist Theory step ties the constituting elements together in a comprehensive theory with boundary, premises, and propositions. The novelty of the resulting pluralist theory is checked against state-of-the-art knowledge and existing theories.

Illustration and Guidelines

To illustrate Pluralist Theory Building and develop practical guidelines for its application, we revisit the politics study in which we analyzed four business units' responses to company-wide efforts to implement process innovations. Our analyses concluded that different types of organizational politics were the underlying mechanisms that gave rise to the observed events, and our goal was therefore to understand and theorize about these mechanisms. The study followed the iterative process of creation and synthesis across empirical description and theory building, as illustrated in Fig. 9.2. First, based on extant theory, we applied Bradshaw-Camball and Murray's (1991) multiperspective framework of organizational politics to create single perspective accounts of political tensions and maneuvering at the case company. This resulted in empirical descriptions (single perspective accounts) of process innovation politics within each of the four business units from each of the applied political perspectives. Second, we synthesized these descriptions into an overall storyline (multiperspective account) for each unit with an explanation of what happened and why. Third, we engaged in "disciplined imagination" (Weick, 1989) to create theory fragments from our synthesized accounts. Specifically, we consistently applied the diverse political perspectives and captured the essence of the synthesized accounts by means of metaphors.

We compared the synthesized accounts in terms of similarities and differences, discerning patterns of process innovation politics and considering both the political perspectives of the Bradshaw-Camball and Murray (1991) framework and extant theory. Fourth, we synthesized these fragments of theory into a pluralist theory by alternating between generalizations and specific instances of observed process innovation politics. The results were a model of process innovation politics and nine propositions related to political responses and counter-responses.

Triangulation, peer feedback, and critical self-reflection were central to our theory-building efforts. During data analysis, we triangulated between data sources; compared evidence from interviews with meeting notes, plans, and other documents; and checked evidence with key stakeholders. We also triangulated at an aggregate level by systematically comparing summary data presented in tables with written accounts of what happened and why. Furthermore, to validate our findings we concluded the data analyses with key stakeholder reviews in which company employees provided feedback on all write-ups and synthesized storylines. Throughout our theory-building process, we relied on peer feedback and engaged in critical self-reflection through continuous discussions of empirical findings and theoretical contributions. In addition to our critical self-reflection, journal reviewers challenged us during consecutive revisions of our paper to clarify how we moved between description and theory in our research. For example, in the first round of reviews, one reviewer commented: "What is the basis and orientation of the theorizing? Is it from theory to theory, or is it from empirical observations to theory?" (Müller et al., 2017). This question relates to the types of generalization we engaged in. In the following subsections, we unfold our process of generalization—moving between empirical description and theory building—by describing each step and the related challenges and activities. These challenges and activities are summarized in Table 9.2 as guidelines for other researchers that want to apply Pluralist Theory Building. Although we present the challenges and activities in logical order, their sequence may vary based on peer feedback and self-reflection. It is also important to note that the challenges in Table 9.2 are specific to Pluralist Theory Building and researchers may face other challenges across the four

Table 9.2 Guidelines for Pluralist Theory Building[a]

Step	Challenges	Activities
Create Perspective Accounts	• Identifying and selecting theoretical perspectives • Understanding concepts underlying theoretical perspectives • Switching between contrasting theoretical perspectives	• Define area of concern • Review literature to identify contrasting theoretical perspectives • Develop research question • Establish case study protocol • Collect rich, multi-dimensional data • Develop coding scheme and data analysis guide • Code and organize data • Assess intra- and inter-coder reliability • Analyze coded data to develop single perspective accounts • Document link between single perspective accounts and theoretical perspectives
Synthesize Multiperspective Account	• Establishing a coherent storyline across contrasting perspectives • Determining combination of perspectives that makes for the most compelling account	• Compare single perspective accounts using data analysis guide • Evaluate explanatory power of single perspective accounts • Assess configuration of theoretical perspectives to develop storyline • Synthesize storyline in the form of a multiperspective account • Document link between single perspective accounts and resulting multiperspective account
Create Theory Fragments	• Contrasting accounts and data as basis for pattern recognition • Identifying comparable theoretical concepts and relationships in extant literature that stimulate imagination and creativity in theorizing	• Define boundary of theory • Identify and analyze patterns in and across accounts • List similarities and differences among accounts • Analyze concepts and relationships found in extant theory • Compare identified fragments to perspective accounts

(continued)

Table 9.2 (continued)

Step	Challenges	Activities
Synthesize Pluralist Theory	• Developing a strong (i.e., nontrivial) theoretical contribution • Theorizing beyond the immediate case but within the limits of the context	• Distinguish between premises and propositions • Formulate relationships as propositions • Evaluate propositions against extant theory • Validate concepts and propositions against data • Re-evaluate theory boundaries through self-reflection • Derive managerial and theoretical implications

[a]Please note that activities in a step may not always occur in the same order in which they are listed in the table

steps that are common to many if not all theorizing efforts based on rich, multi-dimensional data.

Create Perspective Accounts

Initially, we adopted a pluralist approach to review relevant literature streams with a political component. These streams included process innovation, software process improvement, IS implementation, and organizational change. The review led to the identification of knowledge gaps on organizational politics and was critical in developing and articulating a research question focused on uncovering political tensions and maneuvering and revealing their impact on process innovation efforts.

The first challenge was deciding on a combination of contrasting theoretical perspectives to drive our ensuing data analysis. We faced the options of either relying on an existing multiperspective framework of politics (e.g., Lukes' (2005) three dimensions of power or Clegg's (1989) circuits of power) or establishing our own by combing two or more contrasting perspectives that complement each other. In the end, we decided on Bradshaw-Camball and Murray's (1991) multiperspective framework of organizational politics because it facilitates a comprehensive and holistic understanding of politics based on contrasting perspectives such as the surface and deep structures of organizational politics (Bradshaw-Camball & Murray, 1991). Further, it has been applied both in management (ibid.) and IS research (Jasperson et al., 2002).

We subsequently used the framework to analyze process innovation behaviors and outcomes in each business unit. We distinguished between pluralist,[2] rationalist, interpretive, and radical perspectives on politics as representations of different sociological paradigms (Bradshaw-Camball & Murray, 1991) to analyze our data that had been collected from interviews, observations, meetings, archival documents, and process maturity assessments. We analyzed each business unit from all four contrasting theoretical perspectives to create four single perspective accounts of

[2] Note: The use of "pluralist" in this context is based on the categorization of sociological paradigms by Burrell and Morgan (1979), and it differs from the concept of "pluralism" used elsewhere in this chapter, which refers to multiperspective, multimethod approaches to conducting research.

process-innovation politics within each business unit. To develop these distinct political accounts, we went through three stages of data analysis.

First, we developed a case study protocol with a data analysis guide and coding scheme (Tables 9.3 and 9.4 in the Appendix). We developed the analysis guide based on Bradshaw-Camball and Murray's (1991) framework to identify and classify expressions of politics in the data. The guide contains key questions and concepts for each political perspective that helped us apply the four perspectives to understand how politics had shaped process innovation within each business unit. Whereas the questions reflect our interpretation of how the perspectives apply to process-innovation initiatives, the concepts are derived directly from Bradshaw-Camball and Murray's (1991) framework. We also developed a coding scheme based on (Bradshaw-Camball & Murray, 1991), which allowed us to distinguish statements based on who the messenger is (different transformation agents and process users), what the statement is about (intra-organizational level), and what the political nature (perspective) of the statement is (Miles, Huberman, & Saldana, 2014). The analysis guide and coding scheme helped us bring structure to and manage the rich, multi-dimensional data. Also, the process of identifying key concepts for each perspective during the development of the guide (see Table 9.3 in the Appendix) helped us understand the differences across perspectives and reach consensus about how to interpret them during data analysis. Hence, by writing down key concepts for the different theoretical perspectives, discussing what they meant to each of us, and developing the coding scheme, we were able to deal with the challenge of understanding the nuances of the underlying concepts and reach consensus about how we would interpret them in our data.

Second, we used the analysis guide and the coding scheme to code the data using ATLAS.ti. Initially, we identified all expressions of politics using the guide and coded each expression in accordance with the coding scheme. Next, we sorted all expressions of politics according to political perspective, evaluated the resulting categorization with respect to internal consistency and homogeneity, and recoded statements as needed. Through this process, we identified nearly 600 expressions of organizational politics ranging from a single sentence to half a page of transcript. We took various steps to ensure both intra- and intercoder reliability

(Miles et al., 2014), which correspond to investigator triangulation (Denzin, 1978). In terms of intracoder reliability, we wrote memos on all expressions of politics to document coding rationale and provide preliminary interpretations of political content (Neuman, 2014). We also recoded one transcribed audio recording and compared it to the first coding. In terms of intercoder reliability, two authors brought definitional clarity to the coding scheme by engaging in "check coding" (Miles et al., 2014).

Third, we created single perspective accounts (Table 9.5 in the Appendix) that describe the process-innovation behaviors and outcomes in every business unit from each of the four political perspectives. To facilitate perspective-based accounts of how politics influenced behaviors and outcomes within each business unit, we selected all expressions related to a unit and a political perspective and organized these into 16 tables (four perspectives on four units) (Table 9.6 in the Appendix). Each table contains illustrative quotations from process innovation participants at different organizational levels. These tables provide an overview that allowed us to engage in data triangulation (Denzin, 1978) by systematically comparing, contrasting, and relating statements from multiple sources (from, e.g., interviews, meetings, and archival documents). We then described the process-innovation behaviors and outcomes for each business unit and political perspective, yielding 16 distinct accounts of process-innovation politics.

When creating single perspective empirical accounts, we found it challenging to switch between contrasting theoretical perspectives as we analyzed data and created numerous empirical accounts through perspectives grounded in disparate paradigms. In particular, it was challenging to shift our frame of mind as we adopted different philosophical premises and ways of thinking across non-native paradigms. To address this challenge, we relied on the data analysis guide with key concepts and questions for each perspective. Although it was not easy to come up with a comprehensive set of useful questions across Bradshaw-Camball and Murray's (1991) rather abstract perspectives, the guide helped prime our brains and change mindset between successive and iterative analyses. Thus, it helped us engage in theoretical triangulation (Denzin, 1978). In addition, we attached a memo to every piece of coded data to sensitize us to contrasting perspectives and different understandings of the data. Finally, we established four tables for all business units—one for each political perspective—containing empirical evidence linking data and perspectives through the questions in the data analysis guide. Such tables are examples of within-case displays "for drawing and verifying *descriptive* conclusions about the phenomena in a bounded context that makes up a single 'case'" (Miles et al., 2014, p. 90). An example is shown in the Appendix (Table 9.7). This systematic documentation of our coding rationale and preliminary interpretations enabled us to gradually develop and distinguish numerous empirical accounts based on the rich, multi-dimensional data, and it helped us reconcile divergent interpretations among researchers and across theoretical perspectives. It also forced us to deal with the challenge of fully understanding the concepts associated with each theoretical perspective.

Synthesize Multiperspective Account

To establish an overall, coherent storyline, we compared the four single perspective accounts for each business unit and synthesized them into one empirical description of what happened and why in each of the four cases. For that purpose, we relied on our data analysis guide with its emphasis on key concepts and questions for each perspective. In doing so, we moved from single perspective accounts of the data to synthesized multiperspective accounts. A special feature of our study was the fact that we had four embedded cases, which in turn resulted in four separate, synthesized accounts.

Initially, the authors collaboratively interpreted four tables with illustrative quotations for one of the business units (one for each theoretical perspective) and synthesized an overall storyline with an explanation of what happened and why. As a participant in the process innovation project, the first author provided context information as basis for interpreting data and synthesizing the final storyline. The other authors acted as devil's advocates and contributed to developing a coherent storyline based on consistent use of coded data. The first author then developed similar multiperspective accounts for the remaining business units. The other authors reviewed these independently, leading to changes and additions. We discussed issues until reaching consensus on an interpretation or deciding to revisit the data. This approach is another expression of investigator triangulation (Denzin, 1978). We also made comments and observations about how each of the four political perspectives could help explain behaviors and outcomes. Moreover, we established a table for each business unit highlighting what each single perspective account, and thus each theoretical perspective, helped and did not help explain (Table 9.9 in the Appendix). These tables document empirical evidence of the links between the single perspective accounts and the synthesized multiperspective account.

Hence, we carefully considered the four contrasting political explanations of the behaviors and outcomes expressed in the single perspective accounts to arrive at a holistic understanding of how organizational politics had shaped process innovation within each business unit. We documented the explanatory power of each political explanation in a table (Table 9.8 in the Appendix) to support both within-case analysis and cross-case comparisons. Specifically, for each business unit, we evaluated which perspectives offered "minimal", "some", "major", or "dominant" explanations of the observed behaviors and outcomes. The resulting distribution of explanatory power across political perspectives revealed differences across business units and documented that combinations of perspectives were needed to fully account for our findings. These tables (Tables 9.8 and 9.9 in the Appendix) were instrumental in comparing different explanations of observed process innovation behaviors and outcomes within each business unit and determining the combination of perspectives that made for the most compelling account of what happened and why.

It was challenging to compare and synthesize the four perspective accounts for each business unit into their respective multiperspective accounts—each with a coherent storyline across contrasting perspectives—because of the complexity and data richness of the individual perspective accounts. This challenge was exacerbated by differences in vocabularies across the theoretical perspectives. For example, when looking at top management support through different power lenses, the rational view focuses on the transfer of authority, the pluralist and interpretive views focus on power-seeking behavior, and the radical view focuses on opportunities for power gains. The use of the 16 tables (again, four perspectives for each of the four business units) helped us organize the data, systematically compare statements, and synthesize the single perspective accounts of the observed process-innovation behaviors and outcomes into one multiperspective account for each business unit. Consequently, we needed to adopt and combine vocabularies of individual theoretical perspectives to synthesize the accounts. This required extensive discussions among the researchers during which we moved back and forth between data, single perspective accounts, and the emerging multiperspective accounts. In doing so, we had to bridge the theoretical perspectives by going back to the data analysis guide to better understand the differences and similarities across perspectives and accounts. The analysis guide was helpful because it contains the concepts—and therefore the vocabulary associated with each theoretical perspective and thus perspective account—as well as questions that reflect our interpretation of the political perspectives as they relate to process innovation. During this process, we established and documented the trail of evidence from raw data to the different levels of interpretation. This was time-consuming and challenging, and it involved organizing exemplar empirical statements into tables to support our knowledge claims (Tables 9.6 and 9.7 in the Appendix). However, documenting the trail of evidence was valuable in analyzing the rich, multi-dimensional data in a systematic and comprehensive manner and in ensuring that the multiperspective accounts accurately reflected the data and the single perspective accounts.

Create Theory Fragments

During the initial theorizing, we framed and created the basic elements of our theory by identifying and analyzing patterns across the multiperspective accounts and by revisiting extant literature in search of inspiration. A major challenge during this step was to contrast the synthesized multiperspective accounts while taking the underlying perspective accounts and empirical data into account in order to identify and theorize about patterns of politics across the business units. We started by comparing the four synthesized multiperspective accounts using metaphors as descriptive and heuristic devices. The metaphors helped us abstract and generalize political patterns as a first step toward theory. Hence, we expressed the particular characteristics of politics within each unit by relating the final storyline of each business unit to a metaphor that encapsulates the observed political behaviors and outcomes and highlights its key characteristics (Kendall & Kendall, 1993; Morgan, 1996). The four metaphors we used are: *applying-the-hammer*, *struggling-to-engage*, *walking-the-talk*, and *keeping-up-appearances*. In the words of Kendall and Kendall, "metaphors are like the magical incantations of old. By using words that people understand and believe in to make linkages with the new and unfamiliar, the speaker provides the ability to envision the world in a new way" (Kendall & Kendall, 1993, p. 149). As such, the metaphors not only express the sequence of political responses and counter-responses during the process-innovation project within the business units at an aggregate level, but also represent a new understanding of organizational politics that transcends individual political perspectives and facilitates theorizing about process innovation politics.

One journal reviewer questioned whether the metaphors were linked to and limited by political perspectives. This question prompted us to reflect upon the underlying characteristics of and possible mechanisms behind the patterns that we had identified. We responded by arguing that "the metaphors can display variations in degrees of the political perspectives. For example, *walking-the-talk* may make major or even dominant use of supportive deeper perspectives, either interpretive or radical. However, given the different perspectives on a common overarching goal, it is unlikely that rationalist politics could become more dominant in cases where pluralist politics are already strong (i.e., *struggling-to-engage* or *keeping-up-appearances*)" (Müller et al., 2017). Thus, this and other review comments challenged us to revisit the synthesized multiperspective accounts to create theory fragments and develop preliminary propositions regarding process-innovation politics. Through this process we received peer feedback and continued our theorizing through critical self-reflection. As part of the theorizing process, we realized that our metaphors had both strengths and limitations as vehicles for theorizing through cross-case synthesis. Their strengths lie in communicating and abstracting the essence of the synthesized multiperspective accounts to the level of cross-case comparisons (patterns across accounts). However, there is a risk of over-interpreting the cases by attaching too much importance to the metaphors in moving toward theory.

Having described the patterns of process-innovation politics in terms of four distinct metaphors, we faced the challenge of identifying theoretical concepts and relationships in extant literature that could serve as sources of inspiration and means of stimulating imagination and creativity in creating theory fragments. This process corresponds to Rivard's theory-building practice of alternating "between abstractions and specific instances of the explanation of the phenomenon under study" (Rivard, 2014: viii). The new round of literature review was motivated, in part, by suggestions from the review team to look at the IS development, IS implementation failure, and business process transformation literatures. We repeatedly compared existing theoretical models and concepts in these and other literature streams with the characteristics of the four synthesized multiperspective accounts. Through this process, we drew on previous research and state-of-the-art knowledge to create theory fragments from the empirical descriptions. We found inspiration in Keen's (1981) paper "Information Systems and Organizational Change" and adopted his distinctions between implementation and counterimplementation in IS-related organizational change (Keen, 1981) to identify patterns of responses and counter-responses in process-innovation politics across the synthesized multiperspective accounts. We further decided to adopt the vocabulary of process user and transformation agent based on Keen's distinctions between management and users (Keen, 1981). We considered other candidate theories but eventually rejected them. For example, we considered Rahim's conflict management tactics of integrating, obliging, compromising, and dominating (Rahim, 1985, 2002). However, Rahim's theory focuses on how individuals manage conflicts and the difference in unit of analysis compared to our organizational-level investigation of process innovation politics did not allow for analogical reasoning.

Distinguishing between responses and counter-responses allowed us to reinterpret the metaphors as theory fragments of process user responses and transformation agent counter-responses. As such, the identified metaphors constitute exemplars of behavioral patterns in process-innovation politics, although other political patterns may unfold under different circumstances. This led to discussions of the boundaries of our theory and the contexts under which it is valid. For example, we identified circumstances under which the identified patterns would manifest, and which other yet unidentified patterns might manifest under different circumstances. Drawing on Sabherwal and Grover's study of politics in system development projects, we contemplated several other patterns manifesting themselves, for example, *tug-of-war* and *empire-building* (Sabherwal & Grover, 2009). This comparison of patterns and preliminary theorizing confirmed our assumption that process-innovation politics depends on circumstances and context. This realization led us to adopt contingency theory (Iivari, 1992) as an important part of our theorizing and investigate the contingencies at play.

To identify case-specific contingencies, we listed similarities and differences across the four business units. This allowed us to break simplistic frames (Eisenhardt, 1989) and helped us develop possible explanations (Miles et al., 2014). Among other things, we compared the units in terms of business domain, process-innovation plans, alignment of process needs, and process-innovation outcomes (Table 9.10 in the Appendix). By contrasting similarities and differences across the synthesized multi-perspective accounts (i.e., across the business units), we identified goal alignment and goal compliance as important contingencies. In support of this cross-case comparison, we drew on the table highlighting the explanatory power of each political explanation in each of the four cases (Table 9.8 in the Appendix). The table reveals that the degree to which the perspectives explain the process-innovation behaviors and outcomes is contingent upon goal alignment and goal compliance. This pattern helped us synthesize the individual theory fragments into a pluralist theory and develop propositions regarding the contingencies of process-innovation politics.

Another theory fragment that we created through the initial process of theorizing was the role of structure in organizational politics. Early in the process, we responded to a journal reviewer's request that we clarify our use of Bradshaw-Camball and Murray's concept of "deep structures". In realizing the significance of structures, we added an appendix that not only identified structures, processes, and outcomes, but also employed the concepts of deep and surface structures to explain two of our propositions. This led to a meaningful dialogue with reviewers and the senior editor. For example, the senior editor noted: "Issues of goal congruence and process alignment characterize the sort of underlying political structure and influence the friction with which different technologies and implementation strategies are likely to work. It isn't clear why you back off of this and fall back on 'deep structure'. To me this is a core of your findings and analysis and needs more not less forward presence in the discussion" (Müller et al., 2017). This, in turn, encouraged us to develop one proposition dealing specifically with the role of structure.

In creating theory fragments, it was challenging to mobilize and leverage extant IS literature in our theorizing. We compared our empirical findings, that is, the synthesized multiperspective accounts, to existing theory and drew on concepts and relationships documented in the literature. In doing so, we further defined and narrowed down our area of concern and the literature to which we wanted to contribute. This, in turn, translated into the difficult task of determining the boundaries of our theory. As our case was rich and multi-dimensional, we had to decide which parts of the IS, management, and organization science literatures to include, even if it meant excluding a literature stream recommended by the review team. We wrestled with the role of the IS artifact in our research, and our literature searches spanned an exhaustive range of general and specific focal areas. In the end, we decided to concentrate on process innovation as a particular form of organizational change that involves a complex interplay between technology and people (Grover & Markus, 2008). Although it was time-consuming to arrive at this decision, settling on boundaries and establishing a focus allowed us to identify the appropriate terminology, concepts, and relationships to build upon. As such, our experiences suggest that theorizing should start by clearly defining the area of concern that the research is contributing to and identifying existing theories that need to be considered. These theories should in turn be broken down into their constituent components (concepts, relations, and boundaries) to facilitate cross-theory comparisons and comparisons with the empirical accounts that drive the theorizing efforts.

Synthesize Pluralist Theory

Our synthesizing of theory fragments into a pluralist theory was motivated by two goals. First, we wanted to contribute a pluralist theory as an analytical framework for understanding process-innovation politics. Second, we wanted to develop a theory as a practical tool for managing process-innovation politics. The resulting model of process-innovation politics and the associated propositions describe how process users react politically to process-innovation efforts, how transformation agents engage with process users, and the interplay between the two.

In building the theory, we drew on Rivard (2014), Weber (2012), Gregor (2006), Whetten (1989), and Bacharach (1989) to specify the type of theory and define its concepts, relationships, and boundaries. During this process, we developed nine propositions articulating the relationships. Specifically, we drew on extant literature and the identified structures and response–counter-response patterns (theory fragments) in relating key concepts to each other and developing the propositions. In effect, we theorized different types of process-innovation politics as mechanisms that explain process-innovation behaviors and outcomes. As we related key concepts to each other and developed propositions, we constantly revisited the data and synthesized empirical descriptions to explain the observed political behaviors. This process forced us to iteratively reconsider questions about the boundaries of our theory such as "where" the theory applies and to "whom" it applies.

As such, the theory-building step (i.e., Synthesize Pluralist Theory) involved an iterative process of abstraction during which propositions were developed for subsequent empirical investigation and comparison. Through several iterative cycles of self-reflection during which we went back and forth between theoretical statements in extant literature, empirical descriptions of the cases (the synthesized multiperspective accounts), and our propositions, we were able to achieve analytical stability of the theorized mechanisms (Zachariadis, Scott, & Barrett, 2013). The iterative cycles involved extensive discussion and critical self-reflection among the authors and feedback from journal reviewers that challenged our theoretical statements by asking whether they: (1) explain the empirical cases studied; (2) explain alternative, imagined scenarios; and (3) offer a better explanation than extant theory.

Because our aim was to build a contingency theory, our theorizing efforts focused on articulating contingencies, that is, the context and circumstances under which the identified political patterns are likely to manifest. The resulting theory describes patterns of process user responses and transformation agent counter-responses depending on the degree of goal alignment and level of goal compliance. We articulated the transformation agent counter-responses as different types of politics that can be utilized when confronted with varying process innovation challenges. We described these transformation agent counter-responses as reinforcement, persuasion, accommodation, and confrontation politics. They reveal tactics as well as threats and opportunities that can help managers maneuver process-innovation efforts. Further, they reflect underlying surface and deep structures of process-innovation politics. Finally, we summarized and visualized our theory in a model of process-innovation politics (Fig. 2 in (Müller et al., 2017)) and tables of exemplar process user responses and transformation agent counter-responses (Tables 9.8 and 9.9 respectively in (Müller et al., 2017)).

In synthesizing the theory fragments into a pluralist theory, we found it challenging to develop a strong (i.e., nontrivial) theoretical contribution. We specifically wrestled with how to establish propositions that were interesting, testable, and bold. In maturing our thinking through consecutive versions of the manuscript, journal reviewers persistently asked us to avoid "truisms" which "offered no new insights" and to go beyond the empirical analyses to make wider knowledge claims—to transcend trivial observations and develop more universally valid theoretical statements. This presented us with the additional challenge of theorizing beyond the immediate case but within the considered context. In good keeping with the tenets of pluralism, the journal editor encouraged us to define "speculative propositions that were true in your observations and that could be tested in other circumstances and, if robust after adequate testing, could be applied in semi-formulaic manner by those who follow" (Müller et al., 2017). In the subsequent review of the revised manuscript, the senior editor elaborated by stating that "I am not suggesting that the authors project their findings as if they were universally true, but rather to suggest if they are universally true, what would the theoretical propositions say?" (Müller et al., 2017). In doing so, we found it helpful to re-evaluate the theory boundaries through self-reflection and to distinguish between premises and propositions. The premises essentially articulate the boundaries of the theory as the foundation upon which we built our propositions. One premise, for example, posits that process-innovation politics is contingent upon goal alignment and goal compliance. We included this statement as a premise and not a proposition, because it easily can be inferred from extant literature. As expected, it was also confirmed by our study. In making the other candidate propositions more assertive, we found it useful to constantly consider the dreaded "so what" question (Whetten, 1989). Among other things, that meant clarifying the managerial implications of the propositions, as well as deciding what to address with our theory in terms of contribution and focus. In the end, the resulting propositions specify criteria for process-innovation success and the kinds of political responses and counter-responses that are likely to unfold under different circumstances.

Discussion

Based on extant research and grounded in our own experiences (Müller et al., 2017), we have presented Pluralist Theory Building as an approach to empirically based theory building. By describing and showcasing this methodology, we address the lack of research that explores the value and feasibility of pluralist research. In comparison with extant literature, Pluralist Theory Building is unique within the IS field in offering theoretical and practical guidance on how to move from empirical description to theoretical statements through an iterative theory-building process of creation and synthesis. As such, it is aligned with Rivard (2014) advocating the practice of alternating between abstractions and specific instances when developing theoretical statements from empirical data, and it addresses the identified lack of methodological frameworks and guidance on IS theory building in general (Weber, 2003, 2012) and the generalization process specifically (Goeken & Börner, 2012; Seddon & Scheepers, 2012). While others adopt an abstract perspective and offer little help in terms of how to theorize (Eisenhardt, 1989), Pluralist Theory Building describes the process of creation and synthesis across empirical description and theory building, and it prescribes the iterative steps, deliverables, challenges, and activities (Fig. 9.2, Table 9.1 and 9.2) involved in the process. In the following, we discuss these contributions to state-of-the-art knowledge on IS research methodologies.

Importantly, we offer a methodological framework with iterative steps (Fig. 9.1) and specific deliverables (Table 9.1) that builds on and combines the concepts of generalization (Lee & Baskerville, 2003) and pluralism (Mingers, 1997, 1999, 2001). As described earlier, these concepts are debated in the literature with conflicting views on how they may be leveraged in theory building (e.g., Lee & Baskerville, 2012; Tsang & Williams, (2012). Pluralist Theory Building draws on both concepts through the iterative movement between, on the one hand, description and theory (i.e., generalization from empirical to theoretical statements as well as between levels of empirical and theoretical statements), and, on the other hand, between single and multiple perspectives (i.e., a plurality of perspectives on the same phenomena). As detailed in the Illustration and Guidelines section and summarized in Table 9.2, we take a pragmatic stance (Mingers, 2001) in which we are first and foremost concerned with making sense of data rather than with philosophical issues related to conflicting paradigms (Lewis & Grimes, 1999). This pragmatic stance focuses on developing empirical descriptions and theoretical claims (Lee & Baskerville, 2012) rather than on the philosophical discourse about the role and forms of induction in research (Tsang & Williams, (2012). This pragmatic approach bridges the "methodological space that lies between *empiricism* and *interpretivism*" (Zachariadis et al., 2013, p. 856) and it enables us to develop and combine the concepts of generalization and pluralism into practical methodological knowledge that researchers can use to build theory from data.

The ideas behind and design of the methodology, including the pragmatic stance, grew out of our attempts to theorize based on rich, multidimensional data as well as our frustrations over not being able to practice metatriangulation to build new theory (Lewis & Grimes, 1999). Though we owe much in terms of critical self-reflection to the metatriangulation theory-building strategy, we found that Lewis and Grimes (1999) offer little advice in terms of how to explore so-called metaconjectures, how to attain a metaparadigm perspective, and how to identify let alone explore the transition zones between paradigms in any meaningful manner. Hence, while metatriangulation seeks to reconcile the paradigmatic tensions involved in applying contrasting perspectives on a theoretical level, for example through the notion of transition zones, we found it difficult to practice. This experience led us to develop Pluralist Theory Building, which allows for tensions between multiple and irreconcilable perspectives to be resolved at an empirical level by synthesizing empirical accounts. In other words, instead of trying to solve irreconcilable tensions of an ontological and epistemological nature between theoretical perspectives, we show how to leverage pluralism based on contrasting perspectives to establish coherent empirical accounts and to generalize these into parsimonious theoretical statements.

Confronted with the problems of applying metatriangulation, we consulted the research methodology literature. First, we decided to rely on Lee and Baskerville's (2003) generalization framework to navigate the complex relationships and dynamics between different forms of theoretical and empirical statements involved in theory building from data. Second, we found inspiration in Mingers (2001) and his vision of a critical pluralist methodology. His ideas and arguments for using multiple theoretical perspectives resonated with us and helped us deal with the problems encountered in trying to practice metatriangulation. Combining these two sources allowed us to successfully develop and publish our process-innovation politics paper (Müller et al., 2017), which in turn became the foundation for developing the Pluralist Theory Building methodology. In doing so, we went beyond Lee and Baskerville's (2003) types of generalization and drew on pluralism (Mingers, 1997, 1999, 2001) to turn their descriptive framework into a process with steps, deliverables, challenges, and activities for how researchers can iteratively move from data analysis to theory building (Fig. 9.2, Tables 9.1 and 9.2).

Accordingly, Pluralist Theory Building does not offer a solution to the paradigm incommensurability problem. Instead, we accept that the social world is—borrowing Mingers' expression—"ontologically stratified and differentiated" (Mingers, 2001, p. 243), and that we need an approach based on different paradigms to understand and explain complex empirical events. Whereas Mingers focuses primarily on multiple methods (Mingers, 2001, 2003; Mingers & Brocklesby, 1997), we focus primarily on multiple theories, which explains why we decided to present it as a Pluralist Theory Building methodology rather than a multimethod research design. This fundamental positioning of Pluralist Theory Building has allowed us to offer a generic methodology in which certain boundary conditions must be met. First, the research must involve rich, multi-dimensional data that allows for contrasting interpretations. This suggests inclusion of qualitative and possibly quantitative data collected through different types of studies, for example, ethnographies, case studies, action research, or design science. Researchers may use multiple methods, for example, observation, interviews, and surveys to collect the data, and they may consider a mixed-method approach to data collection. Second, the research must involve at least two contrasting (theoretical) perspectives to support synthesizing perspective accounts and developing pluralist theory. Third, the research goal must be to contribute new theory. In addition to these boundary conditions, it is necessary to adapt the methodology depending on research context. For example, in action research and design science, the multiple, contrasting perspectives should be integrated into the problem-solving and design activities.

By detailing the iterative steps, deliverables, challenges, and activities in Pluralist Theory Building, we provide guidance to both experienced and novice researchers. Still, based on our own experiences in applying the methodology, it is not easy to build theory in practice. Pluralist Theory Building demands both structure (by following our proposed methodology) and creativity and imagination (as emphasized by (Weick, 1989) and others). As such, our methodology does not offer a process that ensures successful theorizing. Creativity and imagination are indeed indispensable to the process, which accentuates the need for providing researchers with creativity support and encouragement to improve their theory-building capabilities. Weick (1989) brings attention to the concepts of "disciplined imagination", and while our methodology brings discipline to the process of theory building, it does not in the same way ensure requisite imagination. We therefore invite future research to better understand how complementary imagination capabilities can be supported and encouraged in relation to our methodology.

At this stage of development, Pluralist Theory Building has certain limitations. First, we have developed and exemplified the methodology through an embedded IS case study (Müller et al., 2017). Still, though the IS artifact is central to this case study, it is not part of the methodology, which suggests that Pluralist Theory Building in its current form is a social science research methodology with limited IS specificity. Hence, there is ample opportunity to apply and further develop the methodology related to IS research. Second, we have not tested the methodology based purely on quantitative data. It is conceivable that it can be used based on quantitative data and appropriate statistical techniques (e.g., Hierarchical Linear Modeling) to build theory on multi-level phenomena such as trust or resistance. Other researchers are encouraged to investigate whether such theory building based on quantitative data and techniques is feasible. In its current state of development, we argue that the methodology can be used to theorize based on rich qualitative data, possibly in combination with quantitative data.

In conclusion, Pluralist Theory Building combines a critical realist (Bhaskar, 2008) research approach based on Mingers' pluralist methodology (Mingers, 1997, 2001) with Lee and Baskerville's perspectives on generalization (Lee & Baskerville, 2003) into a practical methodology for building theory. The methodology involves creation and synthesis based on rich, multi-dimensional data and theoretical perspectives through four iterative steps with accompanying deliverables, challenges, and activities. As such, it leverages Mingers' pragmatic approach to pluralism (2001) and extends Lee and Baskerville's (2003) generalization framework to a detailed process for empirically based theory building.

Appendix

In this appendix, we have collected example evidence from our Pluralist Theory Building process behind the Müller et al. (2017) process-innovation politics paper. We conceal the identity of the four business units and refer to them as Alpha, Beta, Gamma, and Delta.

Table 9.3 shows our data analysis guide. The guide contains key questions and concepts that helped us apply the four political perspectives to

Table 9.3 Data analysis guide

Perspective	Questions	Concepts
Pluralist	How are conflicting interests between involved stakeholders expressed and negotiated during the initiative? How do differences in powerbase between stakeholders influence the process and its outcomes?	Stakeholders Interests Powerbases Conflicts Negotiation
Rationalist	How are goals expressed and data collected and used as a basis for evaluating options during the initiative? How are choices between alternative processes and outcomes made based on legitimate and formal authority structures?	Goals Authority Value judgments Decision making
Interpretive	How do actors make sense of the initiative based on past experience and symbolic expressions? How do actors use symbols to socially construct the process and influence its outcomes?	Experiences Symbols Sensemaking Social constructions Organizational culture
Radical	How are actors influenced during the initiative by the ideologies and constraints of the firm's environment? How does the resulting struggle between opposing forces influence the process and its outcomes?	Ideology Constraints Struggle Oppression Emancipation

understand how politics had shaped process-innovation behaviors and outcomes within each business unit at the case company.

Table 9.4 provides an overview of our coding scheme. The scheme includes codes that allowed us to identify all political statements, who made the statement in question, which business unit the statement was made in Müller et al. (2017) to, and the political perspective of the statement. Thus, the coding scheme helped us categorize political statements according to type and sort our data into manageable chunks.

Table 9.5 contains an example of a single perspective account that describes the process-innovation behaviors and outcomes in the Gamma business unit from the rationalist political perspective. All case analyses follow the same structure. The contribution of each actor to the

Table 9.4 Coding scheme

Organizational Unit	Interviewee	Perspective
Corporate	Corporate management	Interpretive
	Corporate PI[a] management	Pluralist
	Corporate PI agent	Radical
Alpha[b]	Alpha management	Rationalist
	Alpha PI management	
	Alpha implementation management	
Beta	Beta management	
	Beta PI management	
	Beta PI agent	
	Beta implementation management	
Gamma	Gamma management	
	Gamma PI management	
	Gamma PI agent	
	Gamma implementation management	
Delta[a]	Delta management	
	Delta PI management	
	Delta implementation management	

[a]PI is short for process innovation
[b]No PI agent was appointed for this business unit

interpretation from the political perspective in question is described in turn. However, for the sake of presentation, the analysis is not only sorted by person (e.g., the CEO and the corporate BPIP manager) but also grouped by organizational level (project level, business unit level, and corporate level). Each analysis is summarized in a "results" subsection.

Table 9.6 contains data coded as expressions of pluralist politics in the Alpha business unit. We selected all expressions related to a unit and a political perspective and organized these into 16 tables (four perspectives on four units).

Table 9.7 contains sample evidence of interpretive politics in Alpha. We created similar tables for each business unit—one for each political perspective—containing empirical evidence linking data and perspectives through the questions in the data analysis guide. Consequently, these tables contain answers to the questions (in the data analysis guide) that we used in interrogating the data.

Table 9.8 shows the qualitative impact of the political perspectives on the overall storylines, that is, the synthesized multiperspective accounts.

Table 9.5 Example of single perspective account

Gamma Project Level
At the project level, the attitude towards the BPIP was positive, and all the project managers offered constructive criticism of various aspects of the BPIP. The new processes were evaluated and implemented, although some project managers required assistance in doing so. From Gamma PM#1's perspective, the introduction of project status meetings was the most apparent change to existing practices coming out of the BPIP. He described this change as a positive initiative. Said Gamma PM#1: "One of the things that unknowingly has become an excellent practice—and that the BPIP has imposed on us—is holding monthly status meetings, project status meetings {8:6} … It is a matter of us doing it so often that we can begin to call it a good habit" {8:8}.
In the opinion of Gamma PM#2, several of the changes that resulted from the BPIP were valuable. He also pointed to project status meetings as one example. Other examples included improvements to Configuration Management and a new template for project presentations that improved the handover of projects from the sales unit (staff function) to the project organization. Based on his value judgment, he decided to press for the adoption of new processes, e.g. the writing of minutes of meetings, among participants in his project. During the BPIP implementation, he relied on both process maturity assessments and the leadership of the Gamma implementation manager and the Gamma roll-out & training manager. Assessments allowed for snapshots in time that revealed trends in the process implementation over time. He saw these assessments as reliable measures of progress, and he put his trust in them. Since he did not read the new process descriptions, he also put his trust in other sources of information about the new process requirements. Thus, he trusted the Gamma implementation manager and the Gamma roll-out & training manager to provide him with the information that he needed. Said Gamma PM#2: "I mean, I have not … the only processes I have read are about project management {46:10} … I guess it was [the Gamma roll-out & training manager] who wielded the baton at the time. [The Gamma implementation manager] has probably been part of it. Anyhow, it was [the Gamma roll-out & training manager] and [the Gamma implementation manager] who held the progress status meetings or whatever they are called" {46:3}.

According to Gamma PM#3, many of the new processes were inapplicable and of little value because his project was at a late stage in its life cycle. Many of the activities addressed by the processes had already taken place. However, like Gamma PM#1 and Gamma PM#2, he regarded the project status meetings as a positive change. Generally speaking, the implementation was managed like any other project with a timetable, requirements, a plan for meeting these requirements, monitoring to ensure progress, and tests, i.e., in this case, process maturity assessments to evaluate the result. The Gamma roll-out & training manager had interpreted the new processes in terms of process requirements and established a list of things to do to ensure process compliance. The implementation projects were confronted with this list which was updated continually. Elaborations were added based on lessons learned from the assessments and meetings among the project managers. These meetings were held every month in order to discuss implementation related issues and progress. Specifically, the projects were required to use JIRA (project management tool) and IFS (ERP system) in addition to holding project status meetings. Implementation progress was evaluated through assessments, and steps were taken to ensure the accuracy of the results. The Gamma implementation manager coached the implementation responsible project managers in how to administer the questionnaires used for the assessments. He did not tell them which answers to give but how to interpret and fill out the questionnaires. Said Gamma PM#3: "One of the things we did before the last assessment in May … [the Gamma implementation manager] instructed people from the whole department in how to understand each question—nothing about how we should answer or anything—but when answering, you should be aware that if you have done so and so, you should answer this; 'do not know', if so and so … so people do not answer arbitrarily" {47:7}.

(*continued*)

Table 9.5 (continued)

From Gamma PM#4's perspective, it was necessary, in addition to the Gamma implementation manager's coaching, to explain to the project participants how to interpret the survey questions in relation to their project practices. These explanations had a positive effect on the assessment results. Within Gamma, the assessment results and accompanying measurement reports were discussed among the project managers and with the Gamma roll-out & training manager and the Gamma implementation manager to ensure follow-up and to identify the reasons for non-compliance within a given process area. Misinterpretations of the survey questions were also discovered during these discussions. All in all, he described the BPIP implementation in Gamma as appropriate, although the short duration of the pilot and its overlap with the broader implementation resulted in the new processes not being evaluated and modified before wider dissemination. The implementation pace was hurried and some processes, e.g. within the process area of Measurement and Analysis, were still on the drawing board at the time when they were supposed to be put into practice. Yet, Gamma PM#4 found it reasonable to define and implement the new processes concurrently. In his capacity as project manager, he analyzed the gaps between existing practices, including plans and other forms of documentation, and the new processes. It was determined that only a few changes were needed, estimated at a few hundred man-hours. Status meetings during the implementation ensured continuous evaluation and progress. Said Gamma PM#4: "I evaluated my project. The first step was an evaluation according to the new processes: Where did I see changes in my project being necessary in order for me to meet the requirements. I saw that I needed to update my management plan; I need to make a data management plan … I suggested that I needed this, this, and this in the project. Then it was planned that in the first period from this date to that date, we work on these process areas … It was [the Gamma roll-out & training manager] and [the Gamma implementation manager] who had already planned such an implementation sequence to keep us on track … then follow-up meetings were held [to determine the outcome]: How did it go? Have you updated the management plan? Have you made the configuration management plan? … It was a sensible approach" {48:2}.

Gamma Unit Level

In the opinion of the Gamma senior VP, getting to CMMI level 2 was "fundamental". The diversity of the projects, i.e. what he termed a "diverse business model", made it necessary to focus on processes to ensure common ground within the business unit. Despite his commitment to the goal of the BPIP, he expressed doubts about the overall maturity of Gamma. In his opinion, Gamma had not reached level 2 yet. By distinguishing between the process maturity of different Gamma departments, he demonstrated a detailed understanding of the status quo. One reason for the variation in process maturity was the relocation of one implementation project from Gamma to Beta due to organizational restructuring. This project was the original implementation flagship of the BPIP and with it disappeared the process knowledge and process-oriented people. It left Gamma Ground Solutions without any process champions to drive the BPIP implementation forward. The other departments had the benefit of the leadership and process push of the Gamma roll-out & training manager and the Gamma implementation manager. The Gamma implementation manager was characterized as dedicated to seeing the BPIP through. Said the Gamma senior VP: "You can say that [the Gamma implementation manager] is extremely biased in this matter. This is his life's blood, right {16:5} … There is no doubt that our process knowledge in the [Gamma Ground Solutions] organization is insufficient. And this means that their ability to help themselves and move on is less than it is here where [the Gamma implementation manager] and [the Gamma roll-out & training manager] can push things. It is undoubtedly a big difference. It is harder for them to move ahead" {16:8}.

(continued)

Table 9.5 (continued)

According to the Gamma roll-out & training manager, not all the new processes were in place and fully defined, e.g. they were not yet available in the TMS when BPIP phase 2 began. As a consequence, the implementation responsible project managers took the initiative to study the CMMI level 2 requirements themselves. Each project manager was then asked to update existing plans etc. to ensure compliance. Despite their initiative, the project managers were confronted by their own uncertainty about the interpretation of the CMMI requirements. Therefore, the Gamma roll-out & training manager and the Gamma implementation manager found it necessary to go through it thoroughly with them and explicate what they needed to do to comply with CMMI level 2. Weaknesses in existing practices were identified, which led to project specific recommendations. In addition, measurements and project status meetings were made mandatory. All in all, the new processes were interpreted, but were put into practice largely unmodified. Also, only minor adjustments were made to existing practices because Gamma was a relatively mature organization to begin with and close to the desired end state. Project Monitoring and Control was the process area impacted the most, with a clear benefit to project follow-up in Gamma. Speaking of project follow-up, three different types of meetings—department meetings, implementation meetings, and project status meetings—served as communication forums that ensured focus on the BPIP within the organization. Implementation meetings were held twice a month with the project managers. Issues were discussed, positive experiences were shared, and status reports were written based on the minutes from these meetings. Department meetings and project status meetings were held to facilitate communication both across and within projects and were well-received. Consequently, even though some people felt that the BPIP had been forced upon them, and although authority had to be exerted, the same people came to appreciate the value of the new processes. Said the Gamma roll-out & training manager: "I guess they feel it has been forced down their throat. I think if you actually asked them, they would say: 'Yes, we have been forced into doing it. Somebody in Gamma decided we should have it in Gamma, and then we had to do whatever they wanted to do' … I believe that along the way—while they were being, you might say, coerced—they were able to see that, well, this is actually quite sensible" {24:6}.

From the Gamma implementation manager's perspective, only Gamma Ground Solutions had a need for a CMMI certificate. At the same time, it was, paradoxically, the least mature part of the organization. Nevertheless, he described Gamma as committed to the BPIP and willing to cooperate with corporate services to see the project through. However, in terms of the actual process implementation, he identified two main obstacles. One was insufficient training in the use of the new processes. The other was a lack of implementation related communication between the two Gamma locations. Training and communication were means to convince people of the value of the new processes and were therefore important for the success of the BPIP. Because both were lacking, Gamma senior management had to exert authority to ensure implementation progress. Said the Gamma implementation manager: "Preferably, people are able to see the point in continuing the process {27:9} … Anyway, I know from Gamma that local management applies massive pressure to ensure that [the BPIP] succeeds … everybody understands that the [BPIP] items put on the agenda by the CEO at the strategy seminar are important. And then people say it is important because it is not beneficial to your career to say something different" {1:13}.

(*continued*)

Table 9.5 (continued)

Corporate Level
In the opinion of the corporate BPIP manager, Gamma was committed to the BPIP, although their support for the project varied over time. For example, at times they expressed their need for a CMMI certificate in no uncertain terms. At other times, the rhetoric was softened. At yet other times, the Gamma senior VP was concerned with value chain analyses as the basis for process improvements and not the CMMI. The overall goal was, however, not challenged. In fact, the CMMI and its emphasis on planning and documentation fitted the Gamma business strategy with its focus on large projects. Despite their commitment to the BPIP, Gamma focused on CMMI compliance rather than TMS process adherence. Whether or not a certain template was being used was of little consequence as long as the process in question was implemented one way or another. The implementation progressed satisfactorily. The implementation began later than expected, but Gamma was the first business unit to start putting the new processes into practice. Constant meetings and discussions about the BPIP preceded the actual implementation. The corporate BPIP manager was satisfied with the implementation progress and the level of activity, although only minor changes were made to existing practices. Gamma's implementation plan was based on the premise that the business unit was already operating close to CMMI level 2. It was just a matter of adding activities to Project Monitoring and Control and establishing a few extra plans. The impact of the BPIP was as high as could be expected. The BPIP was taken seriously, and the implementation was managed "quite sensibly". Their diligence was reflected in their request for being assessed ahead of time in order to better determine the gaps between the new processes and existing practices. Subsequent process maturity assessments were used as a driver in Gamma's improvement efforts. First, the assessment results attracted the attention of the Gamma senior VP because they were on the agenda at the quarterly operational review meetings attended by the CEO. Unsatisfactory results made the Gamma senior VP increase the pressure for better process performance on the Gamma implementation manager and the Gamma roll-out & training manager who in turn increased their pressure on the project managers. Second, the assessments were—like the project status meetings that had been introduced—a vehicle for communication and learning. The assessment results were analyzed and corrective actions were initiated to address identified shortcomings. For example, the project status meetings were introduced because a lack of internal project communication had been identified as a problem through the assessments. Said the corporate BPIP manager: "The only reason [the Gamma senior VP] is angry about not being in the green this time is that it came up during his operational review, and he knows we are interested in it. And when he is angry, [the Gamma roll-out & training manager] and [the Gamma implementation manager] keep a tight rein on [Gamma PM#3] to make sure he does his things {28:10} … I have spoken quite a bit with [the Gamma roll-out & training manager] because a lot of the things in Gamma seemed to have to do with the lack of communication … I believe they have done something about it because [now] they have these weekly meetings with the project managers who have run this [implementation]" {1:7}.

According to corporate SPI agent#2, the BPIP implementation in Gamma was a success, and she attributed it to Gamma management commitment and attention. Said corporate SPI agent#2: "It was as if it was up to the business units to decide if they wanted to participate [in the BPIP] or if they did not want to be part of it, and the extent to which they wanted training and what not. This impacts the way it is implemented within each business unit. That much is evident in Gamma. It is fairly well-implemented there, and that is a great example. It is also because there has been [management] attention over there, right" {31:5}.

Results

At the project level, individual value judgments led project managers to identify positive effects of the BPIP, for example, the impact of regular project status meetings on project progress and outcomes. Having said that, most if not all project managers found that only few changes or minor adjustments to existing practices had taken place as a result of the BPIP. One project manager cited the relatively high maturity at the outset as an explanation for the few changes at the project level. The corporate BPIP manager confirmed this interpretation as did the results of the baseline assessments. In the Gamma senior VP's judgment, the process maturity varied across Gamma departments. He attributed the lack of progress in particularly one department to a lack of people like the Gamma implementation manager and the Gamma roll-out & training manager. By implication, he credited them with the success in other parts of the organization. For his part, the Gamma implementation manager stated that the decision to commit resources to the BPIP was based on a judgment of the value it would contribute to Gamma and that authority was subsequently exerted to ensure the project's success. For her part, the Gamma roll-out & training manager stressed that the processes were too incomplete to ensure implementation success based on the data available at the beginning of the implementation stage of the BPIP. Therefore, it was decided that the implementation responsible project managers needed guidance. Consequently, the Gamma implementation manager and the Gamma roll-out & training manager assessed the gap between the new processes and existing practices to identify implementation requirements and needs for guidance in each project. Implementation plans were developed for the benefit of the project managers, including recommendations for what was needed for CMMI compliance. Project Monitoring and Control was one process area targeted for improvement across all projects. Furthermore, it was determined to organize BPIP status meetings within Gamma on a regular basis. These meetings were mandatory for all project managers as part of the overall implementation strategy within Gamma. More generally, Gamma management decided on organizing careful follow-up on implementation progress to ensure that goals were reached. As a consequence, the BPIP was put on the agenda of three types of meetings, namely implementation, department, and project meetings within Gamma. These

(continued)

Table 9.5 (continued)

meetings served to ensure communication both across and within projects and they were considered an important management vehicle for creating and sharing knowledge about implementation progress in Gamma. There were also close follow-up on and joint discussions of the reports from the corporate led maturity assessments that were conducted several times during the implementation phase of the BPIP. Thus, assessments helped drive the implementation effort in the sense that assessment results attracted management attention and resulted in an organization-wide pressure for change. On account of these efforts, the corporate BPIP manager described Gamma as a proactive organization that was attuned to process innovation and had managed to achieve CMMI level 2 compliance. Because it was a fairly well-run organization to begin with, conditions for successful implementation were present. He described Gamma as proactive because they requested unscheduled assessments and because their implementation plan was ambitious (in terms of the number of projects that were allocated). The corporate BPIP manager attributed the Gamma success to a shared perception of the value of and need for the CMMI. In summary, at the outset, the level of process maturity in Gamma was relatively high compared to the rest of the company. The department was already process oriented and they had previously taken steps to improve work practices. When the BPIP was initiated, they quickly identified the gaps and implemented needed changes to become CMMI level 2 compliant. Through close monitoring and follow-up by the Gamma implementation manager and the Gamma roll-out & training manager at both the project and department level, Gamma was able to meet the BPIP goals. The Gamma project organization was comprised of people who were committed to ensuring the successful implementation of new processes. In particular, there were alignment of interests and a shared understanding of the 'whys' and 'hows' of the project between the local project sponsor and champions, i.e. the Gamma senior VP, the Gamma implementation manager, and the Gamma roll-out & training manager. Dedication and attention to detail were expressions of support from a committed project management team that facilitated successful implementation through communication and careful follow-up. Gamma's success was also attributable to the project managers who were aware of the shortcomings as well as the opportunities offered by the new processes, knew what had to be done to resolve issues, held joint status meetings on a regular basis, were loyal to senior management, and rationally executed the implementation strategy according to plan.

Table 9.6 Example of data on pluralist politics in Alpha

Alpha project level	Alpha unit level	Corporate level
Alpha PM#1[a]	*Alpha senior VP*	*Corporate BPIP manager*
• Implementation easier in production vs. development projects {6:1} • Implementation immense → BPIP not prioritized {6:3} • Lack of IT support {6:4} • No sparring partners → lessons learned not incorporated {6:6} • Generic templates unsuitable {6:7} • One-man army, sidetracked {6:8} • Lack of guidance {6:12}	• Funding conflict {45:8}, {45:11} • Expediters (project manager type) need retraining {45:10} • Funding unresolved → scope change {45:14} • Process misalignment: Alpha vs. special programs {45:16} • Corporate demand used as forcible means {45:17}	• Layoffs + organizational changes → BPIP standstill {5:5} • PM#3 + Alpha senior VP as supporters vs. all others {29:20} • Lack of creativity → PM#2 removal {36:12}
Alpha PM#2	*Alpha implementation manager*	
• Processes unsuitable for production {10:1} • Cookbook ⇔ negotiation of interests → delays {10:5} • Heterogeneity → generic templates not applicable {10:7} • No production template {10:10} • Lack of time → frustration {10:11}, {10:17} • BPIP = corporate dictate {10:13} • No value → annoyance {10:14} • Implementation forced through {10:18} • Message not conveyed, management dictate → dirty work by project managers {10:20}	• No push from middle management vs. self-interest of PM#1 & PM#2 {17:11} • Alpha senior VP as driver & enforcer {17:12} • Pull by implementation responsible persons not project owner {17:13} • Contradiction: standardization vs. BU differences {17:19}	
Alpha PM#3		
• Generic processes and Alpha approach sensible {23:2} • Cookbook adaptations to blueprint possible {23:4} • Young vs. old project manager conflict {23:7} • Standardization inhibits creativity {23:8} • Old technicians opposed {23:11} • Procurement falls short {23:20}, {23:21}		

[a]PM is short for project manager

Table 9.7 Sample evidence of interpretive politics in Alpha

Perspective	Concepts	Illustrative Quotations	Observed Interactions
Interpretive	• Sensemaking • Symbols • Social constructions • Organizational culture • Experiences	• "Our project managers have come to realize that the PI project gave them some tools that were actually useful" (Alpha implementation manager). • "Speaking of the cookbook—at one point in time we realized that we needed to understand all this, and then we established a CMMI guideline for the projects to use. We wrote down what it is all about" (Alpha implementation manager). • "I see it as a leap forward that each development process has been thoroughly defined … it commands greater respect" (Alpha project manager #1). • "My fear is that having this cookbook will stop people from asking: 'What does all this mean to me?' and make them follow it blindly … my belief is that, in Alpha, they don't have the maturity to reflect upon processes" (corporate PI manager).	• *Sensemaking* activities among Alpha managers led them to see the PI project as a solution to the crisis situation in Alpha • The Alpha Senior VP became a *symbol* of decisive action • As part of *social construction*, Alpha managers continuously communicated the PI project as a *symbol* of the unit's future directions • Alpha management decided to adapt the generic processes to Alpha's *organizational culture* based on past *experiences* with project managers' inability to adopt off-the-shelf processes

Table 9.8 Explanatory power of political perspectives

	Political metaphor	Pluralist politics	Rationalist politics	Interpretive politics	Radical politics
Alpha	Applying the hammer	Minimal	Major	Dominant	Minimal
Beta	Struggling to engage	Dominant	Some	Some	Minimal
Gamma	Walking the talk	Minimal	Dominant	Some	Some
Delta	Keeping up appearances	Major	Minimal	Dominant	Minimal

Table 9.9 Four perspectives on process-innovation behaviors and outcomes in Alpha

Political perspective	Helps explain	Does not help explain
The pluralist perspective	• The implementation speed	• The latency of conflicts • The implementation approach
The rationalist perspective	• The support for the BPIP • The implementation approach • The implementation process • The implementation outcome	• The latent conflicts
The interpretive perspective	• The alignment of interests	• The implementation process • The lack of support for competing perceptions of the BPIP
The radical perspective	• The support for the CMMI	• The tailoring of processes

Table 9.9 highlights what each of the four political perspectives helps and does not help explain in terms of process-innovation behaviors and outcomes in the Alpha business unit.

Table 9.10 contains a comparison of four business units at the case company in terms of key factors describing and explaining the process-innovation behaviors and outcomes.

Table 9.10 Four cases of process innovation at the case company

	Alpha	Beta	Gamma	Delta
Business domain	Aerostructures for commercial and military customers	Aerospace technology for military customers	Integrated systems for military customers	Radar systems for civilian customers
BPT plan	Process tailoring through guidelines, checklists, and templates	Generic implementation plan	Generic implementation plan	Management-driven tailoring to suit business unit needs
Software development	Limited	Major	Major	Limited
Goal alignment	A priori low	A priori high	A priori high	A priori low
BPT outcomes	Process maturity increased; not CMMI level 2 compliant; met BPT goals	Process maturity decreased; not CMMI level 2 compliant; did not meet BPT goals	Process maturity increased; CMMI level 2 compliant; met BPT goals	Process maturity decreased; CMMI level 2 compliant; did not meet BPT goals

References

Bacharach, S. (1989). Organizational theories: Some criteria for evaluation. *The Academy of Management Review, 14*(4), 496–515.

Baskerville, R., & Myers, M. (2002). Information systems as a reference discipline. *MIS Quarterly, 26*(1), 1–14.

Benbasat, I., & Zmud, R. (2003). The identity crisis within the is discipline: defining and communicating the discipline's core properties. *MIS Quarterly, 27*(2), 183–194.

Bhaskar, R. (2008). *A realist theory of science* (1st ed.). New York, NY: Routledge.

Bradshaw-Camball, P., & Murray, V. (1991). Illusions and other games: A trifocal view of organizational politics. *Organization Science, 2*(4), 379–398.

Burrell, M., & Morgan, G. (1979). *Sociological Paradigms and Organizational Analysis*. Ashgate.

Bygstad, B. (2010). Generative mechanisms for innovation in information infrastructures. *Information and Organization, 20*(3), 156–168.

Bygstad, B., Munkvold, B., & Volkoff, O. (2016). Identifying generative mechanisms through affordances: A framework for critical realist data analysis. *Journal of Information Technology, 31*(1), 83–96.

Carroll, J., & Swatman, P. (2000). Structured-case: A methodological framework for building theory in information systems research. *European Journal of Information Systems, 9*(4), 235–242.

Checkland, P. (1986). *Systems thinking, systems practice*. Chichester: John Wiley & Sons.

Checkland, P., & Scholes, J. (1990). *Soft systems methodology in action*. Chichester: John Wiley & Sons.

Chiasson, M., & Davidson, E. (2005). Taking industry seriously in information systems research. *MIS Quarterly, 29*(4), 591–605.

Chiasson, M., Germonprez, M., & Mathiassen, L. (2009). Pluralist action research: A review of the information systems literature. *Information Systems Journal, 19*(1), 31–54.

Clegg, S. (1989). *Frameworks of power*. London: SAGE Publications.

Denzin, N. (1978). *The research act: A theoretical introduction to sociological methods*. New York, NY: McGraw-Hill.

Dobson, P. (2001). The philosophy of critical realism—An opportunity for information systems research. *Information Systems Frontiers, 3*(2), 199–210.

Eisenhardt, K. (1989). Building theories from case study research. *The Academy of Management Review, 14*(4), 532–550.

Galliers, R. (1993). Research issues in information systems. *Journal of Information Technology, 8*(2), 92–98.

Gregor, S. (2006). The Nature of Theory in Information Systems. *MIS Quarterly, 30*(3), 611–642.

Goeken, M., & Börner, R. (2012). Generalization in qualitative IS research–approaches and their application to a case study on SOA development. *Australasian Journal of Information Systems, 17*(2), 79–108.

Grover, V., & Lyytinen, K. (2015). New state of play in information systems research: The push to the edges. *MIS Quarterly, 39*(2), 271–296.

Grover, V., Lyytinen, K., Srinivasan, S., & Tan, B. (2008). Contributing to rigorous and forward thinking explanatory theory. *Journal of the Association for Information Systems, 9*(2), 40–47.

Grover, V., & Markus, L. (2008). *Business process transformation*. New York, NY: M. E. Sharpe.

Henfridsson, O., Mathiassen, L., & Svahn, F. (2014). Managing technological change in the digital age: The role of architectural frames. *Journal of Information Technology, 29*, 27–43.

Iivari, J. (1992). The organizational fit of information systems. *Information Systems Journal, 2*(1), 3–29.

Jaccard, J., & Jacoby, J. (2010). *Theory construction and model-building skills: A practical guide for the social scientist* (1st ed.). New York, NY: The Guilford Press.

Jasperson, J., Carte, T., Saunders, C., Butler, B., Croes, H., & Zheng, W. (2002). Power and information technology research: A metatriangulation review. *MIS Quarterly, 26*(4), 397–459.

Keen, P. (1981). Information systems and organizational change. *Communications of the ACM, 24*(1), 24–33.

Kendall, J., & Kendall, K. (1993). Metaphors and methodologies: Living beyond the systems machine. *MIS Quarterly, 17*(2), 149–171.

Klein, H., & Myers, M. (1999). A set of principles for conducting and evaluating interpretive field studies in information systems. *MIS Quarterly, 23*(1), 67–93.

Kuechler, W., & Vaishnavi, V. (2012). A framework for theory development in design science research: Multiple perspectives. *Journal of the Association for Information Systems, 13*(6), 395–423.

Landry, M., & Banville, C. (1992). A disciplined methodological pluralism for MIS research. *Accounting, Management and Information Technologies, 2*(2), 77–97.

Lapointe, L., & Rivard, S. (2005). A Multilevel Model of Resistance to Information Technology Implementation. *MIS Quarterly, 29*(3), 461–491.

Lee, A. (1991). Integrating positivist and interpretive approaches to organizational research. *Organization Science, 2*(4), 342–365.
Lee, A. (2001). Research in information systems: What we haven't learned. *MIS Quarterly, 25*(4), v–xv.
Lee, A., & Baskerville, R. (2003). Generalizing generalizability in information systems research. *Information Systems Research, 14*(3), 221–243.
Lee, A., & Baskerville, R. (2012). Conceptualizing Generalizability: New Contributions and a Reply. *MIS Quarterly, 36*(3), 749–761.
Lewis, M., & Grimes, A. (1999). Metatriangulation: Building theory from multiple paradigms. *The Academy of Management Review, 24*(4), 672–690.
Lukes, S. (2005). *Power: A radical view* (2nd ed.). New York, NY: Palgrave Macmillan.
Mathiassen, L. (2002). Collaborative practice research. *Information Technology & People, 15*(4), 321–345.
Miles, M., Huberman, A., & Saldana, J. (2014). *Qualitative data analysis: A methods sourcebook*. Thousand Oaks, CA: SAGE Publications.
Mingers, J. (1997). Towards critical pluralism. In J. Mingers & A. Gill (Eds.), *Multimethodology: The theory and practice of combining management science methodologies* (1st ed., pp. 407–440). Chichester, UK: Wiley.
Mingers, J. (1999). Synthesising constructivism and critical realism: Towards critical pluralism. In D. Aerts, H. Van Belle, & J. van der Veken (Eds.), *World views and the problem of synthesis. The yellow book of "Einstein Meets Magritte"* (1st ed., pp. 187–204). Amsterdam, The Netherland: Kluwer Academic Publishers.
Mingers, J. (2001). Combining IS research methods: Towards a pluralist methodology. *Information Systems Research, 12*(3), 240–259.
Mingers, J. (2003). The paucity of multimethod research: A review of the information systems literature. *Information Systems Journal, 13*(3), 233–249.
Mingers, J., & Brocklesby, J. (1997). Multimethodology: Towards a framework for mixing methodologies. *Omega, 25*(5), 489–509.
Mingers, J., Mutch, A., & Willcocks, L. (2013). Critical realism in information systems research. *MIS Quarterly, 37*(3), 795–802.
Morgan, G. (1996). *Images of organization*. Thousand Oaks, CA: SAGE Publications.
Müller, S., Mathiassen, L. & Saunders, C. (2020). Pluralist theory building: A methodology for generalizing from data to theory. *Journal of the Association for Information Systems, 21*(1), Article 9, 23–49.

Müller, S., Mathiassen, L., Saunders, C., & Kræmmergaard, P. (2017). Political maneuvering during business process transformation: A pluralist approach. *Journal of the Association for Information Systems, 18*(*3*), 173–205.

Neuman, W. (2014). *Social Research Methods: Qualitative and Quantitative Approaches* (Seventh Ed). Pearson Education Limited.

Orlikowski, W., & Iacono, S. (2001). Research commentary: Desperately seeking "IT" in IT research—A call to theorizing the IT artifact. *Information Systems Research, 12*(2), 121–134.

Rahim, M. (1985). A strategy for managing conflict in complex organizations. *Human Relations, 38*(1), 81–89.

Rahim, M. (2002). Toward a theory of managing organizational conflict. *International Journal of Conflict Management, 13*(3), 206–235.

Remenyi, D., & Williams, B. (1996). The nature of research: Qualitative or quantitative, narrative or paradigmatic? *Information Systems Journal, 6*(2), 131–146.

Rivard, S. (2014). The ions of theory construction. *MIS Quarterly, 38*(2), iii–xiii.

Robey, D. (1996). Research commentary: Diversity in information systems research: Threat, promise, and responsibility. *Information Systems Research, 7*(4), 400–408.

Sabherwal, R., & Grover, V. (2009). A taxonomy of political processes in systems development. *Information Systems Journal, 20*(5), 419–447.

Scott, P., & Briggs, J. (2009). A pragmatist argument for mixed methodology in medical informatics. *Journal of Mixed Methods Research, 3*(3), 223–241.

Seddon, P., & Scheepers, R. (2012). Towards the improved treatment of generalization of knowledge claims in IS research: Drawing general conclusions from samples. *European Journal of Information Systems, 21*(1), 6–21.

Singh, R., Mathiassen, L., & Mishra, A. (2015). Organizational path constitution in technological innovation: Evidence from rural telehealth. *MIS Quarterly, 39*(3), 653–665.

Sutton, R., & Staw, B. (1995). What theory is not. *Administrative Science Quarterly, 40*(3), 371–384.

Tashakkori, A., & Teddlie, C. (1998). Introduction to mixed method and mixed model studies in the social and behavioral sciences. In *Mixed methodology. Combining qualitative and quantitative approaches* (1st ed., p. 200). Thousand Oaks, CA: SAGE Publications.

Tsang, E. (2014). Case studies and generalization in information systems research: A critical realist perspective. *Journal of Strategic Information Systems, 23*(2), 174–186.

Tsang, E., & Williams, J. (2012). Generalization and Induction: Misconceptions, Clarifications, and a Classification of Induction. *MIS Quarterly, 36*(3), 729–748.

Walsham, G. (1995). Interpretive case studies in IS research: Nature and method. *European Journal of Information Systems, 4*(2), 74–81.

Weber, R. (2003). Still desperately seeking the IT artifact. *MIS Quarterly, 27*(2), iii–xi.

Weber, R. (2012). Evaluating and developing theories in the information systems discipline. *Journal of the Association for Information Systems, 13*(1), 1–30.

Weick, K. (1989). Theory construction as disciplined imagination. *The Academy of Management Review, 14*(4), 516–531.

Weick, K. (1995). What theory is not, theorizing is. *Administrative Science Quarterly, 40*(3), 385–390.

Whetten, D. (1989). What Constitutes a Theoretical Contribution? *The Academy of Management Review, 14*(4), 490–495.

Yin, R. (2009). *Case study research: Design and methods* (4th ed.). Thousand Oaks, CA: SAGE Publications.

Zachariadis, M., Scott, S., & Barrett, M. (2013). Methodological implications of critical realism for mixed-methods research. *MIS Quarterly, 37*(3), 855–879.

10

Revitalizing Thoughts on Theory, Theorizing, and Philosophizing in Information Systems

Amir Haj-Bolouri

Introduction

In the past three decades, the view on Information Systems (IS) theories, what their purpose is, and how theorizing is undertaken, has evolved in the field of IS. This evolvement ranges from advancements in examining and conceptualizing the structural nature of theories in IS (Gregor 2006), to evolvement of IS design theories (Walls et al. 1992; Gregor and Jones 2007), to the role of theories in IS (Avison and Malaurent 2014; Gregor 2014; Lee 2014; Bichler et al. 2016), to intense discussions about the presence or absence of a theoretical core in IS (King and Lyytinen 2004; Lyytinen and King 2006; Weber 2006), to the complexity and fruitful nature of the IS-theorizing process (Truex et al. 2006; Gregor 2009; Hovorka and Gregor 2012; Markus and Rowe 2018).

A. Haj-Bolouri (✉)
University West, Trollhättan, Sweden
e-mail: amir.haj-bolouri@hv.se

N. R. Hassan, L. P. Willcocks (eds.), *Advancing Information Systems Theories*, Technology, Work and Globalization, https://doi.org/10.1007/978-3-030-64884-8_10

The latter part (theorizing process) has been advocated and highlighted as a central tenet for determining the end characteristics and quality of theories in IS (Truex et al. 2006; Weber 2012; Gregor 2017). One of the fundamental reasons behind this is because the theorizing process varies from case to case in IS research projects. Consequently, variation on what theorizing approach IS scholars adopt, or how they utilize different strategies, varies because IS phenomena are moving targets, changing rapidly with advances in Information Technology (IT). This continuous movement of advancement makes the theorizing process a complex and divergent one to pin-point or successfully formalize into a single coherent way of developing impactful IS theories (Gregor 2017).

A distinction that persists across viewpoints on theorizing is that between the *context of discovery* and the *context of justification* (Reichenbach 1938), where the first mentioned (context of discovery) refers to the generation of a new idea, hypothesis or theory, whereas the latter mentioned (context of justification) refers to the test or verification of it. In IS, the context of justification has been dominating the discourse of theorizing while the context of discovery has been pushed out to the periphery (Fischer and Gregor 2011; Gregor 2017). Consequently, the IS discipline has developed a tradition of borrowing fruitful and rich theories, which scholars (e.g. Smith and Hitt 2005; Colquitt and Zapata-Phelan 2007) from other reference disciplines (e.g. management) have provided through valuable input, reflections, and considerations. However, as stated in the call for this book chapter series, the goal is to advance IS research beyond this form of borrowed legitimization and derivative research, towards fresh and original IS research that naturally produces its own theories, namely IS theories.

The Purpose of This Chapter

As a response to the call, the purpose of this chapter is to revitalize some thoughts around what theory, theorizing, and philosophizing mean for making a knowledge contribution to a larger theory development process in IS, because novel and native IS theories are most likely produced and tested during a longer period of time and research efforts (e.g. as a result of multiple case studies) (Truex et al. 2006; Gregor 2017). The purpose of this chapter is thus not to define or re-define *IS theories* per se, but rather to provide a broad discussion on the aforementioned topics (e.g. theory, theorizing). Consequently, this will be achieved by fulfilling two objectives: the first objective is to examine what theory and theorizing are, whereas the second objective is to advance the discussion around theorizing in IS by exploring the role of philosophizing and how philosophical concepts, ideas, or theories can be used as *kernel philosophies* to inspire and determine the direction and outcomes of theorizing.

In light of these objectives, I will talk about the outcomes of theorizing as *theoretical implications*, rather than only pure IS theories, because as I indicated previously: a novel and native IS theory of any type (Gregor 2006) is most probably produced through multiple cycles of research studies, meaning that it may take a great deal of time and effort before a theory is actualized as a knowledge contribution that has an impact on the IS field (e.g. published in a top ranked journal, cited by others). Hence, by producing theoretical implications such as concepts, principles, or other crucial components of a theory, one may gradually contribute through a cumulative process to the overall theory development process, as well as addressing and providing increments of the future theory as distinct contributions.

The structure of this chapter is as follows. The next section gives an overview on how different philosophical traditions and perspectives encounter and influence the view on theory. The section ends with a presentation of Gregor's (2006) theory types in IS. After that, the third section emphasizes the process of theorizing. The fourth section then emphasizes philosophy and theorizing. Here, the concept of *kernel philosophies* is introduced and discussed. The fifth section provides a move towards a synthesis of spaces for theory development projects in IS. Finally, the sixth section provides concluding remarks.

What Is a Theory?

Before beginning any further discussions on theorizing and philosophizing, an expose on what *theory* means, and how it is defined, will be undertaken, landing in a pluralistic account (an account that advocate pluralism—e.g. different views, notions, ideas) on theory.

In western thinking, scholars have addressed the term *theory* through different meanings and varying definitions. This includes the ancient notion of theory in Greek philosophy, for example, that of Plato where theory is a statement of how and why particular facts are related (Douglas 2008), whereas other western philosophers have elaborated the ancient view of theory in the realm of science into meanings and definitions such as following:

> *Scientific theories are universal statements. Like all linguistic representations they are systems of signs or symbols. Theories are nets cast to catch what we call "the world"; to rationalize, to explain and to master it. We endeavor to make the mesh even finer and finer.* (Popper 1980, p. 59)
>
> *A scientific theory is an attempt to build together in a systematic fashion the knowledge that one has of some particular aspect of experience.* (Ruse 1995, p. 870)
>
> *Theories provide a representation of how a subset of real-world phenomena should be described. In this light, they can be conceived as specialized ontologies.* (Weber 2012, p. 3)

Aligned with the above-mentioned definitions and meanings of theory, dominant forces of *Logical Positivism*, being a strong form of empiricism that was developed by the Vienna Circle in Europe in the 1920s and later on elaborated in what is known as *Logical Empiricism*, considered theory as a product of a strict deductive process of logical verification (Ayer 1936) and correspondence with empirical facts that eliminate all probabilities of theoretical imagination (Carnap 1932; Suppe 1999). The downfall of this tradition became present along with Willard Van Orman Quine's (1951) seminal work, *Two Dogmas of Empiricism*, which essentially questioned and criticized the very foundation of *Logical Empiricism* as well as their dogmatic view on theoretical reasoning (e.g. the meaning of meaning). Additionally, fundamental thoughts on theory emerged in philosophy and were subsequently synthesized into three separate views (Winther 2016). These views on theory have consequently been present and supported in the field of IS (as shown in Table 10.1).

Table 10.1 Three views on theory (adapted from Winther 2016)

The Syntactic View	Emphasizes theory as a formal system that corresponds to the ideas of logical positivism and the Vienna Circle. Here, theories are essentially sentences of a specific logical domain or language, which are verified deductively through *Verificationism* (Schlick 1931; Ayer 1936). According to Winther (2016), the syntactic view has contemporary defenders that are active in the realms of philosophy, whereas in IS, the syntactic view has more or less been adopted (e.g. Langefors 1966, 1977) as a formalized system that conceives theory as a combination of: (1) theoretical constructs that belong to formalized language such as logic or mathematics, and (2) correspondent indicators in the empirical world (e.g. facts). Implications of adopting this view have inspired the production of theoretical outcomes in IS, such as enterprise ontologies (Dietz 2006, Dietz et al. 2013) or data modeling (Hirschheim et al. 1995)
The Semantic View	Investigates meaning and representation and subsequently argues that in order to analyze the structure of a theory, other means beside formal logic is required (Suppes 1957; van Frassen 1989). According to Suppe (2000), the Semantic View—as opposed to the Syntactic View—avoids problems with correspondence rules and shows how the relationship between theory and empirical experiments is mediated by a non-deductive approach that produces theoretical models (Frigg and Hartmann 2013). Additionally, the Semantic View counters the Syntactic View by providing a strong advocacy of system theory concepts (von Bertalanffy 1950), including open and closed systems (e.g. Dubin 1978)
The Pragmatic View	Explores and treats theory through a combination of internally and externally complex features, including a rich variety of formal and non-formal elements such as analogies, metaphors, values, ontological assumptions, natural kinds and classifications, stylized facts, mathematical concepts, techniques and computer modelling etc. Proponents of this view include philosophers such as Cartwright (1983), Hacking (1983), Thomas Kuhn (1962), as well as neo-pragmatic philosophers such as Putnam (1995, 2001), Rorty (1982), Quine (1985), and Habermas (1976, 1992). Furthermore, the Pragmatic View remains under construction because scholars use and need different kinds of theories for a variety of purposes (Winther 2016)

In contrast to the two first mentioned views in Table 10.1 (The Syntactic and Semantic View) stands a disperse and alternative view of theory that resonates with the pragmatic account. This view does not go under any certain label, but rather, reinforces the diverse, contemplative and imaginative nature of theory (Tate 1953). Examples on utterances that manifests this view are as follows:

> *Philosophical theories or ideas, as points of view, instruments of criticism, may help us gather up what might otherwise pass unregarded by us. Philosophy is the microscope of thought. The theory or idea or system which requires of us the sacrifice of any part of this experience, in consideration of some interest into which we cannot enter, or some abstract theory we have not identified with ourselves, or what is only conventional, has no real claim upon us.* (Pater 1873, p. 219)
>
> *Man is so intelligent that he feels impelled to invent theories to account for what happens in the world. Unfortunately, he is not quite intelligent enough, in most cases, to find correct explanations. So that when he acts on his theories, he behaves very often like a lunatic.* (Huxley 1932, p. 270)
>
> *A theory is the more impressive the greater the simplicity of its premises is, the more different kinds of things it relates, and the more extended is its area of applicability.* (Einstein 1946, p. 33)
>
> *Great theories are expansive; failures mire us in dogmatism and tunnel vision.* (Gould 1993, p. 165)

As opposed to the pure rationalistic and scientific view on theories, the above-mentioned quotes shed an alternative light on the meaning of theory. Subsequently, some of the quotes intertwine the word *theory* with other words such as *system* or *idea* (Pater 1873), whereas other quotes (e.g. Huxley 1932) emphasize a critical mindset towards theories that are formulated based on an insufficient and non-reflective foundation (e.g. lack of a present and vital philosophical underpinning). Huxley's (1932) critique in particular is a profound one, especially because of the irony in his critique, which projects a picture of scientists as individuals with the tendency to treat theory as being a dogma or doctrine of scientism (Sorell 1994; Sheldrake 2012). Consequently, such critical mindset towards theory is in tune with Feyerabend's (1975) anarchistic theory of knowledge, which essentially questions the supremacy of *a* single scientific method.

Having now made an argument for recognition on meaning and definitions of theory, I will now turn to a short passage on IS theories in particular.

Information Systems Theories

In the field of IS, Gregor's (2006) paper on the nature of theory is considered by Gregor (2017) as in agreement with the Pragmatic View because it categorizes theories in terms of their purpose and can thus subsequently be applied to IS research. Consequently, Gregor's (2006) categorization provides a taxonomy of five theory types and definitions. These types and definitions are depicted in Table 10.2.

According to Gregor (2017), contemporary work in IS indicates that the IS discipline may be progressing to integrated bodies of knowledge that take support from theoretical implications of different kind (e.g. concepts, principles), which through a cumulative process contributes with increments that help building novel and native IS theories. Examples on work that point toward this direction are, Wiener et al.'s (2016) expanded theoretical framework for IS project control, Alter's (2013) work system theory, Söllner et al.'s (2016) research on trust and information and communication technology, as well as ongoing discussions around knowledge contributions in IS (e.g. Ågerfalk and Wiberg 2018).

Theoretical Implications as Increments of IS Theories

Based on the gathered insights on what a theory is and what it means for the field of IS so far, it is evident that the term *theory* is many times used in light of a pluralistic account that varies depending on: different philosophical views (Winther 2016), different classifications (Gregor 2006), or contemporary discussions on what direction the IS field is (or should be) going with contributions regard to the future of theory development in IS

Table 10.2 Definition of different theory types (adapted from Gregor 2006)

Theory type	Definition and examples
I. Analysis	This type of theory says what is. It does not extend beyond analysis and description, no causal relationships are specified and no predictions are made. Diverse examples of this type are classifications, schema, frameworks, or taxonomies. Explicit examples of this type in IS research are: Iivari et al.'s (2000) framework for classifying IS development approaches and methodologies, or March and Smith's (1995) and Hevner et al.'s (2004) scheme of Design Science Research (DSR) artifacts
II. Explanation	This type of theory says what is, how, why, when, and where. It provides causal explanations but does not aim to predict with any precision and has no testable propositions. These theories could thus be labelled as theories for understanding, or sense-making, as they influence others worldview. Explicit examples of this type in IS research are: Orlikowski's (1992) structurational model of technology or Hirschheim and Klein's (1989) four paradigms of IS development
III. Predicting	This type says what is and what will be. It has testable propositions but does not give explanations for the propositions. This type of theory is not common in IS research, but rather, examples of this type may be found in economics or finance, such as Moore's (1965) law, relating to improved technology circuits and cost
IV. Explaining and Predicting	This type says what is, how, why, when, where, and what will be. It predicts and has testable propositions and causal explanations. This is the most common and well-recognized type in IS research and may be equated with traditional scientific-type theory. Explicit examples of this type in IS research are: Bhattacherjee and Premkumar's (2004) work on temporal changes in user beliefs and attitudes toward IT usage, or Sarker and Sahay's (2004) work on the implications of space and time for distributed work
V. Design and Action	This type gives guidance on how to do something. It is different in nature from the previous types as it will contain imperative, prescriptive, statements such as, do X in order to achieve Y for Z, or provide a system with the following functions and components to achieve a certain outcome. Explicit examples of this type can be found in the works of Markus et al. (2002) where they propose a design theory for emergent knowledge processes, or Lindgren et al.'s (2004) work where they provide a set of design principles for competence management systems

(Gregor 2017; Ågerfalk and Wiberg 2018). Such pluralistic account might mediate a fruitful approach that allows IS scholars to produce different pieces of theoretical implications as sufficient increments that are integrated and contribute to the development of IS theories. This is at least (as I argue in this chapter) a hypothesis worth scrutinizing further in order to advance a fruitful discussion on the future of IS theories. Subsequently, this hypothesis takes support in discussions that consider IS theories from the two following perspectives:

1. **Native Theories**: Scientific theories are characterized by their ability to make falsifiable or testable predictions (Reichenbach 1938; Popper 1980). Consequently, the relevance and specificity of those predictions determine how potentially useful a theory is or not. One can thus question whether or not the IS discipline has produced any native theories at all (Grover et al. 2012; Straub 2012) that are original for the field and have undergone a rigorous process of falsification or iterative testing, or if the discipline has mostly been possessed with a naive view on theory as being *the* ideal contribution to the IS discipline (Avison and Malaurent 2014; Lee 2014; Bichler et al. 2016).
2. **Theoretical core**: Evidentially, prior studies (e.g. Barkhi and Sheetz 2001; Lim et al. 2013) clearly show that the IS discipline lacks a theoretical core (e.g. does not produce dominant theories) and instead relies heavily on theory employment from reference disciplines (e.g. management). One can subsequently question whether or not this indicates that the IS discipline needs a theoretical core (King and Lyytinen 2004; Lyytinen and King 2006; Weber 2006), or if the absence of a theoretical core instead advocates a pluralistic account on future theories in IS.

Having now provided an outlook on what theory is and what it means for IS, as well as emphasizing *theoretical implications* as increments that may contribute to a larger theory development process, I will now turn to the crucial question of *theorizing*.

What Is Theorizing?

Theorizing is equally important for the IS field to take into further consideration, as is the form and structure of which a theory is ordered and expressed (Gregor 2017). However, I argue that the process of theorizing varies heavily depending on two central issues: (1) the perspectives, meanings, or definitions of theory one incorporates; (2) the approach one undertakes for theorizing. Subsequently, I will elaborate further on these two issues as well as discuss theorizing from other perspectives that have a bearing toward how we in IS tend to think about theorizing.

Concerning the first-mentioned issue: perspectives, meanings, and definitions of theory are inevitably incorporated through philosophical underpinnings that flavor and help justify a theory's underlying assumptions and system of thought. For instance, as mentioned earlier in this book chapter, philosophers of science such as Reichenbach (1938), make a distinction between the *context of discovery* and *context of justification* in theorizing and subsequently argue that theorizing is more often concerned with the latter (context of justification)—as can, for instance, be seen in Popper's (1980) idea of *falsification*. Consequently, there are a number of examples within IS (e.g. Davis 1985; Straub et al. 1995; Dennis and Kinney 1998) that support the context of justification in theorizing, whereas the context of discovery has been cast into the periphery (Gregor 2017).

Concerning the second-mentioned issue: there doesn't seem to exist *a* universal approach for theorizing, but rather, different approaches for different theories (Gregor 2017). With the exception of Grounded Theory in IS (Urquhart et al. 2010; Wiesche et al. 2017), there are very few examples that demonstrate or provide systematic guidance on how to organize and undertake a process of theorizing. There are however some prescriptive frameworks, such as the ones situated within the paradigm of DSR in IS. These prescriptive frameworks provide substantial information on how to develop IS theories of various sort—for example, design theories (Walls et al. 1992; Gregor and Jones 2007), mid-range theories (Kuechler and Vaishnavi 2008, 2012), developing IS theories through induction, abduction, and deduction (Fischer and Gregor 2011), multi-grounded theories (Goldkuhl 2004). In other words, the IS discipline does not rely on a single view on what theorizing means or what it shall lead to. This is also the case within reference disciplines to IS.

Theorizing in Reference Disciplines to IS

In addition to how we think about theorizing in IS, there are plenty of other work in reference disciplines that emphasize a view on the process of theorizing, which subsequently allow IS researchers to adopt these different views. Examples of such work include: Smith and Hitt's (2005) work on influential theorists in management and their reflections on their process of theorizing and their personal accounts of the evolution of their theories; Langley's (1999) strategies for theorizing from process data; Colquitt and Zapata-Phelan's (2007) taxonomy on theory building and theory testing in management; and more (e.g. Jaccard and Jacoby 2009). Smith and Hitt's (2005) and Colquitt and Zapata-Phelan's (2007) in particular, describe the process of theorizing similarly by suggesting that theory building is akin to stages that include tension/phenomena and search, which both are, for many theorists, the starting point of struggling with a conflict or friction between one's initial worldview (e.g. initial assumptions, intuitive judgment) and further observations of the world that cause a change to one's worldview. Subsequently, the search process is triggered to explore and discover pieces of puzzles (e.g. new perspectives of a phenomenon) that help producing new theories through a cumulative and iterative process.

Following the organizational and management tradition on theorizing, Cornelissen (2006) explores Karl Weick's (1989) seminal ideas on theory construction and theorizing. Cornelissen (2006) notes that Karl Weick has sketched an account of organizational theorizing as an ongoing and evolutionary process where researchers themselves actively *construct representations*—representations that form *approximations* of the target subject under consideration and that subsequently provide the groundwork for extended theorizing (e.g. construct specification, development of hypotheses) and research. A detailed account of this process is provided by Weick (1989) in his article on *Theory Construction as Disciplined Imagination*, wherein theorizing is likened to artificial selection as "*[...] theorists are both the source of variation and the source of selection when they construct and select theoretical representations of a certain target subject*" (Weick 1989, p. 520).

This last statement by Weick (1989) implies that the theorizing process is heavily dependent on the researcher's (*theorist*) ability and capacity (*source of variation* and *source of selection*) to develop theories. If this is the case, one may wonder how to inspire or fuel the researcher's ability and capacity to maintain a *disciplined imagination* throughout a process of theorizing? Perhaps philosophy and the act of philosophizing may help in addressing such questions.

Philosophizing and Kernel Philosophies

By stepping back from previous reflections on how scholars from IS and other reference disciplines think about theorizing, it is evident that a process of theorizing incorporates both imagination and rationality. As such, researchers from IS or reference disciplines are not the only ones theorizing. Academic philosophers, in particular, have a long tradition of developing theories and concepts. Philosophy is in that sense foundational for many of the intellectual and theoretical movements in other fields of research, including IS (Hassan et al. 2018). This section will thus move beyond aforementioned views on theorizing alone and explore how *philosophizing* (also known as, *doing philosophy*) can provide an added value to the discussions on theorizing in IS. In connection to this discussion, I will also propose how philosophy knowledge can be used as *kernel philosophies* that help inspire and determine the intentionality (direction, focus) of a theorizing process.

Philosophizing: Doing Philosophy

Plato once said in his work, *Theaetetus*, that "*Philosophy begins in wonder*" (Chappell 2005), a view which is echoed by Aristotle in his work *Metaphysics*: "*It was their wonder, astonishment, that first led men to philosophize and still leads them*" (Cohen 2000). Others such as the father of Phenomenology, Edmund Husserl, once said that he had to philosophize, otherwise, he could not live in this world (Welton 1999). In light of such statements, one may consider the entry point of philosophizing as: a beginning that is filled with simple doubts about an accepted worldview. The initial impulse to philosophize may thus arise from suspicion, for example, that we do not fully understand, and have not fully justified, our most fundamental beliefs and notions about the world (Berlin 1978; Magee 2001).

In addition to these broad perspectives on philosophizing, others such as the contemporary philosopher Daniel Dennett (2013), have tried to shed further light on the process of philosophizing by revealing a collection of philosophical thinking tools, also known as *intuition pumps*. Examples of such tools include *thought experiments*, *Occam's razor*, *Jootsing* (acronym for: *Jumping Out of the System*), *semantic engines and syntactic engines*, *wonder tissues*, *folk psychology*, and *making mistakes*. Consequently, Dennett (2013, p. 59) proposes intuition pumps as not only tools for thinking, but also tools for creation of meaning:

> *[…] I am starting with meaning because it is at the heart of all the tough problems, for a simple reason: these problems don't arise until we start talking about them, to ourselves and others.*

To exemplify how Dennett (2013) elaborates on the idea of meaning creation and theorizing from the perspective of his proposed intuition pumps, let us have a look on what he says about *wonder tissues*:

> *The term wonder tissue is a thinking tool along the lines of a policeman's billy club: you use it to chastise, to persuade others not to engage in illicit theorizing. And, like a billy club, it can be abused. It is a special attachment for the thinking tool Occam's Razor and thus enforces a certain scientific conservatism, which can be myopic.* (Dennett 2013, p. 99)

In contrast to Dennett's (2013) account on philosophizing, other philosophers provide alternative accounts by emphasizing the *space* (or *Khôra*) within philosophy, which is characterized by Plato (1961) as resting between the sensible and the intelligible, through which everything passes but in which nothing is retained (e.g. an image needs to be held by something, just as a mirror will hold a reflection). Consequently, Derrida (1993, p. 89) treats this space as *"[...] an interesting space that at times appears to be neither this nor that, at times both this and that, wavering between the logic of exclusion and that of participation"*. The space is in other words one way of surrounding, but also affording, a curious mode of thinking about the reality and its present or emerging phenomena.

One could also say that the space of philosophizing is polymorphous in the sense that it shifts its mode of intentionality (direction) by *actualizing* different versions of philosophical theories, which the philosopher *projects* (e.g. casts an image of) iteratively into other spaces of thoughts and ideas (e.g. the semiotic or symbolic realms). This act of projection and actualization however depends on certain features that address the objective of philosophical inquiry as a constant questioning search for an increased awareness of the experienced phenomena of a particular domain of reality (Sokolowski 1998). These features include: methods of philosophizing typically oriented toward a number of different things including; a sharp and cohesive line of argumentation (Daly 2010); a critical mindset and ability to produce synthesis of ideas and thoughts (Collingwood 1933; Howell 2013); a capacity for problematization; sense-making and meaning creation (Feinberg 2013; Glymour 2015); and last but not least, the discipline and stamina for weaving back and forth between intuition, imagination, and rationality (Ballard et al. 1958).

In summary, philosophizing is a foundational activity that (1) occurs and is actualized in one to many spaces of thought, ideas, and phenomena; (2) incorporates several intellectual resources in order to produce a system of thought that addresses important elements (e.g. critical mindset, meaning creation) that are relevant towards what Weick's (1989) notion of theorizing as being a disciplined imagination. But how does philosophizing help theorizing in IS.

Philosophizing in IS

In the field of IS, there are many prominent examples of studies that incorporate philosophy and philosophizing. A comprehensive overview on this has been provided by Hassan et al. (2018) in their *European Journal of Information Systems* (EJIS) editorial for a special issue on philosophy in IS. In that editorial, Hassan et al. (2018, p. 2) make a call for future contributions in IS to incorporate philosophy:

> *Not only can philosophy help us understand the basis of our research questions, methodologies, and findings, but also, more importantly, it can help us locate ourselves and our research in a greater context*

Others such as Rowe (2018, p. 1) augment Hassan et al.'s (2018) vision by stating that:

> *[...] philosophising helps us break out of limitations that unduly restrict our theoretical imagination and our theoretical reasoning through metatheories and intuition (theoretical rationality).*

Concrete examples of IS studies that reinforce above quotations, and that have successfully used philosophy knowledge in their research vary, thematically from having focus on: metaphysics or axiology of IT (Introna 2002, 2005; Cheikh-Ammar 2018), to DSR and philosophy (Beynon-Davies 2018; Haj-Bolouri et al. 2016a, b; Haj-Bolouri 2018a, b), to philosophy of technology and rationality (Chiasson et al. 2018), to sense-making of information systems (Lyytinen and Hirschheim 1988; Janson et al. 2000; Janson and Cecez-Kecmanovic 2005), and more (e.g. Ngwenyama 1991; De Moor and Weigand 1996; Ngwenyama and Lee 1997; Lee 2004; Ngwenyama and Klein 2018; Markus and Rowe 2018).

These examples provide strong indications of philosophy and philosophizing as a topic that has gained momentum within IS in general and that there are further possibilities for future work on the role and usage of philosophy knowledge in IS. As a response to such call, I will in the subsequent section provide a consideration on how philosophical concepts, ideas, or theories can be used as so-called *kernel philosophies* to inspire or determine the process of theorizing.

Introducing Kernel Philosophies

I introduce *kernel philosophies* as a concept that is akin, yet not equivalent, to the concept of *kernel theories* in DSR, with the intention of advancing an account on how philosophy can be used to advance theorizing in IS. Before going into further details about this, I find it necessary to say a few words about DSR and kernel theories to begin with.

DSR and Kernel Theories

A conventional view of DSR in IS has been provided through an aggregate of seminal works (e.g. Hevner et al. 2004; Vaishnavi and Kuechler 2007; Peffers et al. 2008; Sein et al. 2011), which essentially address DSR as a pragmatic problem-solving paradigm within IS that has rapidly evolved. A typical DSR project seeks to create innovations that define the underlying ideas, practices, technological capabilities, and products through which the analysis, design, implementation, evaluation, and use of IT artifacts can be efficiently accomplished to solve real-world problems. DSR relies thus on a pragmatic philosophy that stresses the importance of bridging the development of viable artifacts that solve real-world problems with the development of academic knowledge such as principles or theories (Gregor and Hevner 2013).

Kernel theories are in turn defined as "justificatory knowledge" (Gregor and Jones 2007) or "theories from natural or social sciences" that either govern design requirements or the design process of DSR (Walls et al. 1992). Kernel theories are in other words theories that are adopted from other fields to inform aspects of the design activity in DSR and to justify a scientific explanation of the design artifact—which can be seen in several DSR-studies (e.g. Markus et al. 2002; Lindgren et al. 2004).

A Rationale for Kernel Philosophies

As a way of supplementing the concept of kernel theories in DSR, a rationale for kernel philosophies is proposed with the aim of going beyond the local scope of kernel theories in IS (theories that are used to inform design in DSR) and instead put a particular emphasis on how philosophy can be used to support the theorizing process. Here, I consider the following aspects as a rationale for justifying the idea of kernel philosophies:

- **Philosophizing as an integral part**: philosophizing can systematically be chosen as an integral part of the theorizing process by allowing and encouraging IS scholars to build a great system of thought (e.g. premises, intuition pumps) that supports a dynamic confrontation with a chosen theorizing approach or strategy—e.g. questioning the significance of chosen data, questioning the underlying rationale of an approach, detailed accounts on what kind of theories it may be most suitable to produce, a critical consideration of how to make a rational theory choice (e.g. deciding what kind of theory to develop and why).
- **Epistemological boundaries**: deliberately choosing and elucidating (making visible) an underlying epistemology (e.g. rationalism, pragmatism) that reinforces the theorizing process with a worldview on what knowledge means, how it is produced, for what reasons, based on what assumptions, etc. This sets epistemological boundaries for the theorizing process by urging the IS scholar to become aware of what knowledge perspectives he/she intends to support or not, and how such perspectives are compatible with a chosen theorizing approach/strategy (e.g. establishing inner coherence between view on theory and strategy for theorizing).
- **Account on the meaning of truth**: considering how a theory of truth—for example, the correspondence theory of truth (David 2015), the coherence theory of truth (Cornelius 1962), a pragmatic theory of truth (Peirce 1901; James 1909)—may either support the theorizing process or help evaluating one's theoretical implications in light of Reichenbach's (1938) *context of justification*. Without an account on what *truth* means and how that affects one's view on theory, the theorizing may be proceeded by superficial motives or false beliefs.
- **Ethical considerations**: considerations on what is *right, good, bad,* or *wrong* are many times omitted in IS research (Myers and Venable 2014). Instead, it is more likely that IS scholars write in terms of *efficiency, viability,* or *sufficiency,* without taking any ethical account into further consideration. The dual potential of IT (Mason 1986) can

very much be used to enhance or destroy human dignity. IT can improve people's lives, but it also has the potential to make them worse. Therefore, any given IS theory that deals with IT needs to take ethical considerations into further action by discussing the ethical implications of a proposed theory.

In addition to the above-mentioned points, the role of kernel philosophies may vary depending on their context of use. For instance, if one chooses to use kernel philosophies that support design theorizing DSR in IS, then one could, for instance, do this iteratively by adopting a certain model or methodology that supports theorizing as an integral part—for example, Goldkuhl's (2004) multi-grounding of theories model, Venable's (2006) activity framework for theory building, or Sein et al.'s (2011) Action Design Research (ADR) method. Here, one could choose different kernel philosophies for different iterations and thus allow the kernel philosophies to have different roles depending on their function, as long as they are not inherently juxtaposing each other (e.g. ethical considerations that are congruent with each other). For example, during the first iteration, one may choose to use a philosophical theory as an analytic lens for sense making, whereas for the second iteration, one may choose to use a philosophical concept as a source of inspiration for the meta-design of artifacts.

Towards a Conceptualization of Spaces for Actualizing IS Theories

To summarize so far, this chapter has shown different types of theory, described different views of the theorizing process, discussed philosophizing, and introduced the concept of kernel philosophies. In light of all this, this section proposes a move towards a conceptualization of three distinct, yet interrelated, spaces (shown in Fig. 10.1) that interplay with each other to support a larger theory development project in IS. I use the term *space* here in a similar sense as the *chora* of doing philosophy (discussed in Section "Introduction"). Similarly, I express the three proposed spaces to discuss the relationships between different activities, events, and phenomena that have a bearing towards a larger

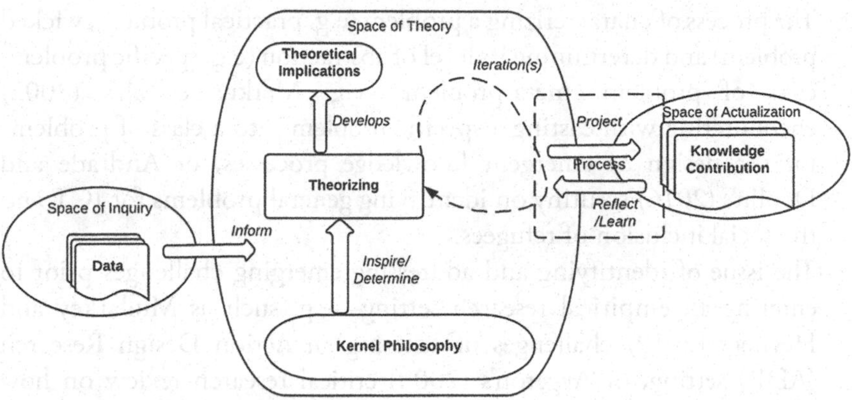

Fig. 10.1 Spaces for theory development projects in IS

encompassing theory development project in IS. The purpose of doing so is to share some creative thoughts that build on previous discussions of this book chapter and which may provoke alternative thoughts on theory development in IS.

The Space of Inquiry

The activities described in this space suggest starting with an investigation on finding out what the object of inquiry is. The investigation may be undertaken together with stakeholders of a given research project or it may be undertaken alone by the researcher within the boundaries of intellectual activities that are prior to entering the empirical context (e.g. literature study). In turn, the object of inquiry may vary from being an idea to being a determined or potential problem, to being a question, or a vision.

General examples that are relevant for the space of inquiry can be to identify and understand a problem in the physical world (e.g. identifying a problem within society) or spotting a "gap" in the literature as a basis for formulating research questions. Concrete IS-related examples that can be framed within the space of inquiry are:

- The process of characterizing a problem (e.g. practical problem, wicked problem) and determining its level of abstraction (e.g. specific problem, class of problem, meta-problem)—e.g. Markus et al.'s (2002) encountering with casting a specific problem into a class of problems for the design of emergent knowledge processes, or Andrade and Doolin's (2016) scrutiny on identifying general problems for ICT and the social inclusion of refugees.
- The issue of identifying and addressing emerging challenges prior to entering an empirical research setting—e.g. such as Mullarkey and Hevner's (2019) challenges of entering an Action Design Research (ADR) setting, or Avgerou's (2008) critical research review on how developing countries have attempted to benefit from the use of information and communication technologies, what challenges they have faced, and how to sufficiently address such challenges before entering future similar projects.

Outcomes from the activities and events that occur in this space result in data, which is used to inform activities and events in the space of theory (as the arrow of this space shows in Fig. 10.1).

The Space of Theory

This space incorporates data from the space of inquiry to inform the theorizing process. Here, the view on what role data has for the theorizing process is similar to Langley's (1999) view for theorizing from process data, which is composed based on various events (e.g. observations, focus group) and used sufficiently according to some particular strategy (e.g. grounded theory strategy, temporal bracketing strategy) for the theorizing process. Choosing a strategy may however depend on what notion of theorizing one intends to adopt.

An IS scholar can adopt different notions of theorizing, varying from Weick's (1989) notion of theorizing as a *disciplined imagination* where the scholar can design, conduct, and interpret imaginary experiments (Cornelissen 2006), to alternative notions such as Colquitt & Zapata-Phelan's or Gregor and Jones's (2007) notions. Either way, the suggestion is that the IS scholar explicitly declares his/her notions in his/her paper so that readers can get a clear sense of which underlying notions (together with its underlying assumptions and premises) govern the theorizing process.

The theorizing process ends when the scholar has reached a point of tentative saturation. I use the term *tentative* here because theorizing is, in the long run, an incremental and iterative process (as the dotted iteration arrow of this space shows in Fig. 10.1) that may take place over a longer period of time (Gregor 2017). Furthermore, the *saturation* part typically refers to the point where an interplay between the scholar's imagination and rationality has taken its toll (Weick 1989). This includes: (1) the scholar reaching a dead end when trying to consume more of his/her vocabulary and images to metaphorically describe a particular phenomenon; (2) the scholar reaching an initial milestone in terms of theoretical implications for a larger theory development project—for example, producing mid-range theories (Gregor and Hevner 2013; Hassan and Lowry 2015) as implications for further reflection that help develop a grand theory.

When it comes to kernel philosophies, the idea is to implement the rationale presented in Sect. 4.3.2 and inspire and determine the intentionality of the theorizing process (as the arrow from kernel philosophies to theorizing in this space shows in Fig. 10.1)—*inspire* in the sense of supporting the context of discovery (Reichenbach 1938) through philosophical tools such as intuition pumps (Dennett 2013) or other rich philosophical concepts or theories (e.g. Wittgenstein's concept of *Language Games*, Heidegger's concept of *Dasein*, Habermas' *Theory of Communicative Action)* and *determine* in the sense of operationalizing the criteria (e.g. epistemological boundaries) of a kernel philosophy's rationale.

The Space of Actualization

Once one has produced theoretical implications in the space of theory, one sufficiently *actualizes* the produced implications by *projecting* them towards a particular knowledge domain in the space of actualization (as the arrow from the space of theory into the space of actualization shows in Fig. 10.1). Here, *projecting* means that a scholar conveys a temporary image or representation of the theorizing outcomes and positions them as a potential knowledge contribution. For instance, if one has produced theoretical implications (e.g. constitutive design principles) for a design theory, then one can project them as a knowledge contribution to the design science paradigm and consequently use Gregor and Hevner's (2013) hierarchy of DSR outputs to motivate the abstraction level of the outcomes as a temporal projection of a larger knowledge contribution.

Another scenario would be when one has done several iterations on refining a research paper according to a set of reviewers' feedback. Here, due to reviewers' comments, each iteration may make changes to the foundation and intentionality of the theorizing process (e.g. the original strategy, kernel philosophy), and thus bring other aspects into light that may benefit or dissolve one's original projection. It is at this point where the knowledge contribution goes from only being *projected* to becoming *actualized*, because of the actual status of the contribution (e.g. from being in progress to becoming finalized). Consequently, different states of the actualization indicate different levels of maturity of one's contribution—a nascent theory versus a mid-range theory (Vaishnavi and Kuechler 2007) versus a general solution concept that supports a class of solutions in DSR (Aken 2004).

Thirdly, when one projects a set of theoretical implications and consequently make an effort to actualize them into a knowledge contribution, one does also project an image of chosen kernel philosophies. This means that the image one projects is an image that represents different parts of the theoretical implications, including their philosophical foundation as well as philosophical view on theory (e.g. Semantic view, Pragmatic view) (Winther 2016). Doing so, the actualization of one's knowledge contribution needs to be witnessed by other scholars in order to make progress—for example, reviewer's comments from a journal or conference, feedback from paper workshops, input from seniors within and across the IS-field. This allows oneself to reflect on the collected feedback and use learning outcomes for the further advancement of the theory development process (as the arrow from the space of actualization to space of theory shows in Fig. 10.1). Consequently, the projection activity becomes a process of its own, which intersects with iterations of theorizing in the theory space and the actualizing of knowledge contributions (as the dotted square shows in Fig. 10.1). Table 10.3 depicts samples of research using philosophical concept, ideas, or theories as kernel philosophies that inspire/determine the theorizing process in order to actualize a knowledge contribution.

Table 10.3 Samples of using kernel philosophies in IS research

Kernel philosophy	Philosophy	Sample research
Philosophical concepts, ideas, or theories that determine epistemological or ontological grounding	– Martin Heidegger's concepts of *Dasein* and *ready-at-hand/ present-to-hand* – Eastern philosophy of Taoism – Pragmatic account on knowledge and truth	– Design epistemologies (e.g. Ehn 1988; Winograd 1995) – Design ontologies (e.g. Stolterman 2008) – Justification of underlying knowledge perspectives for design research (e.g. Niehaves 2007; Sjöström 2010)
Philosophical concepts, ideas, or theories that inspire sense-making of IS phenomena	– Jurgen Habermas' *Theory of Communicative Action* – Phenomenological approaches	– Sense-making of the IS field (e.g. Lyytinen and Hirschheim 1988) – Sense-making of eCommerce systems (e.g. Janson and Cecez-Kecmanovic 2005) – Sense-making of ethics and IT (e.g. Introna 2005)
Philosophical concepts, ideas, or theories that inspire the process of conceptualization	– Jurgen Habermas' *Typology of Social Actions* – Aristoteles' notion of categories of knowledge	– Meta-design of learning platforms (e.g. Haj-Bolouri et al. 2016a, b; Haj-Bolouri 2018a, b) – Conceptualization of practice on the Internet (Heng and De Moor 2003) – Pre-assessment of design priorities (e.g. Nelson and Stolterman 2012)

Concluding Remarks

This chapter has discussed what theory means, showing that different meanings of theory can be traced to their underlying philosophies, discussing different views on theory in philosophy (Winther 2016) and IS (Gregor 2006), what views there are on theorizing, how they relate to philosophy of science (Reichenbach 1938) or management research (e.g. Weick 1989; Smith and Hitt 2005), why to put a focus on *theoretical implications* as a contribution to a larger theory development process in IS, and how philosophy and philosophizing can support the theorizing process through the use of kernel philosophies.

The overall contribution of this chapter is to share revitalizing thoughts concerning theory, theorizing, and philosophizing in IS. As an implication of this, the section prior to this one introduced a move towards a synthesis of spaces for theory development in IS. A limitation of the proposed conceptualization however, is that it does not empirically demonstrate any given theorizing project, but rather, the conceptualization is in its initial mode of progress which can be interpreted as a naive understanding or suggestion for how to advance further discussions around theory and theorizing in IS. This might hopefully be a task for future research to reflect further on.

References

Ågerfalk, P., & Wiberg, M. (2018). Pragmatizing the normative artifact: Design science research in Scandinavia and beyond. *Communications of the Association for Information Systems, 43*(4), 68–77.

Aken, J. E. V. (2004). Management research based on the paradigm of the design sciences: the quest for field-tested and grounded technological rules. *Journal of Management Studies, 41*(2), 219–24.

Alter, S. (2013). Work system theory: Overview of core concepts, extensions, and challenges for the future. *Journal of the Association for Information Systems, 14*(2), 72–121.

Andrade, A. D., & Doolin, B. (2016). Information and communication technology and the social inclusion of refugees. *MIS Quarterly, 40*(2), 405–416.

Avgerou, C. (2008). Information systems in developing countries: A critical research review. *Journal of Information Technology, 23*(3), 133–146.

Avison, D., & Malaurent, J. (2014). Is theory king?: Questioning the theory fetish in information systems. *Journal of Information Technology, 29*(4), 327–336.

Ayer, A. J. (1936). *Language, truth, and logic*. London: Gollancz. (2nd ed., 1946) OCLC 416788667, Reprinted 2001 with a new introduction. London: Penguin. ISBN 978–0–14-118604-7.

Ballard, E. G., Barber, R. L., Feibleman, J. K., Lee, H. N., Morrison, P. G., Reck, A. J., et al. (1958). *The nature of philosophical enterprise*. Dordrecht: Springer.

Barkhi, R., & Sheetz, S. D. (2001). The state of theoretical diversity. *Communications of the Association for Information Systems, 7*(1), 6.

Berlin, I. (1978). *Concepts and categories: Philosophical essays*. Princeton, NJ: Princeton University Press.

Beynon-Davies, P. (2018). What's in a face? Making sense of tangible information systems in terms of Peircean semiotics. *European Journal of Information Systems, 27*(3), 295–314.

Bhattacherjee, A., & Premkumar, G. (2004). Understanding changes in belief and attitude towards information technology usage: A theoretical model and longitudinal test. *MIS Quarterly, 28*(2), 229–254.

Bichler, M., Frank, U., Avison, D., Malaurent, J., Fettke, P., Hovorka, D., et al. (2016). Theories in business and information systems engineering. *Business Information Systems Engineering, 58*(4), 291–319.

Carnap, R. (1932). The elimination of metaphysics through logical analysis of language. *Erkenntnis, 2*. Reprinted in *Logical Positivism*, Alfred Jules Ayer, ed, (New York: Free Press, 1959), pp. 60–81.

Cartwright, N. (1983). *How the laws of physics lie*. New York and St Lucia, QLD: Oxford University Press and University of Queensland Press.

Chappell, S. G. (2005). Plato on knowledge in the theaetetus.

Cheikh-Ammar, M. (2018). The IT artifact and its spirit: A nexus of human values, affordances, symbolic expressions, and IT features. *European Journal of Information Systems, 27*(1), 1–17.

Chiasson, M., Davidson, E., & Winter, J. (2018). Philosophical foundations for informing the future (S) through IS research. *European Journal of Information Systems, 27*(3), 1–13.

Cohen, S. M. (2000). Aristotle's metaphysics.

Collingwood, R. G. (1933). *An essay on philosophical method*. Oxford: Clarendon Press.

Colquitt, J., & Zapata-Phelan, C. (2007). Trends in theory building and theory testing: A five-decade study of the academy of management journal. *Academy of Management Journal, 50*(6), 1281–1303.

Cornelissen, J. P. (2006). Making sense of theory construction: Metaphor and disciplined imagination. *Organization Studies, 27*(11), 1579–1597.

Cornelius, B. A. (1962). Coherence theory of truth. In D. D. Runes (Ed.), *Dictionary of philosophy* (p. 58). Totowa, NJ: Littlefield, Adams, and Company.

Daly, C. (2010). *An introduction to philosophical methods*. Toronto: Broadview Press.

David, M. (2015, 28 May). The correspondence theory of truth. *Stanford Encyclopedia of Philosophy*. Retrieved from http://plato.stanford.edu/entries/truth-correspondence/.

Davis, F. D. (1985). *A technology acceptance model for empirically testing new end-user information systems: Theory and results*. Doctoral dissertation, Massachusetts Institute of Technology.

De Moor, A., & Weigand, H. (1996). The role of social constraints in the Design of Research Network Information Systems. *Proceedings of Eco-Informa, 96*, 4–7.

Dennett, D. C. (2013). *Intuition pumps and other tools for thinking*. New York: WW Norton & Company.

Dennis, A. R., & Kinney, S. T. (1998). Testing media richness theory in the new media: The effects of cues, feedback, and task equivocality. *Information Systems Research, 9*(3), 256–274.

Derrida J. (1993). *Khôra*. Paris: Galilée.

Dietz, J. L. (2006). *What is Enterprise ontology?* (pp. 7–13). Berlin, Heidelberg: Springer.

Dietz, J. L., Hoogervorst, J. A., Albani, A., Aveiro, D., Babkin, E., Barjis, J., …, Mulder, H. (2013). The discipline of enterprise engineering. *International Journal of Organisational Design and Engineering, 3*(1), 86–114.

Douglas, T. (2008). Moral enhancement. *Journal of Applied Philosophy, 25*(3), 228–245.

Dubin, R. (1978). *Theory building*. New York: Free Press.

Ehn, P. (1988). *Work-oriented design of computer artifacts*. Stockholm: Arbetslivscentrum.

Einstein, A. (1946). *Autobiographical notes*.

Feinberg, J. (2013). *Doing philosophy*. Boston, MA: Cengage Learning.

Feyerabend, P. (1975). *Against method: Outline of an anarchist theory of knowledge*. London: New Left Books.

Fischer, C., & Gregor, S. (2011). Forms of reasoning in the design science research process. In H. Jain, A. Sinha, & P. Vitharana (Eds.), *Service-oriented perspectives in design science research (6th DESRIST)* (pp. 17–31). Lecture Notes in Computer Science: Springer.

Frigg, R., & Hartmann, S. (2013). Models in science. In E. Zalta, U. Nodelman, C. Allen, & J. Perry (Eds.), *Stanford encyclopaedia of philosophy*. Retrieved December, 2016, from https://plato.stanford.edu/entries/models- science/.

Glymour, C. (2015). *Thinking things through: An introduction to philosophical issues and achievements* (2nd ed.). A Bradford Book.

Goldkuhl, G. (2004). Design theories in information systems—A need for multi-grounding. *Journal of Information Technology Theory and Application, 6*, 59–72.

Gould, S. J. (1993). *Eight little piggies*.

Gregor, S. (2006). The nature of theory in information systems. *MIS Quarterly, 30*(3), 611–642.

Gregor, S. (2009, May). Building theory in the sciences of the artificial. In *Proceedings of the 4th international conference on design science research in information systems and technology* (p. 4). ACM.

Gregor, S. (2014). Theory—Still king but needing a revolution! *Journal of Information Technology, 29*(4), 337–340.

Gregor, S. (2017). On theory. In R. Galliers & M.-K. Stein (Eds.), *The Routledge companion to management information systems* (pp. 57–72). Routledge.

Gregor, S., & Hevner, A. (2013). Positioning and presenting design science research for maximum impact. *MIS Quarterly, 37*(2), 337–355.

Gregor, S., & Jones, D. (2007). The anatomy of a design theory. *Journal of the Association for Information Systems (JAIS), 8*(5), 312–335.

Grover, V., Lyytinen, K., & Weber, R. (2012, December 16–19). *Panel on native IS theories*. Paper presented at the Special Interest Group on Philosophy and Epistemology in IS (SIGPHIL) Workshop on IS Theory: State of the Art, Orlando, FL.

Habermas, J. (1976). *On the pragmatics of social interaction*.

Habermas, J. (1992). *On the pragmatics of communication*.

Hacking, I. (1983). *Representing and intervening: Introductory topics in the philosophy of natural science*. Cambridge: Cambridge University Press.

Haj-Bolouri, A. (2018a). *Designing for adaptable learning*. Dissertation Thesis. University West, Sweden.

Haj-Bolouri, A. (2018b). Kernel Philosophy: A Way of Inspiring and Mak-ing Sense of Design in Information Systems Re-search?. In AIS SIGPRAG pre-ICIS workshop on" Practice-based Design and Innovation of Digital Artifacts", 2018-12–12, San Francisco, CA, USA (pp. 1-19). American Information Society.

Haj-Bolouri, A., Chandra Kruse, L., Iivari, J., & Flensburg, P. (2016a). How Habermas' philosophy can inspire the design of information systems: the case of designing an open learning platform for social integration. In IRIS39, Information Systems Research Seminar in Scandinavia, Ljungskile, August 7–10, 2016. (Vol. 7, pp. 1–13).

Haj-Bolouri, A., Kruse Chandra, L., Iivari, J., and Flensburg, P. (2016b). *How Habermas' philosophy can inspire the design of information systems: The case of designing an open learning platform for social integration*. Selected Papers of the IRIS, Issue Nr 7. 2.

Hassan, N. R., & Lowry, P. B. (2015, December 13). *Seeking middle-range theories in information systems research.* Paper presented at the International Conference for Information Systems (ICIS).

Hassan, N. R., Mingers, J., & Stahl, B. (2018). Philosophy and information systems: Where are we and where should we go? *European Journal of Information Systems, 27*(3), 263–277.

Heng, M. S. H., & De Moor, A. (2003). From Habermas's communicative theory to practice on the internet. *Information Systems Journal, 13*, 331–352.

Hevner, A. R., March, S. T., & Park, K. (2004). Design research in information systems research. *MIS Quarterly, 28*(1), 76–105.

Hirschheim, R., & Klein, H. K. (1989). Four paradigms of information systems development. *Communications of the ACM, 32*(10), 1199–1216.

Hirschheim, R., Klein, H. K., & Lyytinen, K. (1995). *Information systems development and data modeling: Conceptual and philosophical foundations.* Cambridge: Cambridge University Press.

Hovorka, D. S., & Gregor, S. (2012). 3. Untangling causality in design science theorising. *Information Systems Foundations*, 55.

Howell, K. E. (2013). An introduction to the philosophy of methodology.

Huxley, A. (1932). *Texts and pretexts.*

Iivari, J., Hirschheim, R., & Klein, H. K. (2000). A dynamic framework for classifying information systems development methodologies and approaches. *Journal of Management Information Systems, 17*(3), 179–218.

Introna, L. (2002). The question concerning information technology: Thinking with Heidegger on the essence of information technology. In *Internet Management Issues* (pp. 220–234). Hershey, PA: IGI Publishing.

Introna, L. (2005). Phenomenological approaches to ethics and information technology. *Stanford Encyclopedia of Philosophy*.

Jaccard, J., & Jacoby, J. (2009). *Theory construction and model-building skills: A practical guide for social scientists*. New York: Guilford Press.

James, W. (1909). The meaning of truth.

Janson, M., & Cecez-Kecmanovic, D. (2005). Making sense of E-commerce as social action. *Information Technology & People, 18*(4), 311–342.

Janson, M., Iivari, J., & Oinas-Kukkonen, H. (2000). eCommerce as computer-mediated social action. In *AMCIS 2000 Proceedings*, 68.

King, J. L., & Lyytinen, K. (2004). Reach and grasp. *MIS Quarterly, 28*(4), 539–552.

Kuechler, W., & Vaishnavi, V. (2008). On theory development in design science research: Anatomy of a research project. *European Journal of Information Systems, 17*(5), 1–23.

Kuechler, W., & Vaishnavi, V. (2012). A framework for theory development in design science research: Multiple perspectives. *Journal of the Association for Information Systems, 13*, 6.

Kuhn, T. (1962). *Structure of scientific revolutions*. Chicago: University of Chicago Press.

Langefors, B. (1966). *Theoretical analysis of information systems*. Lund: Studentlitteratur.

Langefors, B. (1977). Information systems theory. *Information Systems, 2*(4), 207–219.

Langley, A. (1999). Strategies for theorizing from process data. *Academy of Management Review, 24*(4), 691–710.

Lee, A. S. (2004). Thinking about social theory and philosophy for information systems. In J. Mingers & L. Willcocks (Eds.), *Social theory and philosophy for information systems* (pp. 1–26). Chichester: Wiley.

Lee, A. S. (2014). Theory is king? But first, what is theory? *Journal of Information Technology, 29*(4), 350–352.

Lim, S., Saldanha, T. J. V., Malladi, S., & Melville, N. P. (2013). Theories used in information systems research: Insights from complex network analysis. *Journal of Information Technology Theory and Application, 14*(2), 5–46.

Lindgren, R., Henfridsson, O., & Schultze, U. (2004). Design principles for competence management systems: A synthesis of an action research study. *MIS Quarterly*, 435–472.

Lyytinen, K., & Hirschheim, R. (1988). Information systems as rational discourse: An application of Habermas' theory of communicative action. *Scandinavian Journal Management, 4*(1/2), 19–30.

Lyytinen, K., & King, J. L. (2006). The theoretical core and academic legitimacy: A response to professor Weber. *Journal of the Association for Information Systems, 7*(11), 714–721.

Magee, B. (2001). *Talking philosophy: Dialogues with fifteen leading philosophers.* New York: Oxford University Press.

March, S., & Smith, G. (1995). Design and natural science research on information technology. *Decision Support Systems, 15*, 251–266.

Markus, M. L., & Rowe, F. (2018). Is IT changing the world? Conceptions of causality for information systems theorizing. *MIS Quarterly, 42*(4), 1255–1280.

Markus, M., Majchrzak, L. A., & Gasser, L. (2002). A design theory for systems that support emergent knowledge processes. *MIS Quarterly, 26*(3), 179–212.

Mason, R. O. (1986). Four ethical issues of the information age. *MIS Quarterly, 10*(1), 5–12.

Moore, G. E. (1965). Cramming more components onto integrated circuits. *Electronics, 38*(8), 114–117.

Mullarkey, M. T., & Hevner, A. R. (2019). An elaborated action design research process model. *European Journal of Information Systems, 28*(1), 6–20.

Myers, M. D., & Venable, J. R. (2014). A set of ethical principles for design science research in information systems. *Information & Management, 51*(6), 801–809.

Nelson, H. G., & Stolterman, E. (2012). *The design way: Intentional change in an unpredictable world.* Cambridge, MA: MIT Press.

Ngwenyama, O. K. (1991). The critical social theory approach to information systems: Problems and challenges. In H.-E. Nissen, H. K. Klein, & R. A. Hirschheim (Eds.), *Information systems research: Contemporary approaches and emergent traditions* (pp. 267–280). Amsterdam.

Ngwenyama, O., & Klein, S. (2018). Phronesis, argumentation and puzzle solving in IS research: Illustrating an approach to phronetic IS research practice. *European Journal of Information Systems, 27*(3).

Ngwenyama, O. K., & Lee, A. S. (1997). Communication richness in electronic mail: Critical social theory and the Contextuality of meaning. *MIS Quarterly, 21*, 145–167.

Niehaves, B. (2007). On epistemological pluralism in design science. *Scandinavian Journal of Information Systems, 19*(2), 7.

Orlikowski, W. J. (1992). The duality of technology: Rethinking the concept of technology in organizations. *Organization Science, 3*(3), 398–427.

Pater, W. (1873). *The renaissance, conclusion.*

Peffers, K., Tuunanen, T., Rothenberger, M., & Chatterjee, S. (2008). A design science research methodology for information systems research. *Journal of Management Information Systems, 24*(3), 45–77.

Peirce, C. S. (1901). "Truth and Falsity and Error" (in part), pp. 716–720 in James Mark Baldwin, ed., *Dictionary of philosophy and psychology*, v. 2. Peirce's section is entitled "*Logical*", beginning on p. 718, column 1, and ending on p. 720 with the initials "(C.S.P.)", see Google Books Eprint. Reprinted, *Collected Papers* v. 5, pp. 565–573.

Plato. (1961). Timaeus. In E. Hamilton & H. Cairns (Eds.), *The collected dialogues of Plato* (pp. 1151–1211). Princeton, NJ: Princeton University Press.

Popper, K. (1980). *The logic of scientific discovery*. London: Unwin Hyman.

Putnam, H. (1995). *Pragmatism: An open question*. Oxford: Blackwell. ISBN 0-631-19343-X.

Putnam, H. (2001). *Enlightenment and pragmatism* (p. 48). Assen: Koninklijke Van Gorcum.

Quine, W. V. (1951). Main trends in recent philosophy: Two dogmas of empiricism. *The philosophical review, 60*, 20–43.

Quine, W. V. O. (1985). *The time of my life: An autobiography*. Cambridge: The MIT Press. ISBN 0–262–17003-5. Harvard Univ. Press.

Reichenbach, H. (1938). *Experience and prediction: An analysis of the foundations and the structure of knowledge*. Chicago: The University of Chicago Press.

Rorty, R. (1982). *Consequences of pragmatism*. Minneapolis: University of Minnesota Press. ISBN 978-0816610631.

Rowe, F. (2018). Being critical is good, but better with philosophy! From digital transformation and values to the future of IS-research. *European Journal of Information Systems, 27*(3), 380–393.

Ruse, M. (1995). Theory. In T. Honderich (Ed.). *The Oxford Companion to Philosophy*. Oxford, UK: Oxford University Press.

Sarker, S., & Sahay, S. (2004). Implications of space and time for distributed work: An interpretive study of US–Norwegian systems development teams. *European Journal of Information Systems, 13*(1), 3–20.
Schlick, M. (1931). Die Kausalität in der gegenwärtigen Physik. *Die Naturwissenschaften, 19*, 145–162. Trans. "Causality in Contemporary Physics" in Schlick 1979b, pp. 176–209.
Sein, M., Henfridsson, O., Purao, S., Rossi, M., & Lindgren, R. (2011). Action design research. *MIS Quarterly*.
Sheldrake, R. (2012). *The science delusion*. Coronet Books.
Sjöström, J. (2010). *Designing information systems: A pragmatic account*. Doctoral dissertation, Uppsala universitet.
Smith, K., & Hitt, M. (Eds.). (2005). *Great minds in management the process of theory development*. Oxford: Oxford University Press.
Sokolowski, R. (1998). The method of philosophy: Making distinctions. *The Review of Metaphysics, 51*(3), 515–532.
Söllner, M., Benbasat, I., Gefen, D., Leimeister, J. M., & Pavlou, P. A. (2016, October 31). Trust. In A. Bush & A. Rai (Eds.), MIS quarterly research curations. Retrieved from http://misq.org/research-curations.
Sorell, T. (1994. *Scientism: Philosophy and the infatuation with science* (p. 1ff). Routledge.
Stolterman, E. (2008). The nature of design practice and implications for interaction design research. *International Journal of Design, 2*(1), 55–65.
Straub, D. (2012). Editorial: Does MIS have native theories. *MIS Quarterly, 36*(2), iii–xii.
Straub, D., Limayem, M., & Karahanna-Evaristo, E. (1995). Measuring system usage: Implications for IS theory testing. *Management Science, 41*(8), 1328–1342.
Suppe, F. (1999). The positivist model of scientific theories. In R. Klee (Ed.), *Scientific inquiry* (pp. 16–24). New York: Oxford University Press.
Suppe, F. (2000). Understanding scientific theories: An assessment of developments, 1969-1998. *Philosophy of Science, 67*(Proceedings), S102–S115.
Suppes, P. (1957). *Introduction to logic*. Princeton: D. Van Nostrand.
Tate, J. (1953). Plato's Theory of Ideas - W. D. Ross: Plato's Theory of Ideas. Pp. 250. Oxford: Clarendon Press, 1951. Cloth, 18s. net. *The Classical Review, 3*(2), 91–94. https://doi.org/10.1017/S0009840X0017581X
Truex, D., Holmström, J., & Keil, M. (2006). Theorizing in information systems research: A reflexive analysis of the adaptation of theory in information systems research. *Journal of the Association for Information Systems, 7*(1), 33.

Urquhart, C., Lehmann, H., & Myers, M. D. (2010). Putting the 'theory'back into grounded theory: Guidelines for grounded theory studies in information systems. *Information Systems Journal, 20*(4), 357–381.

Vaishnavi, V., & Kuechler, W. (2007). *Design science research methods and patterns*. Boca Raton and New York: Auerbach Publications.

van Frassen, B. (1989). *Laws and symmetry*. New York: Oxford University Press.

Venable, J. (2006). The role of theory and theorizing in design science research. *Proceedings DESRIST*.

von Bertalanffy, L. (1950). An outline of general systems theory. *British Journal of Philosophy of Science, 1*, 139–164.

Walls, J. G., Widmeyer, G. R., & El Sawy, O. A. (1992). Building an information system design theory for vigilant EIS. *Information Systems Research, 3*(1), 36–59.

Weber, R. (2006). Reach and grasp in the debate over the IS core: An empty hand? *Journal of the Association for Information Systems, 7*(10), 703–713.

Weber, R. (2012). Evaluating and developing theories in the information systems discipline. *Journal of the Association for Information systems, 13*(1), 2.

Weick, K. E. (1989). Theory construction as disciplined imagination. *Academy of Management Review, 14*(4), 516–531.

Welton, D. (1999). *The essential Husserl: Basic writings in transcendental phenomenology*.

Wiener, M., Mähring, M., Remus, U., & Carol Saunders, C. (2016). Control configuration and control enactment in information systems projects: Review and expanded theoretical framework. *MIS Quarterly, 40*(3), 741–774.

Wiesche, M., Jurisch, M. C., Yetton, P. W., & Krcmar, H. (2017). Grounded theory methodology in information systems research. *MIS Quarterly, 41*(3), 685–701.

Winograd, T. (1995). Heidegger and the design of computer systems. In A. Feenberg & A. Hannay (Eds.), *Technology and the politics of knowledge* (pp. 108–127). Bloomington: Indiana U. Press.

Winther, R. (2016). The Structure of Scientific Theories. In E. Zalta, U. Nodelman, C. Allen, & J. Perry (Eds.), *Stanford encyclopaedia of philosophy*. Retrieved December, 2016, from https://stanford.library.sydney.edu.au/archives/fall2016/entries/structure-scientific-theories/.

11

Reviving the Individual in Information Systems Theorizing

Lars Taxén

Introduction

The IS discipline has been defined as that "which studies the human, social, and technological phenomena associated with the design, construction, implementation, and use of computer-based information systems by individuals, organizations, and societies" (Tarafdar & Davison, 2018, p. 525). Accordingly, central IS phenomena are "information", "Information Technology (IT)", and "Information Systems (IS)". However, the essence of these has been remarkably difficult to elucidate in spite of decades of research. The nature of information "has plagued research on information systems since the very beginning" (Boland, 1987, p. 363). Similarly, debates about what constitutes the IS and the IT artifact do not attain closure—rather the opposite. For example, Alter suggests that the "vastly inconsistent definitions of the term 'the IT

Former Associate Professor, now retired.

L. Taxén (✉)
Linköping University, Linköping, Sweden

N. R. Hassan, L. P. Willcocks (eds.), *Advancing Information Systems Theories*, Technology, Work and Globalization, https://doi.org/10.1007/978-3-030-64884-8_11

artifact' ... demonstrate why it no longer means anything in particular and should be retired from the active IS lexicon" (Alter, 2015, p. 47). Likewise, mutually incompatible definitions of ISs abound (e.g. Alter, 2008). Lee amply summarizes the state of play in IS:

> Virtually all the extant IS literature fails to explicitly specify meaning for the very label that identifies it. This is a vital omission, because without defining what we are talking about, we can hardly know it (Lee, 2010, p. 338).

As a result, serious concerns about the relevance and legitimacy of the IS discipline have been aired (see e.g. Baskerville, 2010; Davison & Tarafadar, 2018; Hassan, 2011; Hirschheim & Klein, 2012; McCubbrey, 2003; Paul, 2010). Nevertheless, the IS field is thriving, as evidenced by annual conferences with many participants, several renowned outlets, and vital professional organizations. This paradoxical situation implies two tendencies: the IS field is becoming increasingly important while being simultaneously discordant about its "core" (e.g. Riemer & Johnston, 2017; Watson, 2014).

In general, attempts to define such a core focus on elucidating the relation between the "social" and "technical/material" as in the socio-technical turn (Mumford, 2006) and the sociomaterial theorising of the field (Cecez-Kecmanovic, Galliers, Henfridsson, Newell, & Vidgen, 2014). As evident from their labels, both foreground the social. The individual is demoted to the background or submerged under the social as, for example, in: "Coordinated *human (social)* and material agencies both represent capacities for action" (Leonardi, 2012, p. 36, our emphasis). Thus, the relation between the individual and the social is regarded as insignificant. This is indeed remarkable, given that the social comes into existence only through individual actions, and human infants become individuals only in social environments.

Accordingly, the purpose of this chapter is to advance IS theorizing by promoting the individual as a foundational element on par with the social and material/technical. To fully understand the IS field we need "to consider the biological adaptations that shaped our innate drives, information processing capabilities, and the information systems we created"

(Watson, 2014, p. 518). However, an increased emphasis on the individual must not lead to methodological individualism where "social science knowledge is best or more appropriately derived through the study of individuals" (Samuels, 1972, p. 249). Likewise, we need to stay clear from the opposite focus on methodological collectivism, where "meaningful social science knowledge is best or more appropriately derived through the study of group organizations, forces, processes and/or problems" (ibid.). We seek to articulate a balanced position, departing from the fundamental assumption that the individual and the social mutually constitute each other—a *dialectical approach* (DA for short):

> Individuals … cannot in principle be understood without taking their developmental environment into account. The opposite is also true … the cultural environment cannot be understood without understanding the individual. The individual, in this context, means first and foremost his or her nervous system, the brain. (Toomela, 2014, p. 325)

The profound insight from this position is that any theorizing of social/material relations needs to include the human, neurobiological system. If not, the analysis will inevitably be incomplete.

The chapter is organized in two main parts. The first one begins with a brief outline of dialectics and praxis, thus providing a philosophical framing of our approach. Praxis is conceptualized as a *communal infrastructure* comprised of individual, *biomechanical* factors, and institutionalized *communal* factors. Next, we describe *communalization*—the dialectical process by which biomechanical and communal factors evolve. Communalization is modeled on the *Theory of Functional Systems*, as elaborated by the Russian biologist Pyotr Anokhin. This theory relates to mental functions by which an individual interprets a particular situation, acts upon it, evaluates the results, and continues so until the motivation for acting is fulfilled. Thus, every action modifies both factors, indicating that communities are always in a process of becoming, only momentarily stabilized. Accordingly, the first part outlines a dialectical "theoretical perspective" (Burton-Jones, McLean, & Monod, 2015) for inquiries into the IS field.

In the second part, the dialectical perspective is used as a guiding lens for rethinking central IS concepts. *Information* is conceived as inherently individual, constituted by integration of previous experiences and sensations emanating from communal factors into an actionable percept. Communalization renders the *IT artifact* into a communally meaningful factor—an *Information System*. Although the IT artifact may be adapted to specific communal needs during communalization, its ontological status as an artifact does not change. Since communalization involves the individual, the "systems" part of Information Systems is comprised of the *individual and the communalized IT artifact*, thus adhering to the view that "Information Systems is Information Technology in Use" (Paul, 2010, p. 379). This means that information, the IT artifact, and the IS cannot be profoundly understood as separate phenomena—they must be seen as inherently related constructs. The practical relevance of the DA is demonstrated by the communalization of an IT artifact in the telecom industry. Further, we indicate how the dialectical perspective may advance mainstream IS thinking concerning communication, sociomateriality, digitalization of artifacts, and the essence of the IS discipline. The key insight of the DA is that the individual needs to be promoted as a prime constituent in researching IS phenomena. We claim that such a move opens up new avenues for advancing the IS field.

Dialectics and Praxis

Dialectics has a long philosophical tradition, from Aristoteles, Hegel, Marx and others (Wan-chi, 2006). The individual–social dialectics of interest in this chapter was originally formulated by Marx in his first thesis on Feuerbach:

> The main defect of all hitherto-existing materialism—that of Feuerbach included—is that the Object [*der Gegenstand*], actuality, sensuousness, are conceived only in the form of the object [*Objekts*], or of contemplation [*Anschauung*], but not as human sensuous activity, practice [*Praxis*], not subjectively. (Marx & Engels, 1998)

The German words are included since these more precisely signify the essence of this thesis. According to Adler (2005, p. 404), Marx refers to *das Objekt as* "simplistic materialism" where the object is merely regarded as a given in the external world. The other form is "pure idealism" in which the object is our mental construction of it; what it means to us [*Anschauung*]. According to Marx, neither of these positions capture the essence of the relation between an actor and the results of acting.

The dialectical synthesis Marx proposes is *Praxis*,[1] in which the object is seen simultaneously as an independently existing, recalcitrant material reality, and a goal or purpose or idea that we have in mind. This stance is referred to as *der Gegenstand* where "*gegen* means against, towards, contrary to, signaling a reality that offers resistance to our efforts and desires, and *der Stand* means category or state of affairs" (Adler, 2005, p. 404, original emphasis).

To clarify the meaning of this exegesis, consider a simple object such as a stop sign on a road. The signpost is clearly a material object [*Objekt*]. An experienced driver has some idea in her mind what the sign means [*Anschauung*]. However, the nature of these phenomena can only be understood in [*Praxis*]. A driver notices the signpost, interprets it and decides how to act—hopefully by stopping the car. The particular materiality of the signpost is irrelevant as long as the driver can perceive it. Thus, the meaning of the object [*der Gegenstand*] is "dependent on the orientation and action of people toward them" (Blumer, 1969, p. 68). The essence of this understanding is that the social reality produced and the producer mutually constitute each other: "The very nature or character of a man is determined by what he does or his *praxis*, and his products are concrete embodiments of this activity" (Bernstein, 1999, p. 44, original emphasis).

The inherent dialectics in praxis has profound implications for how we apprehend parts–whole relationship. Consider the following example

[1] It is necessary to distinguish between 'practice' and 'practices' (in German between *Praxis* and *Praktiken*) (Reckwitz, 2002). 'Practice' (*Praxis*) describes the whole of human action, while a 'practice' (*Praktik*) is a "routinized type of behaviour which consists of several elements, interconnected to one other: forms of bodily activities, forms of mental activities, 'things' and their use, a background knowledge in the form of understanding, know-how, states of emotion and motivational knowledge" (pp. 249–250).

from Levins and Lewontin (1985). A person cannot fly by flapping her arms, no matter how much she tries. Nor can a group of people fly, by all flapping their arms simultaneously. But people do in fact fly as a result of a long cultural–historical process where purposeful human activity over time has produced airplanes, pilots, landing strips, fuel, and all other things necessary to fly. Although our neurobiological constitution remains the same, we have in fact acquired a qualitatively new property as *social* beings—we can fly. Now, we can look down at clouds from above, while before the twentieth century, we could only look up at them. Further, the parts of the airplane can also fly by being parts of the totality of flying. A jet engine cannot be airborne unless it is part of this totality. Thus:

> The ancient debate on emergence, whether indeed wholes may have properties not intrinsic to the parts, is beside the point. The fact is that the parts have properties that are characteristic of them only as they are parts of wholes; the properties come into existence in the interaction that makes the whole. (Levins & Lewontin, 1985, p. 273)

Accordingly, this perspective sees the individual and the social as constituting a dialectical whole in which both acquire properties as parts of this whole. The individual and the social cannot be separately studied without destroying the phenomenon that defines them.

The communal infrastructure

In this section, we outline how the abstract terms praxis and dialectics can be operationalized. Our point of departure is the fact that our brains "evolved to control the activities of bodies in the world … the mind consists of structures that operate on the world via their role in determining action" (Love, 2004, p. 527). Such structures emanate from the innate, neurobiological predispositions for action that the phylogenetic evolution of humankind has brought about. Thus, every healthy human being meets the world endowed with a certain species-specific, mental and physical "infrastructure". We all "share anatomy and common biomechanical and task constraints ... We all discover walking rather than hopping" (Thelen, 1995, p. 91).

However, it is only in praxis that these latent predispositions develop into specific abilities. Biomechanical predispositions do not translate into biomechanical abilities unless there is the opportunity to exercise them. For example, "Doubtless many contemporaries of Julius Caesar had the biomechanical capacity to become pianists, but were never able to develop the corresponding biomechanical ability because the pianoforte had not yet been invented" (Harris, 1996, p. 29).

During the cultural-historical evolution of humankind, praxis is manifested as diverse *communities*, loosely defined as groups of people that bond together by a shared interest, a shared identity and a shared set of norms (Bradshaw, 2008). Communities "develop, change, *and* remain constant as a result of individual actions, and yet the results of this process, the artifacts and ideational contents it creates, constitute an apparently constant or resistant framework for its inhabitants ... and as such it will constitute, for each new individual born into it, a pre-established environment to be discovered and structured" (Boesch, 1991, p. 31, original emphasis).

Archer (1995, 1998, 2000) has articulated this process as a "morphogenetic sequence", which we interpret as follows (see Fig. 11.1):

At T^1, existing biomechanical and communal factors provide *structural conditions* for acting at T^2. *Social interaction* between T^2 and T^3 is conceived as individual action, constrained and enabled by the current state of factors at T^1. Action results in *structural elaboration*, which changes

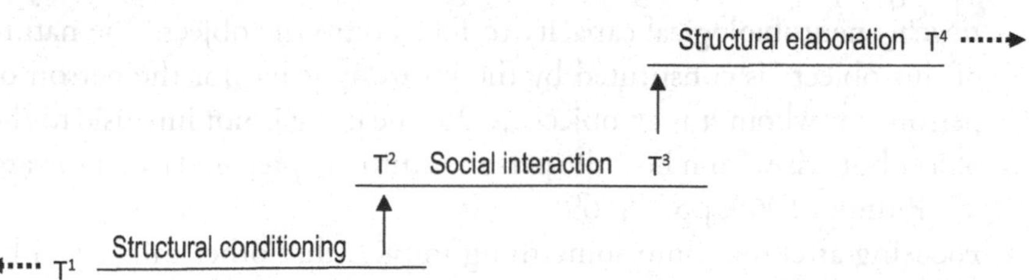

Fig. 11.1 The morphogenetic sequence. Adapted from "Realism and morphogenesis" by M.S. Archer, 1998. In M. Archer, R. Bhaskar, A. Collier, T. Lawson, A. Norrie (Eds.) *Critical Realism: Essential Readings*, p. 375. With permission

both factors into structural conditioning for further actions at T^4. In order to articulate the morphogenetic sequence, we introduce the term *communal infrastructure* for the communal state at T^1. Thus, we understand the sequence from T^1 to T^4 as an evolution of the communal infrastructure.

Importantly, the sequencing in Fig. 11.1 is for analytical purposes. Action is "continuous, cyclical, flow over time: there are no empty spaces where nothing happens, and things do not just begin and end" (Fleetwood, 2005, p. 203). However, a main implication can be drawn—communalization has an intrinsic temporal dimension. Communal changes occur on vastly different timescales, ranging from milliseconds processing by the brain, decades of communal development, and millions of years of biological evolution.

The Dialectics Between Biomechanical and Communal Factors

A crucial step in the elaboration of the communal infrastructure is to conceptualize biomechanical factors required for acting in the world. Since praxis implies that biomechanical and communal factors are dialectically related, we may infer neurobiological capacities by *analyzing the nature of communal factors* involved in action. Accordingly, we posit that at least the following capacities are requisite (albeit not sufficient) to act in a specific situation:

- Acting implies attending "some-thing", an *object*. This entails an *objectivating* neurobiological capacity to focus onto the object. The nature of this object "is constituted by the meaning it has for the person or persons for whom it is an object … this meaning is not intrinsic to the object but arises from how the person is initially prepared to act toward it" (Blumer, 1969, pp. 68–69).
- Focusing attention onto some-thing implies that other things will be unattended. This entails a *contextualizing* capacity to project in the mind a *context* of relevance around the object—a "horizon of meaning" (Gadamer, 1989, p. 383).

- The *spatial* structure of the situation needs to be grasped, which entails a *spatializing* neurobiological capacity. Spatial factors signify "the way we shape the very world that constrains and guides our behavior" (Kirsh, 1995, p. 31).
- A *temporalizing* neurobiological capacity is requisite for anticipating the *temporal* structure of the situation; the sequence of actions towards the object, leading to the fulfillment of the need that motivates the activity in the first place (Toomela, 2014).
- The *normative* structure of the situation, manifested as habits, rules, conventions, traditions, etc., needs to be adhered to, which entails a *habitualizing* neurobiological capacity: "People's thoughts, feelings, and predispositions for action are inherently dynamic, displaying constant change due to internal mechanisms and external forces, but over time the flow of thought and action converges on a narrow range of states—a fixed-point attractor—that provides cognitive, affective, and behavioral stability" (Nowak, Vallacher, & Zochowski, 2005, p. 351).
- When acting in a situation is finished, attention is directed to other situations. A *transition* from one situation to another entails a *transiting* neurobiological capacity to refocus attention in which "the cortical system rapidly breaks functional couplings within one set of areas and establishes new couplings within another set" (Bressler & Kelso, 2016, p. 4).

Hence, we posit that the phylogenetic evolution of humankind has brought about the biomechanical factors *objectivating*, *contextualizing*, *spatializing*, *temporalizing*, *habitualizing*, and *transiting* as requisite neurobiological capacities for acting in the world. These develop into specific neurobiological abilities depending on situations the individual encounters. However, regardless of the endemics of a particular situation, action always necessitates the mental capacity to confer signhood onto communal factors signifying *object*, *context*, *space*, *time*, and *norms*. Transitional factors enable the *transition* from one community to another, in which other objects, spaces, times, and norms are contextualized. Accordingly, we conceive of the dialectical relation between the individual and the social as constituted by these six factors. The gist of this conceptualization is that our neurobiological predispositions are reflected in the way we understand and construct social reality. We cannot make sense of environmental sensations that do not resonate with our neurobiological constitution.

To further articulate this dialectic, we introduce the concepts of *activity modalities* (Taxén, 2003, 2009) for the biomechanical factors and *communal anchors* for communal factors.[2] These terms are coined to provide "conceptual bridges" across to other related disciplines such as neuroscience for biomechanical factors and social sciences for communal factors. Such bridges are required to further sustain the DA. Neuroscience is necessary to ground the activity modalities. To exemplify, neural correlates of the spatializing modality include at least the place cells found in the posterior hippocampus (Jeffery, Hayman, & Chakraborty, 2004; O'Keefe & Nadel, 1978) and the grid cells in the entorhinal cortex (Witter & Moser, 2006). A lesion in any of these cortical zones destroys the ability to navigate spatially in the environment and, consequently, to act.

Concerning social sciences, one implication is that "the cognitive, normative, and regulative structures and activities that provide stability and meaning to social behavior" (Scott, 1995, p. 33) will be structured according to the activity modalities. Every community will contain communal anchors that can be recognized as objects, contexts, spaces, times, norms, and transitions. This provides a point of departure for further investigating social structures from an action point of view.

Joint Actions

Even if the human act is profoundly individual, we mostly collaborate jointly with others. Joint action can be defined as "the larger collective form of action constituted by the fitting together of the lines of behavior of the separate participants. … Joint actions range from a simple collaboration of two individuals to a complex alignment of the acts of huge organizations or institutions" (Blumer, 1969, p. 70). To understand joint action, it is necessary to distinguish between two types of individual actions. When a pianist gives a recital, she performs an *autonomous* act (Clark, H.H., 1996, p. 19). There is no other musician involved. When

[2] The connotation of the term 'anchors' is that these exert a 'gravitational pull' for individual meanings, thus contributing to developing and sustaining a communal world-view.

the same pianist plays in a piano trio, she also preforms an individual act, but now together with other musicians. Such individual acts, "performed only as parts of joint actions", are called *participatory* ones (ibid.). Communal anchors are requisite for both types of actions. However, in joint actions, these anchors become *common identifiers* (Blumer, 1969, p. 71) towards which participatory actions gravitate. Thus, participatory actions are never identical—they are still uniquely individual—but they may be sufficiently aligned to contribute to a common goal, motivated by some social need.

An illustration

To illustrate the dialectics between activity modalities and communal anchors construct, we may use the example of a guitar quartet in concert (Fig. 11.2):

Fig. 11.2 A guitar quartet in concert

The quartet can be seen as a musical community, the rationale of which is to entertain an audience. Communal factors such as the instruments, teaching practices, concert halls, compositions, and much more, have been refined and improved over a long period. Together, these anchors provide a communal infrastructure, which prospective musicians encounter as newcomers to this community. Correspondingly, the biomechanical factors of each musician have evolved over many years of practicing alone with the instrument, and together with the quartet.

A conspicuous communal factor in musical communities is the score, as exemplified in Fig. 11.3:

The evolution of the score into its present form has occurred over hundreds of years (Hoskin, 2004). From the dialectical perspective, this suggests that the score harmonizes with our mental abilities for acting in musical communities as follows:

- The entire score anchors the attentions of musicians to the *object* of the activity, which in this case is the concert.
- Focusing attention onto the score implies that musicians project in their minds a *context* of relevance around the object such as the playing of other musicians, the acoustics of the hall, the audience, and so on. Other things, like the painting on the wall behind in the quartet in Fig. 11.2 is certainly irrelevant for the concert, but may of course be relevant in another situation.
- The sequence of notes from left to right, as well as the shape of note stem flags, anchor a *temporal* structure of the music.

Fig. 11.3 Anchoring a musical activity

- At each moment in time, the position of a note relative to a note line anchors a *spatial* position of the guitar neck, where a certain string shall be pressed down and plucked to sound the note.
- The various music-specific notations on the score, such as the *mf* (mezzo forte) signifying the level of playing, and the symbol signifying the F-clef bass notation, anchor a *habitual* dimension of the activity.

Accordingly, we can infer that the score is a particularly efficient communal anchor since it anchors five of the activity modalities in a single artifact. Regarding the transition modality, the musicians attend other activities when the concert is finished. For example, one guitarist may have noticed a certain jangle from the guitar when playing, and therefore contacts the guitar builder to fix this. This presumes a *transition* between two mutually dependent communities—one in which the guitar is built, and another in which the guitar is played. How such a transition should be carried out requires the development of "an infrastructure for transferring outcomes of disparate activities across social worlds" (Winter & Butler, 2011, p. 102).

Concerning joint action, a musician performs an autonomous act according to her corresponding staff when practicing alone. When playing together, the very same individual act becomes participatory. The score itself remains unchanged, but now it is conceived as a common identifier for the quartet. The alignment of the staffs on top of each other indicates how individual acts should be coordinated as participatory acts. However, regardless of how tightly these acts are fitted together, this "cannot be resolved into a common or same type of behavior on the part of the participants. Each participant necessarily occupies a different position, acts from that position, and engages in a separate and distinctive act" (Blumer, 1969, p. 70).

Further, the dialectical relation between parts and whole can also be seen in the score. Voices are indispensable elements in forming the totality of the music, but voices are also formed by the same totality. If each voice is played as autonomous acts, the music becomes void of meaning. The totality of the piece can only come about as participatory actions. If one voice is missing or badly played, the music makes no sense. Thus, parts have properties only as parts of wholes, and the whole properties only as made up from parts. The whole constitutes the parts, and the parts constitute the whole.

A final observation is that the balance between biomechanical and communal factors may change. One player may learn his voice by heart, in which case the biomechanical factors dominate, and the score becomes irrelevant. Conversely, another player may be entirely dependent on the score, in which case communal factors dominate. However, both factors are always needed.

Communalization

In this section, we describe the dynamics by which the communal infrastructure evolves—a process we call *communalization*. In the literature, two main approaches to this issue can be discerned. The term "socialization" refers to "the process by which somebody, especially a child, learns to behave in a way that is acceptable in their society" (*Oxford Learner's Dictionary of Academic English*). Likewise, the term "institutionalization" refers to the process of making "something become part of an organized system, society or culture, so that it is considered normal" (ibid.). From the dialectical perspective, socialization relates to biomechanical factors, and institutionalization to communal factors. Usually the relationships between socialization and institutionalization are investigated from either a socialization or an institutional perspective. The term "communalization" is deliberately coined to *foreground the relation* between socialization and institutionalization rather than on these two separately.

A suitable point of departure for elaboration of communalization is the concept of *affordances*:

> An affordance cuts across the dichotomy of subjective–objective and helps us to understand its inadequacy. It is equally a fact of the environment and a fact of behaviour. It is both physical and psychical, yet neither. An affordance points both ways, to the environment and to the observer. (Gibson, 1979, p. 129)

The "affordances of the environment are what it *offers* the animal, what it *provides* or *furnishes*, either for good or ill" (Gibson, 1979, p. 127, original emphasis). Even if Gibson focused on natural affordances, he insisted that it is "a mistake to separate the cultural environment from the natural environment, as if there were a world of mental products distinct from the world of material products" (ibid., p. 130). Conceived in this way, affordances are compliant with the dialectics since dualism is "out from the beginning" (Baerentsen & Trettvik, 2002, p. 52).

The Theory of Functional Systems

The essence of affordances is "its foundation in activity" (Baerentsen & Trettvik, 2002, p. 52). However, this aspect remained underdeveloped in Gibson's writings since he focused mainly on the perceptual side of affordances (ibid, p. 53). We suggest that the *theory of functional systems* (TFS), conceived in the 1930s by the Russian biologist Pyotr Anokhin, can be elaborated into a coherent activity concept where "every phenomenon that enters consciousness may mean something to the subject" (ibid., p. 54).

The distinctive feature of TFS is its emphasis on *stability* based on self-regulation principles as the primary characteristic of life processes (Red'ko, Prokhorov, & Burtsev, 2004). In Fig. 11.4, a simplified version of TFS is illustrated:

Two groups of functions are involved, depending on which kind of nerves are actuated: *afferent* ones going from the periphery of the body to the brain, and *efferent* ones going from the brain to effectors such as muscles or glands. Action proceeds in the following stages.[3] In *Afferent synthesis*, sensations from the external world (Situational afferentation), previous experiences retained from memory, and motivation are integrated into a coherent Gestalt of the situation—the perception of a pattern or structure as a whole. Based on this Gestalt, a decision of *what* to do, *how* to do, and *when* to do is taken. *Decision making* involves two functions: anticipation of the expected result (Acceptor of result) and the

[3] These stages are separated for analytical purposes. In reality, they are highly intertwined. For example, perception is guided by anticipation of action as well (Lewis, 2002).

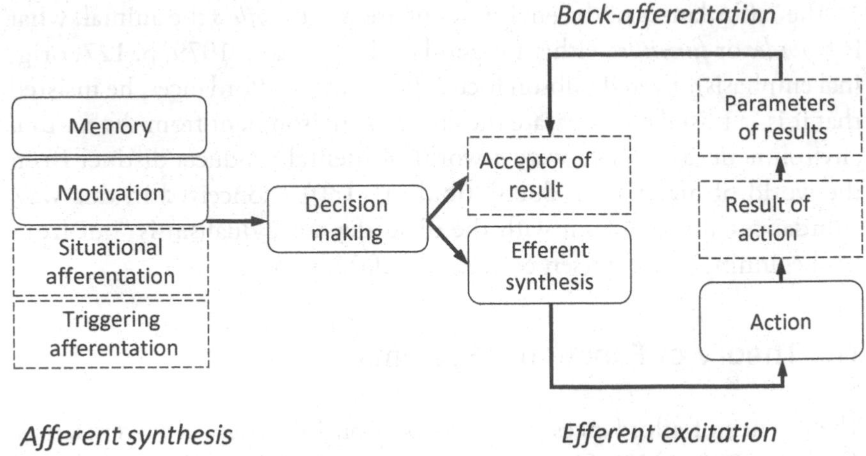

Fig. 11.4 General architecture of an individual functional system. Solid lines mark internal functions, while dotted lines indicate functions that depend on external sensations. Adapted from A. Toomela "Biological Roots of Foresight and Mental Time Travel" (2010). *Integrative Psychological and Behavioral Science, 44*, p. 115. With permission

formation of an action program (Efferent synthesis): "if I act in this way, I assume this will result". Triggering afferentation sets off *Efferent excitation*, in which the action is performed and the result is evaluated against the anticipated result via *Back-afferentation*. Depending on the outcome of the evaluation, the sequence is repeated or stopped. The entire episode is then retained in memory for acting relevantly in future, similar situations.

In this way, the individual perceives and acts upon the affordances provided by the environment, regardless of whether these are natural or cultural-historical in origin. It follows that an affordance can be conceived as having both an *afferent* and an *efferent* facet related to the various stages in the TFS model. Afferent affordances are foregrounded in Afferent synthesis and Back-afferentation stages, while efferent affordances prevail in Efferent excitation. Thus, every affordance has both an inward-directed, informative aspect—"how can I make sense of this?"—and an outward-directed action aspect—"what can I do with this?" This applies equally well to things we conceive as predominantly material, such as a hammer, and things we tend to see as predominantly ephemeral such as an utterance. All in one, the TFS provides a neurobiological model for the dialectics between biomechanical and communal factors.

The Dialectical Approach—A Summary

The dialectics between the individual and the social can be summarized as follows (see Fig. 11.5).

Communalization proceeds from a momentarily stabilized communal infrastructure, comprised of biomechanical and communal factors. The *structural conditioning* phase between T^1 and T^2 corresponds to the *Afferent synthesis* stage in TFS; the result of which is an integrated unitary percept—a Gestalt—comprised of the activity modalities and other neurobiological functions necessary for action. This phase requires that the individual is able to perceive communal factors signifying objects, contexts, spaces, times, and norms.

The *social interaction* phase between T^2 and T^3 corresponds to the *Efferent excitation* stage, in which the individual intervenes in and changes the environment. The *structural elaboration* phase between T^3 and T^4 corresponds to the *Back-afferentation* stage. This is where the result of the action is assessed. This result is compared with the anticipated, and the outcome is stored in memory as modified neurobiologically structures pre-existing the next cycle. Importantly, these structures comprise the

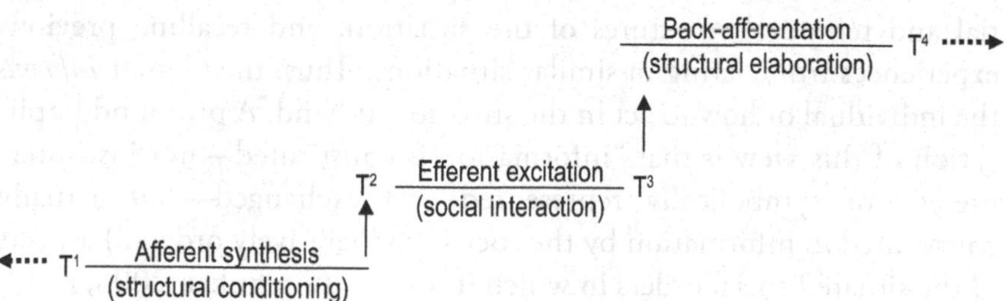

Fig. 11.5 TFS and Archer's morphogenetic sequence. Adapted from "Realism and morphogenesis" by M.S. Archer (1998). In M.S. Archer, R. Bhaskar, A. Collier, T. Lawson, A. Norrie (Eds.) *Critical Realism: Essential Readings*, p. 375. With permission

entire cycle from T^1 to T^4. Thus, what is kept in mind is the totality of the situation.

Accordingly, communalization simultaneously encompasses two opposite tendencies—towards stabilization and towards becoming. Without stability, no becoming and without becoming, no stability. A main advantage of this marriage between the morphogenetic sequence and TFS is that a dialectical bridge is built towards other research fields. The stages in the TFS may be related to neuroscientific findings, and the stages in the morphogenetic sequence to findings in social sciences. This means that a theoretical "kernel" is provided for future research into the dialectics between the individual and social realms.

The Dialectics Between Information, IT and IS

The dialectical perspective provides an alternative conceptualization of information, the IT artifact, and the IS as follows.

Information

Information is conceptualized as the Gestalt resulting from the Afferent synthesis stage in TFS. This Gestalt is structured according to the activity modalities. Stated differently, the decision to act is based on the perception of an integrated pattern comprised of the identification of the object of the action, contextualizing the situation, grasping the spatial and temporal structures of the situation, and recalling previous experiences from acting in similar situations. Thus, the Gestalt *informs* the individual of how to act in the situation at hand. A profound implication of this view is that "information is constituted—not just interpreted—or symbolically represented and exchanged—but actually constituted *as* information by the social (cooperatively ordered) aspects of the situated social orders in which it occurs" (Garfinkel, 2008, p. 13, original emphasis). Thus, the locus of information is the *individual neurobiological system*, the constitution of which is requisite on the input from the external world:

> [Cognitive] systems construed as dynamic systems do not process information transduced from the outside world; they reconfigure themselves in response to an ongoing stream of sensory events. (Lewis, 2005, p. 173)

Consequently, information is seen as inherently associated with the evolution of Homo Sapiens, and not restricted to the era starting with the introduction of computerized information systems in the 1960s. Information in this sense has been requisite for the survival of our species ever since the dawn of humankind (cf. Watson, 2014). Such a view implies that *any* external sensation may contribute to the constitution of relevant information, regardless of whether it originates from an IT artifact or some other source. Information is "knowledge for the purpose of taking effective action" (Mason & Mitroff, 1973, p. 475).

The IT Artifact

An IT artifact is a physical artifact based on technology such as software and hardware, which is intentionally designed to be informative: "This is actually the *most important trait* and what distinguishes it from many other types of technical artefacts" (Goldkuhl, 2013, p. 93, our emphasis). Someone using the IT artifact should be informed about the state of things in the world in order to act relevantly. This implies that the IT artifact should be designed to render *afferent* affordances anchoring objects, contexts, spaces, times, norms, and transitions; thus, contributing to the Afferent synthesis and Back-afferentation stages in the TFS model.

However, most IT artifacts also render *efferent* affordances. You can do something with the artifact besides monitoring it; sending commands, starting a conversation, modifying it, and so on. This means that a "pure" informative IT artifact is an extreme case. Likewise, artifacts that render mainly efferent affordances, such a hammer or a shotgun, also render informative, afferent affordances in the sense that someone using it must recognize what it is and how to use it: "A tool is also a mode of language. For it says something, to those who understand it, about the operations of use and their consequences" (Dewey, 1991, p. 52). Accordingly, there is no sharp borderline between IT artifacts and other classes of artifacts, only a qualitative difference.

The Information System

When individuals encounter an IT artifact, it becomes an object of attention. Knowing an object is to master "sets of sensorimotor skills and possible actions that can be chosen to explore or utilize the object" (Engel, Maye, Kurthen, & König, 2013, p. 206). At first, the IT artifact is what Heidegger calls *present-at-hand* (Riemer & Johnston, 2017). The meaning of the artifact, and what can be done with it, is unclear. By repeated engagement with the artifact, it may turn into *ready-at-hand*, fluently employed in order to achieve the task at hand. This process changes the neurobiological structure of the individual:

> [External] aids or historically formed devices are *essential elements in the establishment of functional connections between individual parts of the brain*, and that by their aid, areas of the brain which previously were independent become the *components of a single functional system*. (Luria, 1973, p. 31, original emphasis)

From an information point of view, we may understand the artifact's progression from present-at-hand to ready-at-hand as the gradual formation of an "information system", comprised of the *individual's neurobiological structure* and *the IT artifact*. In line with Clark's (Clark, H.H., 1969) terminology, we may call such an information system autonomous, since no other individual is involved.

When an IT artifact is introduced in a community, it comes forth as the inception of a communal anchor towards which individuals direct their attention. Individual "information systems" now become participatory, meaning that individual "information systems" gravitate towards the IT artifact. Gradually, the IT artifact is communalized into a meaningful communal anchor—it becomes an institutionalized *Information System*. This means that "Information Systems is Information Technology in Use" as Paul (2010, p. 379) maintained. In this process, the IT artifact emerges as Gegenstand, regarded simultaneously as recalcitrant materiality and something which makes sense in the community.

Importantly, communalization of IT artifact does not mean that it is transmuted into a different kind of artifact. The IS emerges in the eyes of the beholders as communalization proceeds. Hence, the IS will always be an indeterminate, open-ended, never settled, and always changing phenomenon. During communalization, the artifact may be adapted to the needs of the community, but it will *maintain its ontological status as an artifact*. A modified IT artifact is still an artifact, and its "IS status" can only be assessed by investigating the artifact's communalization history. The sociality of the artifacts lies in its function as a communal anchor. Accordingly, we "do not need to put humans inside the boundary of the IT artifact in order to make these artifacts social" (Goldkuhl, 2013, pp. 93–94). The key implication of this conceptualization of the IS is that there will be no separate "IS artifact", unrelated to the IT artifact.

The conceptualization of information, the IT artifact, and the IS as above, implies that these phenomena are dialectically related—they mutually constitute each other. Consequently, studying information, the IT artifact, and the IS as isolated, unrelated phenomena will at best provide incomplete results, and at worst fallacious.

Communalization of an IT Artifact in Practice

The purpose of this section is to substantiate the practical relevance of the DA by recapitulating the communalization of an IT artifact in the telecom industry. In the late 1990s, Ericsson™—a major provider of telecommunication systems worldwide—was developing the third generation of mobile systems. The challenges posed by this endeavor were by that time unprecedented in terms of people, organization, and new technology. For example, one project involved about 1000 persons distributed over 22 subprojects and 18 design units worldwide (Taxén, 2003, 2009).

In order to coordinate the projects involved, extensive IT support was necessary. With the introduction of modern, object-relational databases in the mid-1990s, quite new information management capabilities became available. In 1997, a decision was made to introduce such an IT

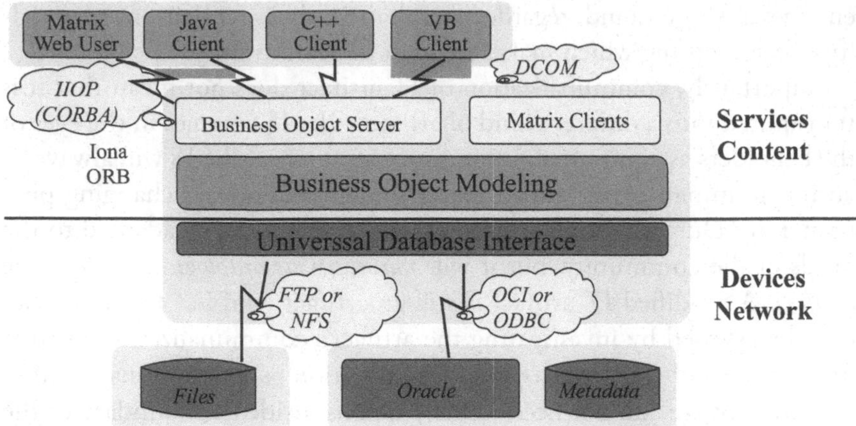

Fig. 11.6 The layered modular architecture of the Matrix IT system. With permission from Technia™

system called Matrix in one of the subprojects. The architecture of Matrix is illustrated in Fig. 11.6:[4]

Matrix was a general-purpose platform offered by the vendor as an infrastructure for organizational-specific adaptations.[5] The network and device layers offer basic functionalities for making the system up-and-running, such as data base interfaces, installation scripts, distribution to physical databases, client–server communication, interconnections to other IT systems, etc. The service and content layers provide functions for defining and managing organizational-specific information, setting up workflows, configuring user interfaces, formatting reports, allocating role responsibilities, and more.

The History

The configuration of Matrix to the Ericsson environment unfolded as follows. The period between 1997 and 1999 was a "trial period", devoted to investigating the capabilities of Matrix before it was used in an

[4] Technia Addnode Group: https://www.technia.com/.

[5] Matrix was later acquired by Dassault Systèmes (https://www.3ds.com), and is now a key component in their 3D experience product.

ongoing subproject. The development task was allocated to a small team consisting of the project manager, key persons in the project, a consultant from the vendor, and this author. The first task after the installation was to device a tentative model signifying which items were to be managed in Matrix. By the end of 1997, this "information model" had evolved into the one shown in Fig. 11.7:

The model is a variant of an Entity-Relationship model (Chen, 1976) in which each box represents an item to be managed. The names of these items are Ericsson-specific, and thus meaningful only in that context. Lines indicate relations between items. In the center, the main object in focus—the "Feature Increment"—is shown. This item represents the work to develop a certain function that can be offered to a customer.

The content and structure of the model were subject to intensive discussions within the team. As soon as there was an "interim consensus" of the model, it was implemented in Matrix. Project data were loaded into the system and the results inspected (see Fig. 11.8):

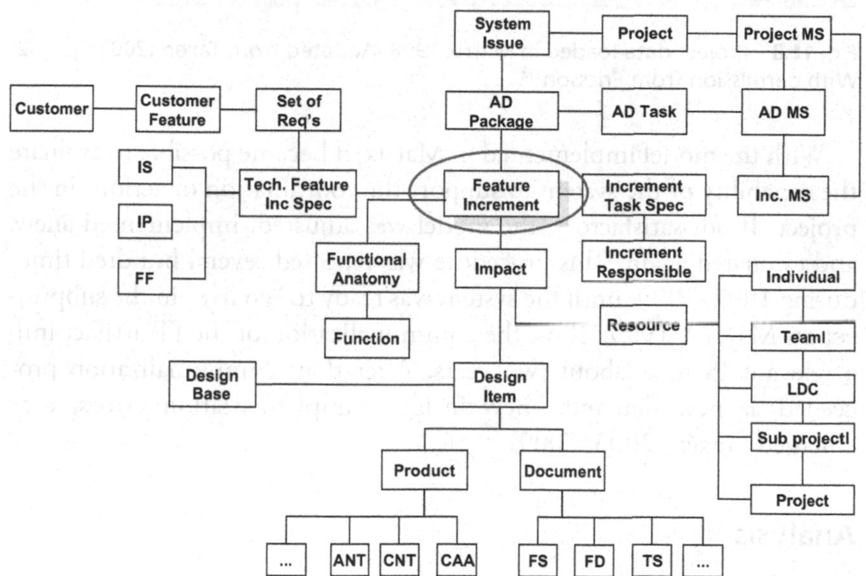

Fig. 11.7 The information model in 1997. Adapted from Taxén (2003), p. 131. With permission from Ericsson™

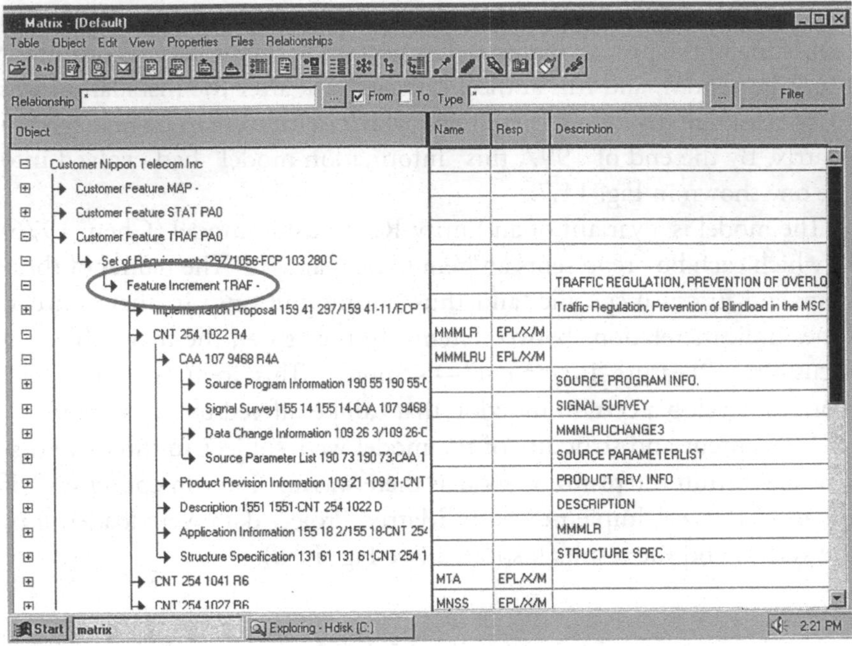

Fig. 11.8 Project data loaded in Matrix 1998. Adapted from Taxén (2003), p. 132. With permission from Ericsson™

With the model implemented in Matrix, it became possible to evaluate the capability of the system to support the coordination of actions in the project. If not satisfactory, the model was adjusted, implemented anew, and evaluated again. This procedure was repeated several hundred times during 1997–1999, until the system was ready to "go live" in the subproject on May 10, 1999. Thus, the communalization of the IT artifact into a relevant IS took about two years. After that, communalization proceeded as new demands, new insights, implementation errors, etc., emerged (Taxén, 2003, 2009).

Analysis

The communalization of Matrix departed from the existing communal infrastructure at Ericsson in 1997. Some communal factors, such as norms for naming and revising products and documents, a company-wide documentation archive, the Ericsson project management model, and more had been communalized into stable, communal factors over many years. Consequently, these factors were not changed. Concerning individual factors at the start of communalization, these were quite disparate, since the Ericsson employees had no experience of Matrix, and the consultant from the vendor had no experience of Ericsson.

Communalization

In the Afferent synthesis stage in TFS, individuals integrate their previous experiences, motivation, and situational sensations into a percept informing each individual about the situation at hand, and pre-dating the decision of how to act. Some of these sensations emanate from the information model in Fig. 11.7. This model is the common identifier around which individual interpretations are fitted together. The Efferent excitation stage is restricted to manipulating the physical manifestation of the model on, for example, paper or as a PowerPoint file, thus, modifying this communal anchor. At the same time, individual, biomechanical factors are elaborated through Back-afferentation as a result of discussing the form and structure of the model. This cycle was repeated until the team was prepared to implement the model in Matrix.

Accordingly, the information model provides predominantly *afferent* affordances, directed inwards and modifying the biomechanical infrastructures of the individuals. The *efferent* affordances for the subproject are, however, negligible; the model itself does not make any impact in the subproject practice. However, it is important to realize that the information model was a critical element in the communalization of Matrix. Without it, individual interpretations of what constitutes community-relevant information would not have converged, thus thwarting the adaptation of Matrix to Ericsson-specific circumstances.

With the information model implemented in Matrix, quite new *efferent* affordances became available for the Efferent excitation stage. Examples of such affordances are creating, revising and deleting items, entering and modifying item properties, producing reports, performing quality assurance procedures, computer-supported communication worldwide, and much more. Thus, the capacity of Matrix to support the subproject could be evaluated. The results from these actions were back-propagated in the sense that the information model, the implementation in Matrix and individual, biomechanical factors were changed. This way of working required that changes in the IT system could be easily made without closing down the database, which was the case with Matrix.

In summary, this example demonstrates the significance of the concept of "communalization". Usually, the introduction of a new IT system in an organization is seen as a technical matter—get the system up and running—augmented with educating users how to work with the system. However, communalization implies that the IT system and users need to coevolve into an integrated unity comprised of biomechanical and communal factors. This emphasizes the importance of attending the individual in the sociomaterial discourse.

The Activity Modalities

According to DA, the dialectical relation between individual and communal factors comprises all the dimensions of the activity modalities. This means that objects, contexts, spaces, times, norms, and transitions should be easily recognized affordances in the communalized information model and the IT artifact Matrix. When inspecting the communal factors involved in the communalization of Matrix, it can be seen that the central *object*—the "Feature Increment"—appears in both the information model (Fig. 11.7) as well as in Matrix (Fig. 11.8). The structure of the information model is in essence *spatial*, since it is comprised of "things' related to each other. *Norms* are shown as Ericsson-specific names of products and documents, such as "CNT 254 1022 R4" that signifies a central processor element in the telecom system. *Time* is signified by the state of items, for example "AGREED", "UNDEFINED", "READY", "PREL" (not shown in the screen dump). The *context* is comprised of the totality of signified elements. Thus, traces of the activity modalities were manifested in various communal artifacts, such as the information model and Matrix. The transition modality is not present in this example.

Advancing IS theory

The dialectical perspective so outlined provides an alternative foundation by which mainstream IS research topics may be analyzed and advanced from its current state. Among the multitude of topics, we have chosen communication, information, the IT artifact and the IS, digitalization, sociomateriality, and the rationale for the IS discipline. Obviously, with the limited space available, we can only briefly touch on how these topics may be elaborated towards the dialectical position. Our suggestions should be seen as "prescience"; as the inception of a discourse "discerning or anticipating what we need to know and, equally important, of influencing the intellectual framing and dialogue about what we need to know" (Corley & Gioia, 2011, p. 13).

Communication

As emphasized by many scholars (e.g. Beynon-Davies, 2010; Beynon-Davies, 2013; Garfinkel, 2008; Mingers & Standing, 2018; Mingers & Willcocks, 2017; Stamper, 2001), communication and information are inevitably related: "Whichever formulation of communication theory one chooses, the concept of information is always a strategic part of it" (Stichweh, 2000, p. 11). Thus, to understand information, we need to understand communication.

In the IS literature, the mainstream communication model is that of an informant sending a message containing the information to an informee (Boell, 2017) (see Fig. 11.9):

The essence of this "transmission" model is that "mental content (conceived variously as concepts, ideas, symbolic representations, etc.) is neatly conveyed intact from the mind of one party to the other" (Orman, 2017, p. 176).

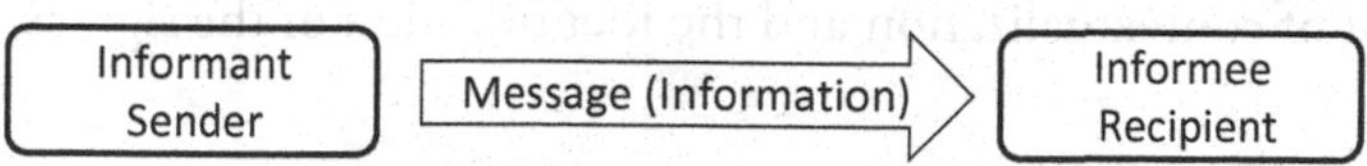

Fig. 11.9 The mainstream communication model in IS

Elaboration

The transmission model is incompatible with the DA stance of information as internally constituted by the individual. Information is not something that can be sent from one person to another. An alternative communication model, compatible with the DA, is *Integrational Linguistics* worked out by Roy Harris, professor of General Linguistics at Oxford (Harris, n.d.; Harris, 1996; IAISLC, 2017).[6] Integrational Linguistics is a radical departure from traditional Western assumptions about language and communication in that

- it abandons the idea of communication as a "sender-receiver" process,
- it rejects code-based and rule-based models of language,
- it questions the existence of any natural or universal distinction between language and non-language.

Communication is seen as an open-ended continuum of integrated activities, shaped by the initiative of individuals. This means that

- there is continuous and simultaneous creation of meaning,
- all signs are products of the communicational situation,
- there are no autonomous, context-free signs,
- signs are created by individuals.

From an integrationist perspective, the "primary function of the sign is to integrate an individual's past, present and (anticipated) future experience. That is an essential prerequisite for making sense of any situation in which we are involved. Without it, there can be no question of communication" (IAISLC, 2017).

Integrational Linguistics concurs with the DA in several ways. Concerning temporalization and contextualization, Integrational Linguistics states that "There are no timeless signs" (Harris, 1996, p. 97). The "act of contextualization and the identification of the sign as sign are

[6] For an extensive overview of Harris' works, see (Love, 2015).

one and the same. We contextualize as a condition of integrating new signs into the temporal dimension of experience" (Harris, 2009, p. 103).

Further, the tenet that there is no universal distinction between language and non-language implies that language can be seen as "complementary to more basic forms of neural processing" (Clark, 2006, p. 372). Thus, language is an

> extension of the physical environment generally, and one that we may perceive (by language comprehension) and act upon (by language production), just as we do with any physical environment. (Andrews, Frank, & Vigliocco, 2014, p. 363)

This means that language has both afferent and efferent affordances (comprehension and production respectively), like any other communal anchor. In essence, communication is conceived as coordination dynamics "in which words and structured linguistic encodings act to stabilize and discipline (or 'anchor') intrinsically fluid and context-sensitive modes of thought and reason" (Clark, 2006, p. 372).

In summary, elaborating communication towards the DA stance requires that the transmission model is replaced by some communication model in line with Integrational Linguistics. This is clearly a strenuous task, given the massive thrust the transmission model has, not only in the IS field, but also in mainstream linguistics.

Information

Regardless of how the IS discipline defines its core matter, the concept of information is of "ultimate interest and value … Lacking a clear conceptualization of information prevents the IS research community from engaging deeply with our true core subject matter—information systems" (Carter, Petter, & Randolph, 2015, p.2). However, in spite of intensive research, information is "a problem that has plagued research on information systems since the very beginning. The problem is the elusive nature of information itself, and the way we as researchers have failed to address the essence of information in our work" (Boland, 1987, p. 363).

Boell (2017) has made an extensive compilation of extant conceptualizations of information in the IS literature. Four different views can be discerned:

- The *physical* stance asserts that information "is a fundamental property of the material world" (Boell, 2017, p. 5), existing independently of human observers.
- The *objective* stance sees information as contained in sign vehicles "that exists independently, outside of humans" (ibid., p. 7).
- The *subjective* stance regards information as "something that is appropriated by a subject" (ibid., p. 7), thus informing the individual by conveying some "information substance", as it were, from the environment into the individual.
- The *sociocultural* stance affirms that "information is specified by a social context determining what is regarded as information" (ibid., p. 9), thus suggesting that information cannot be understood at the level of the individual.

Elaboration

These views comply with the dialectical perspective as follows:

- The *physical* stance is incompatible with the dialectical one, in which information is seen as a mental phenomenon. However, the physical view may be elaborated as "sensations emanating from physical world is requisite for the individual constitution of information".
- The *objective* stance is rejected since its very premise is irreconcilable with the dialectical position. Signs do not "contain" information: "signs, and hence knowledge, arise from creative attempts to integrate the various activities of which human beings are capable" (Harris, 2009, p. 162).
- The *subjective* stance of information as being transported from the outside into the individual is also rejected; there is no "information" in the external stream of sensory events. The subjective view may however be

elaborated into the position that "information exists only in relation the subject, but the external world supplies input to this in-forming".

- The *sociocultural* stance suggests that information cannot be understood at the individual level. This view recognizes that the social is an inherent aspect of information, but fails to see the individual factor in constituting information. Further, the sociocultural view indicates that the context is in some way a pre-existing "container", determining what counts as information. This view is rejected in the DA, in which contextualization and context are seen as dialectically related. However, the sociocultural stance may be elaborated into the position that "communal factors in the sociocultural environment are requisite for informing the individual".

The dialectical view of information has far-reaching implications for the concept of "information sharing" as discussed by Beynon-Davies and Wang (2019). They argue that the central problem with this term is that "it relies on a rather brittle convention of signifying information as stuff that can be manipulated, transmitted, and used in an unproblematic manner both within and between organizations" (ibid., p. 477). Accordingly, in line with the DA stance, Beynon-Davies & Wang suggest that information cannot be shared or transmitted since information "is not *stuff*" (ibid., p 479, italics in original).

As an alternative, they suggest that situations of information sharing "consist of actors, structures, messages, and actions, all taking place within some environment" (ibid.) in one holistic, enacted view. In such situations, an actor's capacity to act

> involves interaction with its external environment, and that such interaction relies on two critical forms of apparatus making up the body of the actor: a sensory apparatus and an effector apparatus. A sensory apparatus consists of a series of sensors which continually monitor differences in the state of the external environment. An effector apparatus consists of a series of effectors that allow the actor to perform instrumental action in relation to this external environment—to manipulate structures within the external environment and through so doing to change the state of the external environment. (ibid., p. 479)

Further, Beynon-Davies & Wang assert that information is subjective because it involves "acts of making sense of certain structural changes in the environment (data structures) by individual actors" (p. 481). This resonates well with the dialectical perspective.

At the same time, however, information is seen as inherently "physical in that it comprises physical differences formed in some substance and that conveys some message as a signal" (ibid.). What is being shared in information sharing is "data structures" in their virtue of being "communicative acts" communicating "intent between one actor and others within some institutional context" (ibid., p. 487). Data structures are "physical entities that have an independent life over and above the actors that created them" (ibid., p. 481).

The description of an actor's capacity to act concurs with the TFS model outlined previously. However, the view of data structures as "communicative acts" needs to be abandoned. Humans communicate—data structures don't. For example, a packaging list—a data structure—in the form [productNo; quantity; weight; size …] (ibid., p. 487), may make sense to an actor depending on whether the list has been communalized or not. This is not about the list "communicating" with the actor, but of the actor interpreting it based on previous experiences. In total, the contribution of Beynon-Davies & Wang is an important step towards understanding information sharing, and as such a point of departure for elaborating it further into the dialectical perspective.

The IT artifact and the IS

The view of the IS as the communalized IT artifact implies that information, the IT artifact, and the IS are distinct ontological relata constituted by the dialectical relation between them. This means that the "artifact" connotation applies to the IT artifact only. Concepts such as "IS artifacts" (Chatterjee, Xiao, Elbanna, & Sarker, 2017; Lee, Thomas, & Baskerville, 2015), "information artifact" (Lee et al., 2015), "social artifact" (ibid.), and the like, are not sustained by the DA.

Likewise, expressions indicating an ontological transmutation of the IT artifact are rejected, such as: "IT artifacts (those bundles of *material and cultural properties packaged* in some socially recognizable form such as hardware and/or software)" (Orlikowski & Iacono, 2001, p. 121, our emphasis); "structures of the organizational domain *are inscribed into the artifact* during its development and use" (Sein, Henfridsson, Purao, Rossi, & Lindgren, 2011, p. 38, our emphasis); "even without a change in the technical components, a change in a *social dimension can produce a very different artifact*" (Silver & Markus, 2013, p. 83, our emphasis).

All these examples suggest the "social" leaves some kind of insignia on the IT artifact when it is used. However, this is a category mistake. Doubtless, the IT artifact is changed during communalization as the Ericsson example illustrates. However, although modified in a social setting, it *remains an artifact*. There is no way to decide whether an IT artifact is "social" or not by merely inspecting it.

Elaboration

The IS community is well prepared to move its present state of play into a more dialectically oriented position adhering to Paul's declaration:

> [After] a nigh-on 20-year search for the answer to the 'What is I.S.?' question, I have come up with a concise and precise answer to the question, which is 'Information Systems is Information Technology in Use'. (Paul, 2010, p. 379)

However, this requires that extant views on information, the IT artifact and the IS are rethought. Such a move would further sustain the relevance of IS research since a dialectical position concurs with that of the practitioner. As the Ericsson example demonstrates, modifying an IT artifact to the needs of a community would be seen just like that, and not as inscribing something undifferenced "social" into the artifact.

Digitalization

The concept of "digital" plays a profound role in the IS discourse. However, the ontological nature of the "digital" is unclear (e.g. Kallinikos, Aaltonen, & Marton, 2013). To clarify this concept, it is helpful to distinguish between two aspects of "digital"—"digitization" and "digitalization".

Digitization is the process of "converting any *thing* into digits (ones and zeros), make it digital. The *thing* here could be analog data, logs, description about some *things*, let's say, equipment, its location, attributes etc. So, any *thing*, description of any *thing*, documents, information available in papers, hard copies etc. Converting them into digital format is called digitization" (Armarnath, 2019, italics in original). *Digitalization* is the "strategy of adopting recent technologies in IT to make the most of the digital resources available in the enterprise" (ibid.).

This means that an artifact is digital when it contains some digitalized part such as computers, special purpose ASICs (Application Specific Integrated Circuit), field-programmable gate arrays (FPGA), microprocessors, analog-digital converters, sensors, and the like. Some affordances provided by a digital artifact may be controlled by software (materialized as load modules containing program instructions). However, in any digitized artifact, there are ultimately also non-digitized, analogue elements. Otherwise, we could not interact with it. For example, the central digitized artifact of our times—the cell phone—would be useless if there were no analogue/digital converters transforming speech (analogue soundwaves) into ones and zeros. Ontologically, a digital artifact remains an artifact: a "material thing whose meaning is established only in and through the activity of individuals in social practice using the artifact" (Blunden, 2007, p. 264).

Elaboration

To illustrate digitalization, we may consider the musical score in Fig. 11.3. Artifacts like these have for long been realized in analog form on paper or, in historical times, on papyrus or parchment (Wikipedia, 2018). As such, the affordance of the score is mainly afferent. The efferent aspect—how the score can be manipulated—is limited to making notations on the paper.

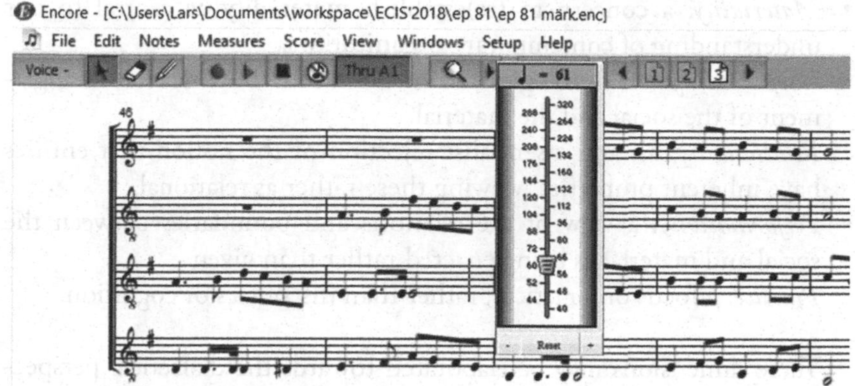

Fig. 11.10 A digitized score

However, the same score digitized in an IT artifact enables quite new affordances (see Fig. 11.10).

The score can be played back, easily modified, electronically distributed, and so on. These affordances make the convergence of the cycle in TFS more efficient, mainly with respect to the Efferent synthesis and Back-afferentation stages.

Importantly, however—the layout of the score *is not changed* when digitalized. A key insight is thus that the afferent-informative affordances of the score *are retained* in the transition from analog to digital medium. It makes no sense to introduce quite a new notation system just because the score is digitized. Thus, although obvious, we need to bear in mind that digitalization of an activity does not impact our evolutionary developed, neurobiological constitution. We are not transformed into some "digitized" species just because our technology evolves.

Sociomateriality

Sociomateriality has been an influential theoretical trend in IS research for more than a decade (e.g. Cecez-Kecmanovic et al., 2014). According to Jones (2014), sociomateriality (SM) has five characteristic dimensions:

- *Materiality*: a concern to (re)establish materiality as central to our understanding of contemporary organizations.
- *Inseparability*: an ontological claim about the inextricable entanglement of the social and the material.
- *Relationality*: an anti-essentialist rejection of the notion that entities have inherent properties, viewing these rather as relational.
- *Performativity*: a view of the relations and boundaries between the social and material as being enacted rather than given.
- *Practice*: a focus on practices, rather than discourses or cognition.

These dimensions may be elaborated towards the dialectical perspectives in the following way.

Materiality

The concept of "materiality" has surfaced to the forefront in IS discourse (see e.g. Leonardi, Nardi, & Kallinikos, 2012). However, as Kallinikos, Leonardi, and Nardi (2012) point out, the problem

> is that the term "materiality" and other related terms that scholars have begun to use, such as "material," "material property," "material consequence," "materialize," "materialism," "sociomaterial," and "sociomateriality," are neither well defined nor consistently used … Moreover, their relationship to concepts in regular use in the social sciences such as "technology," "form," "function," "artifact," and "socio-technical system" is not yet clear. (p. 5)

In line with Pentland and Singh (2012) we focus on how materiality, in whatever apparition it may come forth, constrains and empowers action: "Something is material insofar as it has consequences in a particular context" (p. 292). As discussed previously, action is pre-dated by a momentarily stabilized communal infrastructure comprised of both biomechanical and communal factors. A particular action may be triggered from internal impulses, such as hunger or thirst, or from external sensations emanating from the environment. Thus, materiality has both an individual and social quality.

A consequence of this conceptualization is that materiality "relates to what is relevant or pertinent to a given situation" (Cooren, Fairhurst, & Huet, 2012, p. 301). Regardless of which materiality is involved in the communalization process—be that drawings, models, IT artifacts, speech, and so on—physical sensations emanating from them need to be sensed by the individual to integrate them into an actionable Gestalt.

Such a conceptualization of materiality has implications for current IS discussion of the "nonmaterial nature of many artifacts, particularly those associated with ICT" (Faulkner & Runde, 2013, p. 811). Some examples given of nonmaterial objects are "articles, sales reports, employment contracts, product designs, musical compositions, and bitstrings such as computer files" (ibid., p. 806). However, these are always material in the sense that we can attend them, talk about them, and act towards them. For example, our understanding of bitstrings (ones and zeros) are fundamental for the entire technical evolution of the computer. Touching a physical "computer", seeing a 3D model of a "computer", reading the word "computer", or the hearing someone uttering "computer", are all meaningful for someone communalized into computerized communities. They all make "a difference in the current situation" (Leonardi, 2010) as do the terms ones and zeros.

Accordingly, the qualitative step for elaborating extant discussions of materiality toward the DA stance is to acknowledge the biomechanical ability of the individual to perceive sensations: "each and every organism with a mind relates mentally with the world only through senses. There is no exception to this principle" (Toomela, 2016, p. 102). This means that "materiality is an observable quality of the world. It might not always be tangible (e.g., our thoughts) but nonetheless we can always trace it back to some form of material presence" (Kedzior, 2014, p. 6). Accordingly, the term "non-materiality" should be replaced by the term "non-tangible".

Inseparability

SM states that "the social and the material are considered to be inextricably related—there is no social that is not also material, and no material that is not also social" (Orlikowski, 2007, p. 1437). In the DA, inseparability can emerge in the dialectical relation between an *individual* and the social/material. A pertinent example is the following statement from the famous cello player Mstislav Rostropovič:

> There no longer exist relations between us. Some time ago I lost my sense of the border between us.... I experience no difficulty in playing sounds ... The cello is my tool no more (Mstislav Rostropovič, quoted in Zinchenko, 1996, p. 295).

Importantly, the experience of inseparability is entirely an *individual experience*. Rostropovič and the cello remain distinct ontological phenomena as relata in the dialectical relation. They are not metamorphosed into some qualitatively new existence, as SM suggests. The very same cello played by someone else than Rostropovič would not sound the same. Thus, the SM stance of inseparability between the social and material need to be elaborated into a stance including the individual, which concur with everyday human experience.

Relationality

SM suggests that people and things only exist in relation to each other:

> [Entities] (whether humans or technologies) have no inherent properties, but acquire form, attributes, and capabilities through their interpenetration. (Orlikowski & Scott, 2008, pp. 455–456)

This complies with the dialectical perspective in the sense that parts acquire properties that are characteristic of them only as parts of wholes. Consider Rostropovič and his cello in the example above. Taken as separate entities, Rostropovič is just a specific human being, and the cello merely a silent, nicely designed material artifact consisting of wood, strings, tuning screws, and so on. However, in their "interpenetration", playing brings out properties of both, which cannot be actuated in separation—the biomechanical ability of Rostropovič to play the cello and the ability of the cello to produce sounds.

Contrary to the SM position, it is evident that both Rostropovič and the cello do exist as prerequisites before their "interpenetration" in playing. There "is always something that exists first as a given, as an issue, as a problem" (Latour, 2008, p. 5). As in Archer's (1995) morphogenetic cycle, the concept of communalization implies that existing structural conditions, that is, communal and biomechanical factors, predate actions that in turn modify these factors. Indeed, this means that both humans and technology change form, attributes, and capabilities as a result of communalization, but this cannot happen unless there are some momentarily stabilized, pre-existing conditions predating communalization.

Performativity and Practices

Performativity implies "an orientation to the practices through which the relations and boundaries between the social and the material are enacted" (Jones, 2014, p. 897). In the DA, practices are seen as communal infrastructures, momentarily stabilized prerequisites for actions. Performativity is understood as the communalization process by which biomechanical and communal factors are enacted.

The IS Discipline

The question of what constitutes the IS core has been a recurrent topic in the IS discipline. Myers (2003, p. 585) describes how the search for a core has meandered between various focal areas such as: *conceptual modeling and databases* (Weber, 2003); *representation* (Weber, 2003); *systems development* (Nunamaker, Chen, & Purdin, 1990-1991); *business processes*; the *orderly provision of data and information* within an organization using IT (Checkland & Howell, 1998); the *IT artifact* (Benbasat & Zmud, 2003); *systems in organizations* (Alter, 2003). However, the discussion has not been brought to closure. The IS field still does not have a commonly agreed core: "On this point, everyone agrees" (Myers, 2003, p. 584).

Elaboration

Disciplines are formed around a central idea, topic, or problem of interest. Over time, contextualized research programmes evolve from a "hard core" of theoretical assumptions that cannot be abandoned or altered without giving up the programme altogether (Lakatos, 1970). This means that other disciplines, which are potentially relevant for the hard core, become less attended if not outright ignored. As a result, the scientific landscape comes forth as isolated islands made up of "crusted lava", as it were, without due consideration of the dependencies between them.

We propose that the dialectical interaction of information, the IT artifact and the IS, as outlined previously, provide a foundational kernel from which the IS field can be advanced. The main argument is that the dialectical perspective comprises human, social and material/technical aspects concurring with the ambitions of the field. In line with this view, we suggest that the rationale of the IS discipline is to *research the relationship between disciplines*. This view concurs with several IS scholars (e.g. Baskerville & Myers, 2002; Beath, Berente, Gallivan, & Lyytinen, 2013; Cecez-Kecmanovic, 2002; Tarafdar & Davison, 2018). Lee (2001) summarizes this ambition:

> [Research] in the information systems field examines more than just the technological system, or just the social system, or even the two side by side; in addition, it investigates the phenomena that emerge when the two interact. This embodies both a research perspective and a subject matter that *differentiate the academic field of information systems from other disciplines.* (p. iii, our emphasis)

Seen from the dialectical perspective, parts (disciplines) acquire their properties by being parts of a whole; properties that come into existence in the interaction that make up the whole. At present, however, there is no "whole" constituted by disciplines relevant for the IS context, such as sociology, social psychology, organization theory, semiotics, linguistics, computer science, neuroscience, etc.

However, a dialectical reorientation requires a different mindset, providing

> perspectives for examining disciplinary concepts in a relational way, through the particularity of their positions within a complex net of interrelations … that characterize the research problem. (Tarafdar & Davison, 2018, p. 528)

In such a perspective, the IS discipline has "a tremendous opportunity to take a more prominent, leading role within the larger community of scholars interested in the development, use, and impact of information technology and systems in broadly defined social and organizational settings" (Baskerville & Myers, 2002, p. 8).

Some implications of a dialectical, underlying kernel for the IS discipline can be discerned. IS have been comprehended as a field, which relies on theories from "reference" or "contributing" disciplines for grounding IS research (Baskerville & Myers, 2002; Lee, 2001). With a dialectical kernel, this would be turned on its head. The IS field would become the backbone by which other fields can be integrated, and yet maintaining their particular "hard cores".

The advantage is twofold. First, we may scrutinize our own discipline and sort out what contributes or does not to a dialectical perspective. In the long run, this might consolidate what is now a diversity of unrelated and sometimes incongruent findings, such as the many different definitions of the IT artifact and the IS. Second, our own discipline may be corroborated and in turn corroborate contributing disciplines. For example, the DA view of individual, innate prerequisites for action—the activity modalities—needs to be further sustained in neuroscientific findings. Conversely, neuroscience may profit from a perspective where individual neurobiological abilities evolve dialectically in relation to the social; thus, avoiding the mereological fallacy of allocating powers to the brain that can only be ascribed to the individual as a whole, such as believing, making decisions, interpreting, and so on (Bennett & Hacker, 2003).

Such an interdisciplinary cooperation between disciplines has been inchoated by the recently established NeuroIS subfield of IS, which "relies on neuroscience and neurophysiological knowledge and tools to better understand the development, use, and impact of information and communication technologies" (Wikipedia, 2019). Focusing on the intersection between neuroscience and IT artifacts may eventually impact the design of these artifacts and consequently design science theories.

In summary, with a dialectical kernel as a foundation, the focus of IS research will be on the "simultaneous attention to both the technical and the human (social) side of IT in its organizational context" (Beath et al., 2013, p. ii). It is "precisely this combination that gives IS research its distinctive value" (ibid., p. v). This would create "a unique domain that by far exceeds the individual discipline domains of any of Business/ Management, Social Sciences, Engineering or Computer Science" (Cecez-Kecmanovic, 2002, p. 1700). This opportunity has in fact been latent all along the history of the IS discipline, since it is the only discipline that aims to span the full complexity of the individual, social, and material spectrum. A profound implication is that the IS discipline would become highly relevant, simply because there is no other discipline that can take on this research area.

Conclusion

The IS discipline struggles with central problems obstructing the advancement of IS theorizing. This contribution proposes an escape from this stalemate situation, based on the fundamental assumption that the individual and the social are dialectically related. Thus, the mainstream focus on the social—material/technical is augmented to a trinity comprised of the individual, the social, and the material/technical.

Such a position has been advocated by several IS scholars (e.g. Beynon-Davies, 2013; Mingers, 2001; Mingers & Willcocks, 2014; Mingers & Willcocks, 2017; Ramiller, 2016), which in turn are inspired by the works of Merleau-Ponty (1962) and Maturana and Varela (1980). The gist of this thinking is that the

> body is a nexus for the interaction of both the individual and society, and action and cognition, and is, therefore, of *central importance* both for developing more effective information-based systems, and for observing the effects of such systems on people and society. (Mingers, 2001, p. 124, our emphasis)

We developed this position into a dialectical kernel where individual abilities, conceptualized as activity modalities, and corresponding communal factors, mutually constitute each other. The central idea of this position is that the social structure, as experienced by the individual, reflects evolutionary developed human abilities. This is the ground zero for human action. If we cannot make sense of what is out there, we cannot act.

Based on such a foundation, we discussed how main IS research topics can be elaborated towards the dialectical position. Information, the IT artifact, and the IS is reconceptualized as dialectically related; neither can be defined in isolation from the others. We also proposed an alternative model for communication—Integrational Linguistics—which concurs with the dialectical kernel. Further, we outlined implications for the digitalization of artifacts and how sociomateriality may be rethought to concur with the dialectical perspective. Last but not least, we suggested to reconceptualize the IS discipline as the integrating backbone of other disciplines.

From such a detour from the murky territory of dialectics, we may return to the solid ground of mainstream IS landscape, however now with an alternative mindset where familiar IS phenomena are seen in a new light. It goes without saying that the dialectical perspective has to be discussed, critically scrutinized and elaborated. However, in times of an increasingly, turbulent, interconnected and ubiquitous evolution of IT, we need some stable terra firma for inquiry. This is provided by our neurobiological predispositions, which have not changed significantly since the dawn of our species. If this position is adopted by the IS community, a smorgasbord of opportunities for future research emerges, thus elevating the IS discipline to the central position it deserves.

References

Adler, P. (2005). The evolving object of software development. *Organization, 12*(3), 401–435.

Alter, S. (2003). Sidestepping the IT artifact, scrapping the IS silo, and laying claim to "Systems in Organizations". *Communications of the Association for Information Systems, 12*, 30.

Alter, S. (2008). Defining information systems as work systems: Implications for the IS field. *European Journal of Information Systems, 17*(5), 448–469.

Alter, S. (2015). The concept of 'IT artifact' has outlived its usefulness and should be retired now. *Information Systems Journal, 25*(1), 47–60.

Andrews, M., Frank, S. L., & Vigliocco, G. (2014). Reconciling embodied and distributional accounts of meaning in language. *Topics in Cognitive Science, 6*, 359–370.

Archer, M. S. (1995). *Realist social theory: The morphogenetic approach.* Cambridge, UK: Cambridge University Press.

Archer, M. S. (1998). Realism and morphogenesis. In M. Archer, R. Bhaskar, A. Collier, T. Lawson, & A. Norrie (Eds.), *Critical realism: Essential readings* (pp. 356–382). London: Routledge.

Archer, M. S. (2000). *Being human: The problem of agency.* Cambridge: Cambridge University Press.

Armarnath, B. R. (2019). How to win "Digitization vs Digitalization" debate?—A boring post. Retrieved September 12, 2019, from https://www.linkedin.com/pulse/how-win-digitization-vs-digitalization-debate-boring-r-a/

Bærentsen, K. B., & Trettvik, J. (2002). An activity theory approach to affordance. The *Second Nordic Conference on Human-Computer Interaction. NORDICHI 2002. Tradition and Transcendence* (pp. 51-60). Aarhus, Denmark, October 19-23, 2002.

Baskerville, R. L. (2010). Knowledge lost and found: A commentary on Allen Lee's 'retrospect and prospect'. *Journal of Information Technology, 25*(4), 350–351.

Baskerville, R. L., & Myers, M. D. (2002). Information systems as a reference discipline. *MIS Quarterly, 26*(1), 1–14.

Beath, C., Berente, N., Gallivan, M. J., & Lyytinen, K. (2013). Expanding the frontiers of information systems research: Introduction to the special issue. *Journal of the Association for Information Systems, 14*(4), 4.

Benbasat, I., & Zmud, R. W. (2003). The identity crisis within the IS discipline: Defining and communicating the discipline's core. *MIS Quarterly, 27*(2), 183–194.

Bennett, M. R., & Hacker, P. M. S. (2003). *Philosophical foundations of neuroscience.* Malden, MA: Blackwell Pub.

Bernstein, R. (1999). *Praxis and Action.* Philadelphia: University of Pennsylvania Press.

Beynon-Davies, P. (2010). The enactment of significance: A unified conception of information, systems and technology. *European Journal of Information Systems, 19*, 389–408.

Beynon-Davies, P. (2013). Making faces: Information does not exist. *Communications of the Association for Information Systems, 33*(19), 340–350.

Beynon-Davies, P., & Wang, Y. (2019). Deconstructing information sharing. *Journal of the Association for Information Systems, 20*(4), 476–498. https://doi.org/10.17705/1.jais.00541

Blumer, H. (1969). *Symbolic interactionism: Perspective and method.* Englewood Cliffs, NJ: Prentice-Hall.

Blunden, A. (2007). Modernity, the individual, and the foundations of cultural–historical activity theory. *Mind, Culture, and Activity, 14*(4), 253–265.

Boell, S. K. (2017). Information: Fundamental positions and their implications for information systems research, education and practice. *Information and Organization, 27*(2017), 1–16.

Boesch, E. (1991). *Symbolic action theory and cultural psychology.* Berlin: Springer.

Boland, R. J. (1987). The in-formation of information systems. In R. J. Boland & R. A. Hirschheim (Eds.), *Critical issues in information systems research* (pp. 363–379). New York, NY: John Wiley.

Bradshaw, T. K. (2008). The post-place community: Contributions to the debate about the definition of community. *Community Development, 39*(1), 5–16.

Bressler, S., & Kelso, S. (2016). Coordination dynamics in cognitive neuroscience. *Frontiers in Neuroscience, 10*(September 2016), 1–7. https://doi.org/10.3389/fnins.2016.0039

Burton-Jones, A., McLean, E. R., & Monod, E. (2015). Theoretical perspectives in IS research: From variance and process to conceptual latitude and conceptual fit. *European Journal of Information Systems, 2015, 24*, 664–679.

Carter, M., Petter, S., & Randolph, A. (2015). Desperately seeking information in information systems research. *ICIS 2015 Proceedings.*

Cecez-Kecmanovic, D. (2002). The discipline of information systems: Issues and challenges. *AMCIS 2002 Proceedings*. 232.

Cecez-Kecmanovic, D., Galliers, R. D., Henfridsson, O., Newell, S., & Vidgen, R. (2014). The sociomateriality of information systems: Current status, future directions. *MIS Quarterly, 38*(3), 809–830.

Chatterjee, S., Xiao, X., Elbanna, A., & Sarker, S. (2017). The information systems artifact: A conceptualization based on general systems theory. *In Proceedings of the 50th Hawaii International Conference on System Sciences,* 2017, 5717-5726.

Checkland, P., & Howell, S. (1998). *Information, systems, and information systems: Making sense of the field*. Chichester: John Wiley & Sons.

Chen, P. (1976). The entity-relationship model—toward a unified view of data. *ACM Transactions on Database Systems, 1*(1), 9–36.

Clark, A. (2006). Language, embodiment, and the cognitive niche. *TRENDS in Cognitive Sciences, 10*(8), 370–374.

Clark, H. H. (1996). *Using language*. Cambridge: Cambridge University Press.

Cooren, F., Fairhurst, G. T., & Huet, R. (2012). Why matter always matters in (Organizational) communication. In P. M. Leonardi, B. A. Nardi, & J. Kallinikos (Eds.), *Materiality and organizing: Social interaction in a technological world* (pp. 296–314). Oxford: Oxford University Press.

Corley, K. G., & Gioia, D. E. (2011). Building theory about theory building: What constitutes a theoretical contribution? *Academy of Management Review, 36*(1), 12–32.

Davison, R. M., & Tarafadar, M. (2018). Shifting baselines in information systems research threaten our future relevance. *Information Systems Journal, 28*(4), 587–591.

Dewey, J. (1991). In J. A. Boydston (Ed.), *'Logic', The theory of enquiry. The later works of John Dewey* (Vol. 12). Carbondale and Edwardsville: Southern Illinois University Press. Originally published in 1938.

Engel, A. K., Maye, A., Kurthen, M., & König, P. (2013). Where's the action? The pragmatic turn in cognitive science. *Trends in Cognitive Sciences, 17*, 202–209.

Faulkner, P., & Runde, J. (2013). Technological objects, social positions, and the transformational model of social activity. *MIS Quarterly, 37*(3), 803–818.

Fleetwood, S. (2005). Ontology in organization and management studies: A critical realist perspective. *Organization, 12*(2), 197–222. https://doi.org/10.1177/1350508405051188

Gadamer, H.-G. (1989). *Truth and method*. London: Sheed and Ward.

Garfinkel, H. (2008). *Toward a sociological theory of information*. Boulder, CO: Paradigm Publishers.

Gibson, J. (1979). *The ecological approach lo visual perception*. Boston: Houghton Mifflin.

Goldkuhl, G. (2013). The IT artefact: An ensemble of the social and the technical?—A rejoinder. *Systems, Signs & Actions, 7*(1), 90–99.

Harris. (n.d.). *Integrationism: A very brief introduction*. Retrieved September 12, 2019, from http://www.royharrisonline.com/integrational_linguistics/integrationism_introduction.html

Harris, R. (1996). *Signs, language, and communication: Integrational and segregational approaches*. London: Routledge.
Harris, R. (2009). *After epistemology*. Gamlingay: Bright Pen.
Hassan, N. R. (2011). Is information systems a discipline? Foucauldian and Toulminian insights. *European Journal of Information Systems, 20*(4), 456–476.
Hirschheim, R., & Klein, H. K. (2012). A glorious and not-so-short history of the information systems field. *Journal of the Association for Information Systems, 13*(4), 5.
Hoskin, K. (2004). Spacing, timing and the invention of management. *Organization, 11*(6), 743–757.
IAISLC. (2017). What is Integrationism? Retrieved September 12, 2019, from http://www.integrationists.com/Integrationism.html
Jeffery, K., Hayman, R., & Chakraborty, S. (2004). A proposed architecture for the neural representation of spatial context. *Neuroscience & Biobehavioral Reviews, 28*(2004), 201–218.
Jones, M. (2014). A matter of life and death: Exploring conceptualizations of sociomateriality in the context of critical care. *MIS Quarterly, 38*(3), 895–925.
Kallinikos, J., Aaltonen, A., & Marton, A. (2013). The ambivalent ontology of digital artifacts. *MIS Quarterly, 37*(2), 357–370.
Kallinikos, J., Leonardi, P. M., & Nardi, B. N. (2012). The challenge of materiality: Origins, scope, and prospects. In P. M. Leonardi, B. A. Nardi, & J. Kallinikos (Eds.), *Materiality and organizing: Social interaction in a technological world* (pp. 3–22). Oxford: Oxford University Press.
Kedzior, R. (2014). *How digital worlds become material: An ethnographic and netnographic investigation in second life*. Doctoral theses Economics and Society—281, Hanken School of Economics, Department of Marketing, Marketing.
Kirsh, D. (1995). The intelligent use of space. *Artificial Intelligence, 73*(1-2), 31–68.
Lakatos, I. (1970). *Falsification and the methodology of scientific research programs*. In I. Lakatos & A. Musgrave (Eds.), *Falsification and the methodology of scientific research programs* (pp. 91–132). New York, NY: Cambridge University Press.
Latour, B. (2008). A cautious Prometheus? A few steps toward a philosophy of design (with special attention to Peter Sloterdijk). In F. Hackne, J. Glynne, & V. Minto (Eds.), *Proceedings of the 2008 annual international conference of the design history society* (pp. 2–10). Falmouth: Universal Publishers.
Lee, A. S. (2001). Editor's comments. *MIS Quarterly, 25*(1), iii–vii.

Lee, A. S. (2010). Retrospect and prospect: Information systems research in the last and next 25 years. *Journal of Information Technology, 25*(4), 336–348.

Lee, A. S., Thomas, M., & Baskerville, R. L. (2015). Going back to basics in design science: From the information technology artifact to the information systems artifact. *Information Systems Journal, 25*(1), 5–21.

Leonardi, P. M. (2010). Digital materiality? How artifacts without matter, matter. *First Monday, 15*(6). Retrieved September 12, 2019, from http://firstmonday.org/article/view/3036/2567

Leonardi, P. M. (2012). Materiality, sociomateriality, and socio-technical systems: What do these terms mean? How are they related? Do we need them? In P. M. Leonardi, B. A. Nardi, & J. Kallinikos (Eds.), *Materiality and organizing: Social interaction in a technological world* (pp. 25–48). Oxford: Oxford University Press.

Leonardi, P. M., Nardi, B. A., & Kallinikos, J. (2012). *Materiality and organizing: Social interaction in a technological world*. Oxford: Oxford University Press.

Levins, R., & Lewontin, R. C. (1985). *The dialectical biologist*. Cambridge, MA: Harvard University Press.

Lewis, M. D. (2002). The dialogical brain: Contributions of emotional neurobiology to understanding the dialogical self. *Theory & Psychology, 12*(2), 175–190.

Lewis, M. D. (2005). Bridging emotion theory and neurobiology through dynamic systems modeling. *Behavioral and Brain Sciences, 28*, 169–194.

Love, N. (2004). Cognition and the language myth. *Language Sciences, 26*(6), 525–544.

Love, N. (2015). Roy Harris (1931–2015). *Language & Communication, 42*(2015), iii–iv.

Luria, A. R. (1973). *The working brain*. London: Penguin Books.

Marx, K., & Engels, F. (1998). *The German ideology: Including theses on feuerbach and introduction to the critique of political economy*. Amherst, NY: Prometheus Books. Originally published in 1845.

Mason, R. O., & Mitroff, I. I. (1973). A program for research on management information systems. *Management Science, 19*(5), 475–487.

Maturana, H. R., & Varela, F. J. (1980). *Autopoiesis and cognition: The realization of living*. Dordrecht, Holland: D. Reidel Publishing Company.

McCubbrey, D. J. (2003). The IS Core—IV: IS Research: A third way. *Communications of the AIS, 12*(34), 553–556.

Merleau-Ponty, M. (1962). *Phenomenology of perception*. London: Routledge.

Mingers, J. (2001). Embodying information systems: The contribution of phenomenology. *Information and Organization, 11*(2001), 103–128.

Mingers, J., & Standing, C. (2018). What is information? Toward a theory of information as objective and veridical. *Journal of Information Technology, 33*(2), 85–104.

Mingers, J., & Willcocks, L. (2014). An integrative semiotic framework for information systems: The social, personal and material worlds. *Information and Organization, 24*(1), 48–70.

Mingers, J., & Willcocks, L. (2017). An integrative semiotic methodology for IS research. *Information and Organization, 27*(2017), 17–36.

Mumford, M. (2006). The story of socio-technical design: Reflections on its successes, failures and potential. *Information Systems Journal, 16*(4), 317–342.

Myers, M. D. (2003). The IS Core—VIII: Defining the core properties of the is discipline: Not yet, not now. *Communications of the AIS, 12*(38), 582–587.

Nowak, A., Vallacher, R. R., & Zochowski, M. (2005). The emergence of personality: Dynamic foundations of individual variation. *Developmental Review, 25*(2005), 351–385.

Nunamaker, J. F., Chen, M., & Purdin, T. D. M. (1990-1991). System development in information systems research. *Journal of Management Information Systems, 7*(3), 99–106.

O'Keefe, J., & Nadel, L. (1978). *The hippocampus as a cognitive map*. Oxford: Oxford University Press.

Orlikowski, W. J. (2007). Sociomaterial practices: Exploring technology at work. *Organization Studies, 28*(9), 1435–1448.

Orlikowski, W. J., & Iacono, C. S. (2001). Research commentary: Desperately seeking the "IT" in IT research—A call to theorizing the IT artifact. *Information Systems Research, 12*(2), 121–134.

Orlikowski, W. J., & Scott, S. V. (2008). Sociomateriality: Challenging the separation of technology, work and organization. *The Academy of Management Annals, 2*(1), 433–474.

Orman, J. (2017). Explanation and theory in linguistic inquiry. *Empedocles: European Journal for the Philosophy of Communication, 8*(2), 167–186.

Paul, R. (2010). Loose change. *European Journal of Information Systems, 19*, 379–381.

Pentland, B. T., & Singh, H. (2012). Materiality: What are the consequences? In P. M. Leonardi, B. A. Nardi, & J. Kallinikos (Eds.), *Materiality and organizing: Social interaction in a technological world* (pp. 287–295). Oxford: Oxford University Press.

Ramiller, N. (2016). Editorial: New technology and the post-human self: Rethinking appropriation and resistance. *ACM SIGMIS Database: The DATABASE for advances in information systems, 47*(4), 23–33.

Reckwitz, A. (2002). Toward a theory of social practices. A development in culturalist theorizing. *European Journal of Social Theory, 5*(2), 243–263.

Red'ko, V. G., Prokhorov, D. V., & Burtsev, M. B. (2004). Theory of functional systems, adaptive critics and neural networks. In *Proceedings of international joint conference on neural networks* (pp. 1787–1792). Budapest: ICANN.

Riemer, K., & Johnston, R. B. (2017). Clarifying ontological inseparability with Heidegger's analysis of equipment. *MIS Quarterly, 41*(4), 1059–1081.

Samuels, W. J. (1972). The scope of economics historically considered. *Land Economics, 48*(3), 248–268.

Scott, W. R. (1995). *Institutions and organizations*. Thousand Oaks: Sage.

Sein, M., Henfridsson, O., Purao, S., Rossi, M., & Lindgren, R. (2011). Action design research. *Management Information Systems Quarterly, 35*(1), 37–56.

Silver, M. S., & Markus, M. L. (2013). Conceptualizing the SocioTechnical (ST) artifact. *Systems, Signs & Actions, 7*(1), 82–89.

Stamper, R. K. (2001). Organisational semiotics. Informatics without the computer. In K. Liu, R. J. Clarke, P. B. Andersen, & R. K. Stamper (Eds.), *Information, organisation and technology. Studies in organisational semiotics* (pp. 115–171). Boston: Kluwer Academic Press.

Stichweh, R. (2000). Systems theory as an alternative to action theory? The rise of 'communication' as a theoretical option. *Acta Sociologica, 43*(1), 5–14.

Tarafdar, M., & Davison, R. M. (2018). Research in information systems: Intradisciplinary and inter-disciplinary approaches. *Journal of the Association for Information Systems, 19*(6), 523–551. https://doi.org/10.17705/1jais.00500

Taxén, L. (2003). *A framework for the coordination of complex systems' development*. Dissertation No. 800. Linköping University, Dep. of Computer & Information Science, 2003. Retrieved September 12, 2019, from http://liu.diva-portal.org/smash/record.jsf?searchId=1&pid=diva2:20897

Taxén, L. (2009). *Using activity domain theory for managing complex systems*. Information Science Reference. Hershey PA: Information Science Reference (IGI Global). ISBN: 978-1-60566-192-6.

Thelen, E. (1995). Motor development: A new synthesis. *American Psychologist, 50*(2), 79–95.

Toomela, A. (2010). Biological roots of foresight and mental time travel. *Integrative Psychological and Behavioral Science, 44*, 97–125.

Toomela, A. (2014). There can be no cultural-historical psychology without neuropsychology. And vice versa. In A. Yasnitsky, R. van der Veer, & M. Ferrari (Eds.), *The cambridge handbook of cultural-historical psychology* (pp. 315–349). Cambridge: Cambridge University Press.

Toomela, A. (2016). What are higher psychological functions? *Integrative Psychological & Behavioral Science, 50*(1), 91–121.

Wan-chi, W. (2006). Understanding dialectical thinking from a cultural-historical perspective. *Philosophical Psychology, 19*(2), 239–260. https://doi.org/10.1080/09515080500462420

Watson, R. T. (2014). A personal perspective on a conceptual foundation for information systems. *Journal of the Association for Information Systems, 15*(8), 515–535.

Weber, R. (2003). Still desperately seeking the IT-artifact. (Editor's Comments). *MIS Quarterly, 27*(2), iii–xi.

Wikipedia. (2018). Sheet music. Retrieved September 12, 2019, from, https://en.wikipedia.org/wiki/Sheet_music

Wikipedia. (2019). Neuro-Information-Systems. Retrieved September 12, 2019, from, https://en.wikipedia.org/wiki/Neuro-Information-Systems

Winter, S. J., & Butler, B. S. (2011). Creating bigger problems: Grand challenges as boundary objects and the legitimacy of the information systems field. *Journal of Information Technology, 26*, 99–108.

Witter, M. P., & Moser, E. I. (2006). Spatial representation and the architecture of the entorhinal cortex. *Trends in Neurosciences, 29*(12), 671–678.

Zinchenko, V. (1996). Developing activity theory: The zone of proximal development and beyond. In B. Nardi (Ed.), *Context and consciousness, activity theory and human-computer interaction* (pp. 283–324). Cambridge, MA: MIT Press.

Index[1]

[1] Note: Page numbers followed by 'n' refer to notes.

N. R. Hassan, L. P. Willcocks (eds.), *Advancing Information Systems Theories*, Technology, Work and Globalization, https://doi.org/10.1007/978-3-030-64884-8

Printed in the United States
by Baker & Taylor Publisher Services

Printed in the United States
by Baker & Taylor Publisher Services